and "Neuroinformatics" (1993–1994: Ma 842/6-1). I have received much inspiration from teaching in a mathematical graduate program on "Complex Systems" (supported by the DFG) and an economic program on "Nonlinearity in Economics and Management" at the University of Augsburg. In 1991 and 1993, the Scientific Center of Northrhine-Westphalia (Düsseldorf) invited me to two international conferences on the cultural effects of computer technology, neurobiology, and neurophilosophy.

Last but not least, I would especially like to thank J. Andrew Ross (Springer-Verlag) for carefully reading and correcting the book as a native speaker, and Katja E. Hüther and Jutta Janßen (University of Augsburg) for typing the text.

Augsburg, June 1994   *Klaus Mainzer*

Preface to the First Edition

Complexity and nonlinearity are prominent features in the evolution of matter, life, and human society. Even our mind seems to be governed by the nonlinear dynamics of the complex networks in our brain. This book considers complex systems in the physical and biological sciences, cognitive and computer sciences, social and economic sciences, and philosophy and history of science. An interdisciplinary methodology is introduced to explain the emergence of order in nature and mind and in the economy and society by common principles.

These methods are sometimes said to foreshadow the new sciences of complexity characterizing the scientific development of the 21st century. The book critically analyzes the successes and limits of this approach, its systematic foundations, and its historical and philosophical background. An epilogue discusses new standards of ethical behavior which are demanded by the complex problems of nature and mind, economy and society.

The "nucleus" of this book was a paper contributed to a conference on complex nonlinear systems which was organized by Hermann Haken and Alexander Mikhailov at the Center for Interdisciplinary Studies in Bielefeld, in October 1992. In December 1992, Dr. Angela M. Lahee (Springer-Verlag) suggested that I elaborate the topics of my paper into a book. Thus, I would like to express my gratitude to Dr. Lahee for her kind and efficient support and to Hermann Haken for his cooperation in several projects on complex systems and synergetics. I also wish to thank the German Research Foundation (DFG) for the support of my projects on "Computer, Chaos and Self-organization" (1990–1992: Ma 842/4-1)

Klaus
Mainzer

Thinking
in Complexity

The Complex Dynamics
of Matter,
Mind, and Mankind

Third Revised and Enlarged Edition
With 126 Figures

 Springer

Professor Dr. Klaus Mainzer

Lehrstuhl für Philosophie und Wissenschaftstheorie
Universität Augsburg
Universitätsstrasse 10
D-86135 Augsburg, Germany

ISBN 3-540-62555-0 3rd Edition
Springer-Verlag Berlin Heidelberg New York

ISBN 3-540-60637-8 2nd Edition
Springer-Verlag Berlin Heidelberg New York

Library of Congress Cataloging-in-Publication Data applied for

Die Deutsche Bibliothek - CIP-Einheitsaufnahme
Mainzer, Klaus: Thinking in complexity : the complex dynamics of
matter, mind, and mankind / Klaus Mainzer. - 3., rev. and enl. ed. -
Berlin ; Heidelberg ; New York ; Barcelona ; Budapest ; Hong Kong ;
London ; Milan ; Paris ; Santa Clara ; Singapore ; Tokyo :
Springer, 1997
ISBN 3-540-62555-0

Cover design: Erich Kirchner, Heidelberg
Typesetting: Data conversion by Kurt Mattes, Heidelberg

SPIN: 10567630   55/3144 - 5 4 3 2 1 0 - Printed on acid-free paper

*Das Ganze aber ist nur das durch seine Entwicklung
sich vollendende Wesen.* *

G. W. F. Hegel: Phänomenologie des Geistes (1807)

* *The whole, however, is merely the essential nature reaching its
completeness through the process of its own development.*

G. W. F. Hegel: The Phenomenology of Mind (1807)

Preface to the Third Edition

The second edition of "Thinking in Complexity", like the first edition, was also sold out in less than one year. Meanwhile, Japanese and Chinese translations of the second editon have been published. Once more I have taken the opportunity provided by a new edition to revise and extend the text.

A new Sect. 2.5 "Complex Systems and the Self-Construction of Materials" is included, in order to analyze the role of complex systems in the dramatic success of supramolecular chemistry, nanotechnology, and the technology of smart ("intelligent") materials. These topics lie at the boundary between materials science and life science. In recent years, life science and computer science have been growing together in a common field of research called "artificial life". A further new Sect. 5.5 "From Artificial Intelligence to Artificial Life" has been added, in which the role of complex systems in the field of artificial life is discussed. I also use the opportunity of the new edition to make some remarks about the relationship between the Santa Fe approach to complex systems and the methods of synergetics and order parameters which are key concepts in this book.

Research into complex systems continues worldwide. I have to thank the readers who have written friendly and inspiring letters from all over the world. Some months ago, a German Society of Complex Systems and Nonlinear Dynamics was founded. The honorable German Academy of Natural Scientists Leopoldina invited me to give a lecture on complexity for which I express my gratitude. Last but not least, I would again like

to thank Wolf Beiglböck of Springer-Verlag for initiating
and supporting this new edition.

Augsburg, November 1996   *Klaus Mainzer*

Preface to the Second Edition

The first edition of "Thinking in Complexity" was sold out in less than one year. Obviously, complexity and non-linearity are "hot" topics of interdisciplinary interest in the natural and social sciences. The situation is well summarized by a quotation of Ian Stewart (Mathematics Institute, University of Warwick) who wrote a nice review of my book under the title "Emerging new science" [Nature **374**, 834 (1995)]: "Nonlinearity is not a universal answer, but it is often a better way of thinking about the problem".

I have taken the opportunity provided by a second edition to revise and extend the text. In Sect. 2.4 a supplement is included on the recent importance of conservative self-organization in supramolecular chemistry and the material sciences. Some references are given to the recent discussion of self-organization in alternative cosmologies. Some remarks are made about new results on dissipative self-organization in living cells (Sect. 3.3). The success and limitations of adaptive neural prostheses in neurotechnology are analyzed in more detail (Sect. 5.4). The last chapter is extended into an "Epilogue on Future, Science, and Ethics": After a short introduction to traditional forecasting methods, their limitations and new procedures are discussed under the constraints of nonlinearity and complexity in the natural and social sciences. In particular, the possibilities of predicting and modeling scientific and technological growth are extremely interesting for the contemporary debates on human future and ethics.

General methods of nonlinear complex systems must be developed in cooperation with the natural and social sciences under their particular observational, experimental, and theoretical conditions. Thus, I want to thank some colleagues for their helpful advice: Rolf Eckmiller (Dept. of Neuroinformatics, University of Bonn), Hans-Jörg Fahr and Wolf Priester (Dept. of Astrophysics and Max-Planck Institute for Radioastronomy, Bonn), Hermann Haken (Institute of Theoretical Physics and Synergetics, Stuttgart), Benno Hess (Max-Planck Institute for Medical Research, Heidelberg), S. P. Kurdyumov (Keldysh Institute of Applied Mathematics, Moscow), Renate Mayntz (Max-Planck Institute for Social Sciences, Cologne), Achim Müller (Dept. of Inorganic Chemistry, University of Bielefeld). Last but not least, I would like to thank Wolf Beiglböck (Springer-Verlag) for initiating and supporting this new edition.

Augsburg, November 1995   *Klaus Mainzer*

Contents

1. Introduction: From Linear to Nonlinear Thinking

The theory of nonlinear complex systems has become a successful problem solving approach in the natural sciences – from laser physics, quantum chaos, and metereology to molecular modeling in chemistry and computer-assisted simulations of cellular growth in biology. On the other hand, the social sciences are recognizing that the main problems of mankind are global, complex, and nonlinear, too. Local changes in the ecological, economic, or political system can cause a global crisis. Linear thinking and the belief that the whole is only the sum of its parts are evidently obsolete. One of the most exciting topics of present scientific and public interest is the idea that even our mind is governed by the nonlinear dynamics of complex systems. If this thesis of computational neuroscience is correct, then indeed we have a powerful mathematical strategy to handle interdisciplinary problems of natural sciences, social sciences, and humanities.

What is the reason behind these successful interdisciplinary applications? The book shows that the theory of nonlinear complex systems cannot be reduced to special natural laws of physics, although its mathematical principles were discovered and at first successfully applied in physics. Thus it is no kind of traditional "physicalism" to explain the dynamics of laser, ecological populations, or our brain by similar structural laws. It is an interdisciplinary methodology to explain the emergence of certain macroscopic phenomena via the nonlinear interactions of microscopic elements in complex systems. Macroscopic phenomena may be forms of light waves, fluids, clouds, chemical waves, plants, animals, populations, markets, and cerebral cell assemblies which are characterized by order parameters. They are not reduced to the microscopic level of atoms, molecules, cells, organisms, etc., of complex systems. Actually, they represent properties of real macroscopic phenomena, such as field potentials, social or economical power, feelings or even thoughts. Who will deny that feelings and thoughts can change the world?

In history the concepts of the social sciences and humanities have often been influenced by physical theories. In the age of mechanization Thomas Hobbes described the state as a machine ("Leviathan") with its citizens as cog wheels. For Lamettrie the human soul was reduced to the gear drive of an automaton. Adam Smith explained the mechanism of the market by an "invisible" force like Newton's gravitation. In classical mechanics causality is deterministic in the sense of the Newtonian or Hamiltonian equations of motion. A conservative system is characterized by its reversibility (i.e., symmetry or invariance) in time and the conservation of energy. Celestial mechanics and the pendulum without friction are prominent examples. Dissipative systems are irreversible, like Newton's force with a friction term, for instance.

But, in principle, nature was regarded as a huge conservative and deterministic system the causal events of which can be forecast and traced back for each point of time in the future and past if the initial state is well known ("Laplace's demon"). It was Henri Poincaré who recognized that celestial mechanics is no completely calculable clockwork even with the restrictions of conservation and determinism. The causal interactions of all planets, stars, and celestial bodies are nonlinear in the sense that their mutual effects can lead to chaotic trajectories (e.g., the 3-body problem). Nearly sixty years after Poincaré's discovery, A.N. Kolmogorov (1954), V.I. Arnold (1963), and J.K. Moser proved the so-called KAM theorem: Trajectories in the phase space of classical mechanics are neither completely regular nor completely irregular, but they depend very sensitively on the chosen initial states. Tiny fluctuations can cause chaotic developments (the "butterfly effect").

In this century quantum mechanics has become the fundamental theory of physics [1.1]. In Schrödinger's wave mechanics the quantum world is believed to be conservative and linear. In the first quantization classical systems described by a Hamiltonian function are replaced by quantum systems (for instance electrons or photons) described by a Hamiltonian operator. These systems are assumed to be conservative, i.e., non-dissipative and invariant with respect to time reversal and thus satisfy the conservation law of energy. States of a quantum system are described by vectors (wave functions) of a Hilbert space spanned by the eigenvectors of its Hamiltonian operator. The causal dynamics of quantum states is determined by a deterministic differential equation (the Schrödinger equation) which is linear in the sense of the superposition principle, i.e., solutions of this equation (wave functions or state vectors) can be superposed like in classical optics. The superposition or linearity principle of quantum mechanics delivers correlated ("entangled") states of combined systems which are highly confirmed by the EPR experiments (A. Aspect 1981). In an entangled pure quantum state of superposition an observable can only have indefinite eigenvalues. It follows that the entangled state of a quantum system and a measuring apparatus can only have indefinite eigenvalues. But in the laboratory the measuring apparatus shows definite measurement values. Thus, linear quantum dynamics cannot explain the measurement process.

In the Copenhagen interpretation of Bohr, Heisenberg, et al., the measurement process is explained by the so-called "collapse of the wave-packet", i.e., splitting up of the superposition state into two separated states of measurement apparatus and measured quantum system with definite eigenvalues. Obviously, we must distinguish the linear dynamics of quantum systems from the nonlinear act of measurement. This nonlinearity in the world is sometimes explained by the emergence of human consciousness. Eugene Wigner (1961) suggested that the linearity of Schrödinger's equation might fail for conscious observers, and be replaced by some nonlinear procedure according to which either one or the other alternative would be resolved out. But Wigner's interpretation forces us to believe that the linear quantum superpositions would be resolved into separated parts only in those corners of the universe where human or human-like consciousness emerges. In the history of science anthropic or teleological arguments often showed that there were weaknesses and failures of explanation in science. Thus, some scientists, like Roger Penrose, suppose that the linear dynamics of quantum mechanics is not appropriate to explain cosmic evolution with the emergence of consciousness. He argues that a unified theory of linear quantum mechanics and nonlinear general relativity could at least explain the separated states of macroscopic systems in the world. A measuring apparatus is a macroscopic system, and the measurement process is irreversible far from thermal equilibrium. Thus, an explanation could only succeed in a unified nonlinear theory. Even the generalization of Schrödinger's wave mechanics to quantum field theory is already nonlinear. In quantum field theory, field functions are replaced by field operators in the so-called second quantization. The quantum field equation with a two-particle potential, for instance, contains a nonlinear term corresponding to pair creation of elementary particles. In general the reactions of elementary particles in quantum field theory are essentially nonlinear phenomena. The interactions of an elementary particle cause its quantum states to have only a finite duration and thereby to violate the reversibility of time. Thus even the quantum world itself is neither conservative nor linear in general. In system theory, complexity means not only nonlinearity but a huge number of elements with many degrees of freedom [1.2]. All macroscopic systems like stones or planets, clouds or fluids, plants or animals, animal populations or human societies consist of component elements like atoms, molecules, cells or organisms. The behaviour of single elements in complex systems with huge numbers of degrees of freedom can neither be forecast nor traced back. The deterministic description of single elements must be replaced by the evolution of probabilistic distributions.

The second chapter analyzes *Complex Systems and the Evolution of Matter*. Since the presocratics it has been a fundamental problem of natural philosophy to discover how order arises from complex, irregular, and chaotic states of matter. Heraclitus believed in an ordering force of energy (*logos*) harmonizing irregular interactions and creating order states of matter. Modern thermodynamics describes the emergence of order by the mathematical concepts of statistical mechanics. We distinguish two kinds of phase transition (self-organization) for order states: conservative self-organization means the phase transition of reversible structures in thermal equilibrium. Typical examples are the growth of snow crystals or the emergence of magnetisation in a ferromagnet by annealing the system to a critical value of temperature. Conservative self-organization mainly creates order structures with low energy at low temperatures, which are described by a Boltzmann distribution. Dissipative self-organization is the phase transition of irreversible structures far from thermal equilibrium [1.3]. Macroscopic patterns arise from the complex nonlinear cooperation of microscopic elements when the energetic interaction of the dissipative ("open") system with its environment reaches some critical value. Philosophically speaking, the stability of the emergent structures is guaranteed by some balance of nonlinearity and dissipation. Too much nonlinear interaction or dissipation would destroy the structure.

As the conditions of dissipative phase transition are very general, there is a broad variety of interdisciplinary applications. A typical physical example is the laser. In chemistry, the concentric rings or moving spirals in the Belousov-Zhabotinski (BZ) reaction arise when specific chemicals are poured together with a critical value. The competition of the separated ring waves show the nonlinearity of these phenomena very clearly, because in the case of a superposition principle the ring waves would penetrate each other like optical waves.

The phase transitions of nonlinear dissipative complex systems are explained by synergetics. In a more qualitative way we may say that old structures become unstable and break down by changing control parameters. On the microscopic level the stable modes of the old states are dominated by unstable modes (Haken's "slaving principle") [1.4]. They determine order parameters which describe the macroscopic structure and patterns of systems. There are different final patterns of phase transitions corresponding to different attractors. Different attractors may be pictured as a stream, the velocity of which is accelerated step by step. At the first level a homogeneous state of equilibrium is shown ("fixed point"). At a higher level of velocity the bifurcation of two or more vortices can be observed corresponding to periodic and quasi-periodic attractors. Finally the order decays into deterministic chaos as a fractal attractor of complex systems. Philosophically, I want to underline that in synergetics the microscopic description of matter is distinguished from the macroscopic order states. Thus the synergetic concept of order reminds me of Heraclitus" "logos" or Aristotle's "form" which produces the order states of nature in a transformative process of matter. But, of course, in antiquity a mathematical description was excluded.

In a more mathematical way, the microscopic view of a complex system is described by the evolution equation of a state vector where each component depends on space and time and where the components may denote the velocity components of a fluid, its temperature field, or in the case of chemical reactions, concentrations of chemicals. The slaving principle of synergetics allows us to eliminate the degrees of freedom which refer to the stable modes. In the leading approximation the evolution equation can be transformed into a specific form for the nonlinearity which applies to those systems where a competition between patterns occurs. The amplitudes of the leading terms of unstable modes are called order parameters. Their evolution equation describes the emergence of macroscopic patterns. The final patterns ("attractors") are reached by a transition which can be understood as a kind of symmetry breaking [1.5]. Philosophically speaking, the evolution of matter is caused by a process of symmetry breaking which was earlier mentioned by Heraclitus.

The third chapter analyzes *Complex Systems and the Evolution of Life*. In the history of science and philosophy, people believed in a sharp difference between "dead" and "living" matter. Aristotle interpreted life as a power of self-organization (entelechy) driving the growth of plants and animals to their final form. A living system is able to move by itself, while a dead system can only be moved from outside. Life was explained by teleology, i.e., by non-causal ("vital") forces aiming at some goals in nature. In the 18th century Kant showed that self-organization of living organisms cannot be explained by a mechanical system of Newtonian physics. In a famous quotation he said that the Newton for explaining a blade of grass is still lacking. In the 19th century the second law of thermodynamics describes the irreversible movement of closed systems toward a state of maximal entropy or disorder. But how can one explain the emergence of order in Darwin's evolution of life? Boltzmann stressed that living organisms are open dissipative systems in exchange with their environment which do not violate the second law of closed systems. But nevertheless in the statistical interpretation from Boltzmann to Monod the emergence of life can only be a contingent event, a local cosmic fluctuation "at the boundary of universe".

In the framework of complex systems the emergence of life is not contingent, but necessary and lawful in the sense of dissipative self-organization. Only the conditions for the emergence of life (for instance on the planet Earth) may be contingent in the universe. In general, biology distinguishes ontogenesis (the growth of organisms) from phylogenesis (the evolution of species). In any case we have complex dissipative systems the development of which can be explained by the evolution of (macroscopic) order parameters caused by nonlinear (microscopic) interactions of molecules, cells, etc., in phase transitions far from thermal equilibrium. Forms of biological systems (plants, animals, etc.) are described by order parameters. Aristotle's teleology of goals in nature is interpreted in terms of attractors in phase transitions. But no special "vital" or "teleological" forces are necessary. Philosophically, the emergence of life can be explained in the framework of nonlinear causality and dissipative self-organization, although it may be described in a teleological language for heuristic reasons.

I remind the reader that the prebiological evolution of biomolecules was analyzed and simulated by Manfred Eigen et al. Spencer's idea that the evolution of life is characterized by increasing complexity can be made precise in the context of dissipative self-organization. It is well known that Turing analyzed a mathematical model of organisms represented as complex cellular systems. Gerisch, Meinhardt, et al. described the growth of an organism (e.g., a slime mould) by evolution equations for the aggregation of cells. The nonlinear interactions of amebas cause the emergence of a macroscopic organism like a slime mould when some critical value of cellular nutrition in the environment is reached. The evolution of the order parameter corresponds to the aggregation forms during the phase transition of the macroscopic organism. The mature multicellular body can be interpreted as the "goal" or (better) "attractor" of organic growth.

Even the ecological growth of biological populations may be simulated using the concepts of synergetics. Ecological systems are complex dissipative systems of plants or animals with mutual nonlinear metabolic interactions with each other and with their environment. The symbiosis of two populations with their source of nutrition can be described by three coupled differential equations which were already used by Edward Lorenz to describe the development of weather in meteorology. In the 19th century the Italian mathematicians Lotka und Volterra described the development of two populations in ecological competition. The nonlinear interactions of the two complex populations are determined by two coupled differential equations of prey and predator species. The evolution of the coupled systems have stationary points of equilibrium. The attractors of evolution are periodic oscillations (limit cycles).

The theory of complex systems allows us to analyze the nonlinear causality of ecological systems in nature. Since the industrial revolution human society has become more and more involved in the ecological cycles of nature. But the complex balance of natural equilibria is highly endangered by the linear mode of traditional industrial production. People assumed that nature contains endless sources of energy, water, air, etc., which can be used without disturbing the natural balance. Industry produces an endless flood of goods without considering their synergetic effects like the ozone hole or waste utilization. The evolution of life is transformed into the evolution of human society.

Perhaps the most speculative interdisciplinary application of complex systems is discussed in the fourth chapter, *Complex Systems and the Evolution of Mind-Brain*. In the history of philosophy and science there have been many different suggestions for solutions to the mind-body problem. Materialistic philosophers like Democritus, Lamettrie, et al., proposed to reduce mind to atomic interactions. Idealists like Plato, Penrose, et al. emphasized that mind is completely independent of matter and brain. For Descartes, Eccles, et al. mind and matter are separate substances interacting with each other. Leibniz believed in a metaphysical parallelism of mind and matter because they cannot interact physically. According to Leibniz mind and matter are supposed to exist in "pre-established harmony" like two synchronized clocks. Modern philosophers of mind like Searle defend a kind of evolutionary naturalism. Searle argues that mind is characterized by intentional mental states which are intrinsic features of the human brain's biochemistry and which therefore cannot be simulated by computers.

But the theory of complex systems cannot be reduced to these more or less one-sided positions. The complex system approach is an interdisciplinary methodology to deal with nonlinear complex systems like the cellular organ known as the brain. The emergence of mental states (for instance pattern recognition, feelings, thoughts) is explained by the evolution of (macroscopic) order parameters of cerebral assemblies which are caused by nonlinear (microscopic) interactions of neural cells in learning strategies far from thermal equilibrium. Cell assemblies with mental states are interpreted as attractors (fixed points, periodic, quasi-periodic, or chaotic) of phase transitions.

If the brain is regarded as a complex system of neural cells, then its dynamics is assumed to be described by the nonlinear mathematics of neural networks. Pattern recognition, for instance, is interpreted as a kind of phase transition by analogy with the evolution equations which are used for pattern emergence in physics, chemistry, and biology. Philosophically, we get an interdisciplinary research program that should allow us to explain neurocomputational self-organization as a natural consequence of physical, chemical, and neurobiological evolution by common principles. As in the case of pattern formation, a specific pattern of recognition (for instance a prototype face) is described by order parameters to which a specific set of features belongs. Once some of the features which belong to the order parameter are given (for instance a part of a face), the order parameter will complement these with the other features so that the whole system acts as an associative memory (for instance the reconstruction of a stored prototype face from an initially given part of that face). According to Haken's slaving principle the features of a recognized pattern correspond to the enslaved subsystems during pattern formation.

But what about the emergence of consciousness, self-consciousness, and intentionality? In synergetics we have to distinguish between external and internal states of the brain. In external states of perception and recognition, order parameters correspond to neural cell assemblies representing patterns of the external world. Internal states of the brain are nothing other than self-referential states, i.e., mental states referring to mental states and not to external states of the world. In the traditional language of philosophy we say that humans are able to reflect on themselves (self-reflection) and to refer external situations of the world to their own internal state of feeling and intentions (intentionality). In recent neurobiological inquiries, scientists speculate that the emergence of consciousness and self-consciousness depends on a critical value of the production rate for "meta-cell-assemblies", i.e., cell-assemblies representing cell-assemblies which again represent cell-assemblies, etc., as neural realization of self-reflection. But this hypothesis (if successful) could only explain the structure of emergent features like consciousness. Of course, mathematical evolution equations of cell assemblies do not enable us to feel like our neighbour. In this sense – this is the negative message – science is blind. But otherwise – this is the positive message – personal subjectivity is saved: Calculations and computer-assisted simulations of nonlinear dynamics are limited in principle.

Anyway, the complex system approach solves an old metaphysical puzzle which was described by Leibniz in the following picture: If we imagine the brain as a big machine which we may enter like the internal machinery of a mill, we shall only find its single parts like the cog wheels of the mill and never the mind, not to mention the human soul. Of course, on the microscopic level we can only describe the development of neurons as cerebral parts of the brain. But, on the macroscopic level, the nonlinear interactions in the complex neural system cause the emergence of cell assemblies referring to order parameters which cannot be identified with the states of single cerebral cells. The whole is not the sum of its parts.

It is obvious that the complex system approach delivers solutions of the mind-body problem which are beyond the traditional philosophical answers of idealism, materialism, physicalism, dualism, interactionism, etc. Concerning the distinction between so-called natural and artificial intelligence, it is important to see that the principles of nonlinear complex systems do not depend on the biochemistry of the human brain. The human brain is a "natural" model of these principles in the sense that the cerebral complex system is a product of physical and biological evolution. But other ("artificial") models produced by human technology are possible, although there will be technical and ethical limits to their realization.

In Chap. 5 we discuss *Complex Systems and the Evolution of Artificial Intelligence*. The development of neural and synergetic computers is described and contrasted with Turing machines and knowledged-based systems. In synergetic computers, the order parameter equations allow a new kind of (non-Hebbian) learning, namely a strategy to minimize the number of synapses. In contrast to neurocomputers of the spin-glass type (for instance Hopfield systems), the neurons are not threshold elements but rather perform simple algebraic manipulations like multiplication and addition. Beside deterministic homogeneous Hopfield networks there are so-called Boltzmann machines with a stochastic network architecture of non-deterministic processor elements and a distributed knowledge representation which is described mathematically by an energy function. While Hopfield systems use a Hebbian learning strategy, Boltzmann machines favour a backpropagation strategy (Widrow-Hoff rule) with hidden neurons in a many-layered network.

In general it is the aim of a learning algorithm to diminish the information-theoretic measure of the discrepancy between the brain's internal model of the world and the real environment via self-organization. The recent revival of interest in the field of neural networks is mainly inspired by the successful technical applications of statistical mechanics and nonlinear dynamics to solid state physics, spin glass physics, chemical parallel computers, optical parallel computers, and – in the case of synergetic computers – to laser systems. Other reasons are the recent development of computing resources and the level of technology which make a computational treatment of nonlinear systems more and more feasible. Philosophically, traditional topics of epistemology like perception, imagination, and recognition may be discussed in the interdisciplinary framework of complex systems.

An important application of the complex system approach is neurobionics and medicine. The human brain is not only a cerebral computer as a product of natural evolution, but a central organ of our body which needs medical treatment, healing, and curing. Neurosurgery, for instance, is a medical discipline responsible for maintaining the health of the biological medium of the human mind. The future well-being of the brain-mind entity is an essential aim of neurobionics. In recent years new diagnostic procedures and technical facilities for transplantation have been introduced which are based on new insights into the brain from complex dynamical systems. In this context a change of clinical treatment is unavoidable. Neural and psychic forms of illness can be interpreted as complex states in a nonlinear system of high sensitivity. Even medical treatments must consider the high sensitivity of this complex organ. Another more speculative aspect of the new technology is cyberspace. Perception, feeling, intuition, and fantasy as products of artificial neural networks? Virtual reality has become a keyword in modern philosophy of culture.

After the movement through matter, life, mind-brain and artificial intelligence the book finishes in a Hegelian grand synthesis with the sixth chapter, *Complex Systems and the Evolution of Human Society*. In social sciences one usually distinguishes strictly between biological evolution and the history of human society. The reason is that the development of nations, markets, and cultures is assumed to be guided by the intentional behavior of humans, i.e., human decisions based on intentions, values, etc. From a microscopic viewpoint we may, of course, observe single individuals with their intentions, beliefs, etc. But from a macroscopic view the development of nations, markets, and cultures is not only the sum of its parts. Mono-causality in politics and history is, as we all know, a false and dangerous way of linear thinking. Synergetics seems to be a successful strategy to deal with complex systems even in the humanities. Obviously it is not necessary to reduce cultural history to biological evolution in order to apply synergetics interdisciplinarily. Contrary to any reductionistic kind of naturalism and physicalism we recognize the characteristic intentional features of human societies. Thus the complex system approach may be a method of bridging the gap between the natural sciences and the humanities that was criticized in Snow's famous "two cultures".

In the framework of complex systems the behaviour of human populations is explained by the evolution of (macroscopic) order parameters which is caused by nonlinear (microscopic) interactions of humans or human subgroups (states, institutions, etc.). Social or economic order is interpreted by attractors of phase transitions. Allen et al. analyze the growth of urban regions. From a microscopic point of view the evolution of populations in single urban regions is mathematically described by coupled differential equations with terms and functions referring to the capacity, economic production, etc., of each region. The macroscopic development of the whole system is illustrated by computer-assisted graphics with changing centers of industrialization, recreation, etc., caused by nonlinear interactions of individual urban regions (for instance advantages and disadvantages of far and near connections of transport, communication, etc.). An essential result of the synergetic model is that urban development cannot be explained by the free will of single persons. Although people of local regions are acting with their individual intentions, plans, etc., the tendency of the global development is the result of nonlinear interactions.

Another example of the interdisciplinary application of synergetics is Weidlich's model of migration. He distinguishes the micro-level of individual decisions and the macro-level of dynamical collective processes in a society. The probabilistic macro-processes with stochastic fluctuations are described by the master equation of human socioconfigurations. Each component of a socioconfiguration refers to a subpopulation with a characteristic vector of behavior. The macroscopic development of migration in a society could be illustrated by computer-assisted graphics with changing centers of mixtures, ghettos, wandering, and chaos which are caused by nonlinear interactions of social subpopulations. The differences between human and non-human complex systems are obvious in this model. On the microscopic level human migration is intentional (i.e., guided by considerations of utility) and nonlinear (i.e., dependent on individual and collective interactions). A main result of synergetics is again that the effects of national and international migration cannot be explained by the free will of single persons. I think migration is a very dramatic topic today, and demonstrates how dangerous linear and mono-causal thinking may be. It is not sufficient to have good intentions without considering the nonlinear effects of single decisions. Linear thinking and acting may provoke global chaos, although we locally act with the best intentions.

It is a pity to say that in economics linear models are still dominant. From a qualitative point of view, Adam Smith's model of a free market can already be explained by self-organization. Smith underlined that the good or bad intentions of individuals are not essential. In contrast to a centralized economical system, the equilibrium of supply and demand is not directed by a program-controlled central processor, but is the effect of an "invisible hand" (Smith), i.e., nothing but the nonlinear interaction of consumers and producers. The recent interest of economists in nonlinear dissipative systems is inspired by the growth of knowledge-based high-tech industries with positive feedback (i.e., increasing production depending on increasing know-how, such as electronics, computer industries, etc.) in contrast to the old industries with negative feedback (i.e., decreasing production depending on limited resources like coal or steel). In general, economic processes are very complex and demand nonlinear dissipative models. Recall the different attractors from economic cycles to financial chaos which can only be explained as synergetic effects by nonlinear interactions of consumers and producers, fiscal policy, stock market, unemployment, etc. Even in management possible complex models are discussed in order to support creativity and innovation by nonlinear cooperation at all levels of management and production. The synergetic analysis shows that economic processes are to be embedded in the ecological cycles of nature. It must be the main intention of our politics to achieve a nonlinear complex system of economics and ecology that maintains a balance between human society and nature.

Evidently there are some successful strategies to handle nonlinear complex systems. We shall discuss examples of applications in quantum physics, hydrodynamics, chemistry, and biology, as well as economics, sociology, neurology, and AI. What is the reason behind the successful applications in the natural sciences and humanities? The complex system approach is not reduced to special natural laws of physics, although its mathematical principles were discovered and at first successfully applied in physics (for instance to the laser). Thus, it is an interdisciplinary methodology to explain the emergence of certain macroscopic phenomena via the nonlinear interactions of microscopic elements in complex systems. Macroscopic phenomena may be forms of light waves, fluids, clouds, chemical waves, biomolecules, plants, animals, populations, markets, and cerebral cell assemblies which are characterized by order parameters (Table 1.1).

Table 1.1. Interdisciplinary applications of nonlinear complex systems

DISCIPLINE	SYSTEM	ELEMENTS	DYNAMICS	ORDER PARAMETER
quantum physics	laser	atoms (photons)	phase transition	form of lightwaves
hydrodynamics	fluids	molecules	phase transition	form of fluids
meteorology	weather	molecules	phase transition	form of clouds
geology	lava	molecules	phase transition	hexagonal form (Bénard cells)
chemistry	BZ-reaction	molecules	phase transition	forms of spiral or rings (chemical waves)
biology	biomolecules	molecules	phase transition	structural form
	organisms	cells	organic growth	organic forms (plants, animals)
	population	organisms	evolution of populations	form of population (interactional form)
economics	economical systems	consumer, producers etc.	mechanism of market (f.i. supply and demand)	form of market (interactional form)
sociology	societies	humans, institutions etc.	history	interactional form
neurology (psychology)	brain	neurons	recognition (learning)	forms of neural cell assemblies representing external or internal ('self-referential') states
artificial intelligence (AI)	neural AI-networks	AI-neurons	learning algorithms	forms of neural AI-cell assemblies representing external or internal ('self-referential') states

Philosophically it is important to see that order parameters are not reduced to the microscopic level of atoms, molecules, cells, organisms, etc., of complex systems. In some cases they are measurable quantities (for instance the field potential of a laser). In other cases they are qualitative properties (for instance geometrical forms of patterns). Nevertheless, order parameters are not mere theoretical concepts of mathematics without any reference to reality. Actually they represent properties of real macroscopic phenomena, such as field potentials, social or economical power, feelings or even thoughts. Who will deny that feelings and thoughts can change the world? But the complex system approach is not a metaphysical process ontology. Synergetic principles (among others) provide a heuristic scheme to construct models of nonlinear complex systems in the natural sciences and the humanities. If these models can be mathematized and their properties quantified, then we get empirical models which may or may not fit the data. The slaving principle shows another advantage. As it diminishes the high number of degrees of freedom in a complex system, synergetics is not only heuristic, mathematical, empirical and testable, but economical too. Namely, it satisfies the famous principle of Ockham's razor which tells us to cut away superfluous entities.

Our approach suggests that physical, social, and mental reality is nonlinear and complex. This essential result of synergetic epistemology demands severe consequences of our behavior. As we underlined, linear thinking may be dangerous in a nonlinear complex reality. Recall, as one example, the demand for a well-balanced complex system of ecology and economics. Our physicians and psychologists must learn to consider humans as complex nonlinear entities of mind and body. Linear thinking may fail to yield a successful diagnosis. Local, isolated, and "linear" therapies of medical treatment may cause negative synergetic effects. In politics and history, we must remember that mono-causality may lead to dogmatism, intolerance, and fanaticism. As the ecological, economic, and political problems of mankind have become global, complex, and nonlinear, the traditional concept of individual responsibility is questionable. We need new models of collective behavior depending on the different degrees of our individual faculties and insights. In short: The complex system approach demands new consequences in *epistemology* and *ethics*. Finally, it offers a chance to prevent chaos in a nonlinear complex world and to utilize the creative possibilities of synergetic effects.

2. Complex Systems and the Evolution of Matter

How can order arise from complex, irregular, and chaotic states of matter? In classical antiquity philosophers tried to take the complexity of natural phenomena back to first principles. Astronomers suggested mathematical models in order to reduce the irregular and complex planetary orbits as they are experienced to regular and simple movements of spheres. Simplicity was understood, still for Copernicus, as a feature of truth (Sect. 2.1). With Newton and Leibniz something new was added to the theory of kinetic models. The calculus allows scientists to compute the instaneous velocity of a body and to visualize it as the tangent vector of the body's trajectory. The velocity vector field has become one of the basic concepts in dynamical systems theory. The cosmic theories of Newton and Einstein have been described by dynamical models which are completely deterministic (Sect. 2.2).

But Poincaré discovered that those models may be non-computable in the long run (the many-body-problem). Even in a fully deterministic world, the assumption of a Laplacean demon which can calculate the universe in the long run was exposed as an illusionary fiction. Chaos can arise not only in heaven, but also in the quantum world (as quantum chaos) (Sect. 2.3). From a methodological point of view, nonlinearity is a necessary but not sufficient condition of chaos. It also allows the emergence of order. In the framework of modern physics, the emergence of the structural variety in the universe from elementary particles to stars and living organisms is modeled by phase transitions and symmetry breaking of equilibrium states (Sect. 2.4). But even after Hawking we do not have a complete theory explaining the evolution of matter with its increasing complexity. The presocratic wondering that "there is something and not nothing" is not dissolved.

2.1 Aristotle's Cosmos and the Logos of Heraclitus

Since the presocratics it has been a fundamental problem of natural philosophy to discover how order arises from complex, irregular, and chaotic states of matter [2.1]. What the presocratic philosophers did was to take the complexity of natural phenomena as it is experienced back to "first origins" (ἀρχή), "principles" or a certain order. Let us look at some examples. Thales of Miletus (625–545 B.C.), who is said to have proven the first geometric theorems, is also the first philosopher of nature to believe that only material primary causes could be the original causes of all things. Thales assumes water, or the wet, as the first cause. His argument points to the observation that nourishment and the seeds of all beings are wet and the natural substratum for wet things is water.

Anaximander (610–545 B.C.), who is characterized as Thales' student and companion, extends Thales' philosophy of nature. Why should water be the first cause of all this? It is only one of many forms of matter that exist in uninterrupted tensions and opposites: heat versus cold and wetness versus dryness ... Therefore Anaximander assumes that the "origin and first cause of the existing things" is a "boundlessly indeterminable" original matter (ἄπειρον) out of which the opposed forms of matter have arisen. Accordingly we have to imagine the "boundlessly indeterminable" as the primordial state in which matter was boundless, without opposites, and, therefore, everywhere of the same character. Consequently, it was an initial state of complete homogeneity and symmetry. The condition of symmetry is followed by symmetry breaking, from which the world arises with all its observable opposites and tensions:

> The everlasting generative matter split apart in the creation of our world and out of it a sphere of flame grew around the air surrounding the earth like the bark around a tree; then, when it tore apart and bunched up into definite circles, the sun, moon and stars took its place. [2.2]

The ensuing states of matter that Anaximander described in his cosmogeny were therefore by no means chaotic; instead they were determined by new partial orders. The fascination with Anaximander increases when one reads his early ideas of biological evolution. He assumes that the first human beings were born from sea animals whose young are quickly able to sustain themselves, as he had observed in the case of certain kinds of sharks. A century later searches were already being made for fossils of sea animals as evidence of the rise of humans from the sea. The third famous Milesian philosopher of nature is Anaximenes (†525 B.C.), who is thought to have been a companion of Anaximander. He regards change as the effect of the external forces of condensation and rarefaction. In his view, every form of matter can serve as basic. He chooses air (ἀέρα):

> And rarefied, it became fire; condensed, wind; then cloud; further, by still stronger condensation, water; then earth; then stones; but everything else originated by these. He, too, assumed eternal motion as the origin of transformation. – What contracts and

condenses matter, he said is (the) cold; by contrast, what thins and slackens is (the) warm. [2.3]

Thus Anaximenes assumes external forces by which the various states of matter were produced out of a common original matter and were transformed into each other.

Heraclitus of Ephesus (ca. 500 B.C.), "the dark one", as he was often called, is of towering significance for our theme. His language is indeed esoteric, more phrophetic than soberly scientific, and full of penetrating metaphors. He took over from Anaximander the doctrine of struggle and the tension of opposites in nature. The original matter, the source of everything, is itself change and therefore is identified with fire:

> The ray of lightning (i.e., fire) guides the All. – This world order which is the same for all was created neither by one of the gods nor by one of the humans, but it was always, and is, and will be eternally living fire, glimmering and extinguishing according to measures. [2.4]

Heraclitus elaborated further on how all states of matter can be understood as distinguishable forms of the original matter, fire. In our time the physicist Werner Heisenberg declared:

> At this point we can interpose that in a certain way modern physics comes extraordinarily close to the teaching of Heraclitus. If one substitutes the word "fire", one can view Heraclitus' pronouncements almost word for word as an expression of our modern conception. Energy is indeed the material of which all the elementary particles, all atoms and therefore all things in general are made, and at the same time energy is also that which is moved .... Energy can be transformed into movement, heat, light and tension. Energy can be regarded as the cause of all changes in the world. [2.5]

To be sure, the material world consists of opposite conditions and tendencies which, nevertheless, are held in unity by hidden harmony: "What is opposite strives toward union, out of the diverse there arises the most beautiful harmony (ἁρμονία), and the struggle makes everything come about in this way." [2.6] The hidden harmony of opposites is thus Heraclitus' cosmic law, which he called "logos" (λόγος).

What happens when the struggle of opposites comes to an end? According to Heraclitus, then the world comes to a final state of absolute equilibrium. Parmenides of Elea (ca. 500 B.C.) described this state of matter, in which there are no longer changes and motions in (empty) spaces. Matter is then distributed everywhere equally (homogeneously) and without any preferred direction for possible motion (isotropically). It is noteworthy that infinity is thought to be imperfection and therefore a finite distribution of matter is assumed. In this way Parmenides arrived at the image of a world that represents a solid, finite, uniform material sphere without time, motion or change. The Eleatic philosophy of unchanging being was, indeed, intended as a critique of the Heraclitean philosophy of constant change, which is put aside as mere illusion of the senses. And the later historical impact of the Eleatic philosophy in Plato appears in his critique of the deceptive changes that take place in sensory perception in contrast to the true world of unchangeable being of the Ideas. But from the point of view of philosophy of nature, the world Parmenides described was not necessarily opposite to the teaching of Heraclitus; in his cosmogeny it can be understood entirely as a singular end state of the highest symmetry.

After water, air, and fire were designated as original elements, it was easy to conceive of them as raw materials of the world. Empedocles (492–430 B.C.) took that step and added earth as the fourth element to fire, water, and air. These elements are free to mix and bind in varying proportions, and to dissolve and separate. What, now, according to Empedocles, were the enduring principles behind the constant changes and movements of nature? First there were the four elements, which he thought arose from nature and chance (τύχη), not from any conscious intention. Changes were caused by reciprocal effects among these elements, that is, mixing and separation: "I shall proclaim to you another thing: there is no birth with any of the material things, neither there is an ending in ruinous death. There is only one thing: mixture and exchange of what is mixed" [2.7]. Two basic energies were responsible for these reciprocal effects among the elements; he called them "love" (φιλία) for attraction and "hatred" (νεῖκος) for repulsion. There is an analogy in the yin–yang dualism of Chinese philosophy. Empedocles taught a constant process of transformation, i.e., combination and separation of the elements, in which the elements were preserved. He did not imagine these transformation processes to be at all mechanical (as the later atomists did), but rather physiological, in that he carried over processes of metabolism in organisms to inanimate nature.

In his medical theories, equilibrium is understood to be a genuinely proportional relationship. Thus, health means a particular balance between the opposite components and illness arises as soon as one of them gets the upper hand. If we think of modern bacteriology with its understanding of the antibodies in the human body, then this view of Empedocles proves to be surprisingly apt.

Anaxagoras (499–426 B.C.) advocated what was in many regards a refinement of his predecessors' teaching. Like Empedocles he developed a mixing theory of matter. But he replaced Empedocles' four elements with an unlimited number of substances that were composed of seed particles (σπέρματα) or equal-sized particles (ὁμοιομερῆ (σώματα)). They were unlimited in their number and smallness, i.e., matter was assumed to be infinitely divisible. The idea of a granulated continuum comes forceably to mind. Anaxagoras also tried to explain mixtures of colors in this way, when he said that snow is, to a certain degree, black, although the whiteness predominates. Everything was contained in each thing, and there were only predominances in the mixing relationships.

More distinctly than some of his predecessors, Anaxagoras tried in his philosophy of nature to give physical explanations for the celestial appearances and motions that were described only kinematically in the mathematical astronomy of the Greeks. So in his cosmology he proceeded from a singular initial state: a homogeneous mixture of matter. An immaterial original power, which Anaxagoras called "spirit" ($vo\tilde{\upsilon}\varsigma$), set this mixture into a whirling motion which brought about a separation of the various things depending on the speed of each of them. Earth clumped together in the middle of the vortex, while heavier pieces of stone were hurled outward and formed the stars. Their light was explained by the glow of their masses, which was attributed to their fast speed. Anaxagoras' vortex theory appears again in modern times with Descartes, and then in more refined form in the Kant–Laplace theory of the mechanical origin of the planetary system.

In modern natural sciences atomism has proved to be an extremely successful research program. In the history of philosophy the atomic theory of Democritus is often presented as a consequence of Heraclitus' philosophy of change and Parmenides' principle of unchanching being. The Democritean distinction between the "full" and the "empty", the smallest indestructable atoms ($\mathring{\alpha}\tau o\mu o\varsigma$) and empty space, corresponded to Parmenides' distinction between "being" and "not-being". Heraclitean complexity and change was derived from distinguishable reconfigurations of the atoms. Empty space was supposed to be homogeneous and isotropic.

Atoms differ in their form ($\mu o\rho\varphi\acute{\eta}$), their position ($\vartheta\acute{\varepsilon}\sigma\iota\varsigma$), and their diverse configurations ($\tau\acute{\alpha}\xi\iota\varsigma$) in material combinations. The configuration of the atoms for the purpose of designation is compared with the sequence of letters in words, which has led to the presumption that atomistic ideas were developed only in cultures with phonetic alphabets. In fact, in China, where there was no phonetic alphabet but instead ideographic characters, the particle idea was unknown and a field-and-wave conception of the natural processes prevailed. The Democritean atoms move according to necessity ($\mathring{\alpha}\nu\acute{\alpha}\gamma\kappa\eta$) in a constant whirl ($\delta\tilde{\iota}\nu o\varsigma$ or $\delta\acute{\iota}\nu\eta$). Here, by contrast with later Aristotelian notions, motion means only change of location in empty space. All phenomena, all becoming and perishing, result from combination ($\sigma\acute{\upsilon}\gamma\kappa\rho\iota\sigma\iota\varsigma$) and separation ($\delta\iota\acute{\alpha}\kappa\rho\iota\sigma\iota\varsigma$). Aggregate states of matter, such as gaseous, liquid, or solid, are explained by the atoms' differing densities and potentialities for motion. In view of today's crystallography, the Democritean idea that even atoms in solid bodies carry out oscillations in place is noteworthy.

Plato, in his dialogue *Timaeus*, introduced the first mathematical model of atomism. The changes, mixings, and separations on earth of which the pre-socratics had spoken were to be traced to unchangeable mathematical regularities. In Empedocles' four elements, namely fire, air, water and earth, a classification was at hand that was immediately accessible to experience. Theatetus made a complete determination of all the regular bodies that are possible in 3-dimensional (Euclidian) space: tetrahedra, octahedra, icosahedra, cubes and dodecahedra. Therefore what Plato proposed to do amounted to interpreting Empedocles' four elements with these geometric building blocks.

Plato consciously avoided the Democritean designation "atom" for his elements, since they can be decomposed into separate plan figures. Thus tetrahedra, octahedra and icosahedra have faces consisting of equilateral triangles which, when they are bisected, yield right-angled triangles with sidelenghts 1, 2 and $\sqrt{3}$, while the square faces of cubes, when bisected, yield right-angled triangles with side lengths 1, 1 and $\sqrt{2}$. A consequence is that "fluids" like water, air and fire can combine with each other whereas a solid made of earth building blocks, because of its different triangles, can only be converted into another solid.

Then Plato developed a kind of elementary particle physics in which the specific elements are transformed into each other and "reciprocal effects" can take place with the "elementary particles" (i.e., the corresponding component triangles) according to geometric laws. Transformation of the elements results, for example, from their being cut open along the edges. Plato made this possibility dependent on the acuteness of the angles of the solid. The more acute plane angles can cleave polyhedra which have a regular angle. Thus, in sequence, every tetrahedron, every cube, every octahedron, every icosahedron can, in each case, cleave the following polyhedron, but not the previous one or polyhedra of the same sort. The conclusion for the philosophy of nature is that fire can separate or dissolve all elements; earth, only air and water; air, only water.

Plato stated categorically that the elements are not all of the same size. For instance, in order to be able to explain that fire can cause water in solid form to change into water in liquid form, he maintained that in the liquid state the elements are smaller and more mobile, while in the solid state they are larger.

The escape from fire is called cooling and the state after eradicating fire, solidification. Fire and air can pass through the gaps in earth building blocks (cubes) without hindrance, without dissolution of the earth elements. Condensed air cannot be dissolved without destroying the element. Condensed air, namely, means an accumulation of octahedra under the best surface configurations possible. Even fire would not be able to penetrate into the necessarily remaining gaps, whose plane angles are smaller than those of all elements, without destroying the octahedra. In the case of water, only fire is capable of breaking the strongest condensation. The gaps between adjacent icosahedra form plane angles which do not admit penetration by either earth or air. Only fire (tetrahedra) can penetrate and dissolve the combination.

Indeed, Plato developed an internally consistent mathematical model by which various aggregate states and reciprocal effects of substances could be explained if one accepted his – albeit more or less arbitrary – initial conditions for interpretation of the elements. Naturally, a number of the consequences for the philosophy of nature are strange and ridiculous. And yet we have here the first attempt in the history of sciences to explain matter and its states by simple geometric laws. A high point up to now in this developement is modern elementary particle physics. Heisenberg made this observation about it: "... The elementary particles have the form Plato ascribed to them because it is the mathematically most beautiful and simplest form. Therefore the ultimate root of phenomena is not matter, but instead mathematical law, symmetry, mathematical form" [2.8]. In Antiquity and the Middle Ages Plato's mathematical atomism gained little support. The basic problem, for his successors, in his geometric theory of matter was already evident in the dialogue *Timaeus*. How are the functions of living organisms to be explained? The suggestion that certain corporeal forms are as they are in order to fulfill certain physiological purposes (e.g., the funnel shape of the gullet for assimilation of food) cannot, in any case, be derived from the theory of regular solids. In addition, the idea of explaining the changing and pulsating processes of life on the basis of the "rigid" and "dead" figures of geometry must have seemed thoroughly unnatural, speculative, and farfetched to the contemporaries of that time. Contemporaries of our time still have difficulties understanding the detour that today's scientific explanations take through complicated and abstract mathematical theories. This is where the Aristotelian philosophy of nature begins.

Aristotle formulated his concept of a balance or "equilibrium" in nature chiefly on the basis of the ways in which living organisms such as plants and animals function. The process and courses of life are known to us from everyday experience. What is more obvious than to compare and explain the rest of the world, which is unknown and strange, with the familiar? According to Aristotle, the task of science is to explain the principles and functions of nature's complexity and changes. This was a criticism of those philosophers of nature who identified their principles with individual substances. The individual plant or the individual animal was not simply the sum of its material building blocks. Aristotle called the general, which made the individual being what it was, form ($εἶδος$). What was shaped by form was called matter ($ὕλη$). Yet form and matter did not exist in themselves, but were instead principles of nature derived by abstraction. Therefore Matter was also characterized as the potential ($δύναμις$) for being formed. Not until matter is formed does reality ($ἐνέργεια$) come into being.

The real living creatures that we observe undergo constant change. Here Heraclitus was right and Parmenides, for whom changes were illusory, was wrong. Changes are real. Yet according to Aristotle, Heraclitus was wrong in identifying changes with a particular substance (fire). Aristotle explained those changes by a third principle along with matter and form, namely, the lack of form ($\sigma\tau\acute{\epsilon}\rho\eta\sigma\iota\varsigma$), which was to be nullified by an adequate change. The young plant and the child are small, weak and immature. They grow because in accordance with their natural tendencies (form), they were meant to become big, strong, and mature. Therefore it was determined that movement ($\kappa\acute{\iota}\nu\eta\sigma\iota\varsigma$) in general was change, transition from possibility to reality, "actualization of potential" (as people in the Middle Ages were to say). The task of physics was to investigate movement in nature in this comprehensive sense. Nature ($\varphi\acute{\upsilon}\sigma\iota\varsigma$) – in contrast to a work of art produced by man, or a technical tool – was understood to be everything that carried the principle of movement within itself. If the Aristotelian designations make us think, first of all, of the life processes of plants, animals, and people as they present themselves to us in everyday experience, these designations seem to us to be thoroughly plausible and apposite. Nature is not a stone quarry from which one can break loose individual pieces at will. Nature itself was imagined to be a rational organism whose movements were both necessary and purposeful. Aristotle distinguished two sorts of movement, namely quantitative change by increase or decrease in magnitude, qualitative change by alteration of characteristics, and spatial change by change of location. Aristotle designated four aspects of causality as the causes of changes. Why does a plant grow? It grows (1) because its material components make growth possible (*causa materialis*), (2) because its physiological functions determine growth (*causa formalis*), (3) because external circumstances (nutrients in the earth, water, sunlight, etc.) impel growth (*causa efficiens*), (4) because, in accordance with its final purpose, it is meant to ripen out into the perfect form (*causa finalis*).

Aristotle then employed these same principles, which are obviously derived from the life cycles of plants, animals, and humans, to explain matter in the narrower sense, that is, what was later called the inorganic part of nature. Here too Aristotle proceeded from immediate experience. What we meet with is not so and so many elements as isolated building blocks of nature. Instead we experience characteristics such as warmth and cold, wetness and dryness. Combination of these yield the following pairs of characteristics which determine the elements: warm–dry (fire), warm–wet (air), cold–wet (water), cold–dry (earth). Warm–cold and wet–dry are excluded as simultaneous conditions. Therefore there are only four elements. This derivation was later criticized as arbitrary, but it shows the Aristotelian method, namely to proceed not from abstract mathematical models, but instead directly from experience. Fire, air, water, and earth are contained more or less, more intensively or less intensively, in real bodies and they are involved in constant transformation. According to Aristotle, eliminating the coldness of water by means of warmth results in air, and eliminating the wetness of the air results in fire. The changes of nature are interpreted as maturational and transformational processes.

How could such a predominantly organic philosophy of nature deliver physical explanations for mathematical natural science, insofar as it was extant at that time? There were only two elementary spatial motions – those that proceeded in a straight line and those that proceeded in a circle. Therefore there had to be certain elements to which these elementary motions come naturally. The motions of the other bodies were determined by these elements and their natural motions, depending on which motion predominated with each of them. The most perfect motion was circular motion. It alone could go on without end, which was why it had to be assigned to the imperishable element. This was the fifth element (quintessence), which made up the unchangeable celestial spheres and the stars. The continual changes within the earthly (sublunar) world were to be set off from the unchangeable regularity of the celestial (superlunar) world. These transformational processes were associated with the four elements to which straight-line motion is peculiar, and specifically the straight-line motion toward the center of the world, toward which the heavy elements earth and water strive as their natural locus, and the straight-line motion toward the periphery of the lunar sphere, toward which the light elements strive upwards as their natural locus.

Among the natural motions there was also free fall. But Aristotle did not start out from isolated motions in idealized experimental situations as Galilei did. A falling body is observed in its complex environment without abstraction of frictional ("dissipating") forces. During its free fall a body is sinking in the medium of air like a stone in water. Thus, Aristotle imagines free fall as a hydrodynamical process and not as an acceleration in vacuum. He assumes a constant speed of falling v, which was directly proportional to the weight p of the body and inversely to the density d of the medium (e.g., air), thus in modern notation $v \sim p/d$. This equation of proportionality at the same time provided Aristotle with an argument against the void of atomists. In a vacuum with the density $d = 0$, all bodies would have to fall infinitely fast, which obviously did not happen.

A typical example for a (humanly) forced motion is throwing, which, again, is regarded in its complex environment of "dissipative" forces. According to Aristotle a nonliving body can move only as a result of a constant external cause of motion. Think of a cart on a bumpy road in Greece, which comes to a stop when the donkey (or the slave) stops pulling or pushing. But why does a stone keep moving when the hand throwing it lets go? According to Aristotle, there could be no action at a distance in empty space. Therefore, said Aristotle, the thrower imparts a movement to the continuous medium of the stone's surroundings, and this propels the stone farther. For the velocity v of a pulling or pushing motion, Aristotle asserted the proportionality $v \sim K/p$ with the applied force K. Of course, these are not mathematical equations relating measured quantities, but instead proportionalities of qualitative determinants, which first emerged in this algebraic notation in the peripatetic physics of the Middle Ages. Thus in Aristotelian dynamics, in contrast to Galilean–Newtonian dynamics, every (straight-line) change of position required a cause of motion (force). The medieval theory of impetus altered Aristotelian dynamics by attributing the cause of motion to an "impetus" within the thrown body, rather than to transmission by an external medium.

How did peripatetic dynamics explain the cosmic laws of heaven? The central symmetry of the cosmological model was based on the (unforced) circular motion of the spheres, which was considered natural for the "celestial" element, and on the theory of the natural locus in the centerpoint of the cosmos. Ptolemy was still to account for the position of the earth on the basis of the isotropy of the model and by a kind of syllogism of sufficient reason. Given complete equivalence of all directions, there was no reason why the earth should move in one direction or another.

It was Aristotle's teacher Plato who presented a centrally-symmetrical model with the earth in the center; around it the whole sky turns to the right around the celestial axis, which goes through the earth. Sun, Moon, and planets turn to the left on spheres that have different distances from the earth in the sequence Moon, Mercury, Venus, Sun, Mars, Jupiter, and Saturn. The outermost shell carries the sphere of the fixed stars. According to the Platonic–Pythagorean conception, the rotational periods are related to each other by whole numbers. There is a common multiple of all rotational times, at the end of which all the planets are exactly in the same place again. The motion of each one produces a sound, so that the tunes of the movements of the spheres jointly form a harmony of the spheres in the sense of a well-ordered musical scale. Geometry, arithmetic, and aesthetic symmetries of the cosmos ring through the universe in a harmonious music of the spheres. Soon this emphatically symmetrical model of the cosmos was called into question by exact observations. A difficult problem was presented by the irregular planetary orbits, especially their retrograde movements. The irregularities in the sky were disquieting, especially for philosophers in the Pythagorean tradition, who were accustomed to comprehending the heaven – in contrast to the earth – as the realm of eternal symmetry and harmony.

Plato posed a famous question in order to reduce the complexity of motions in the heaven: by means of what regular, ordered circular movements could the phenomena of the planets be "saved", i.e., kinematically explained? An exact model of the observed curves was achieved when Apollonius of Perga (ca. 210 B.C.) recommended that the common center of the spheres be given up. But the spherical form of planetary movement and the equal speed of the spheres were to be retained. According to this proposal, the planets rotate uniformly on spheres (epicycles), whose imagined centers move uniformly on great circles (deferents) around the centerpoint (the earth). By appropriately proportioning the speed and diameter of the two circular motions and by varying their directions of motion, it was possible to produce an unanticipated potential for curves, and these found partial application in astronomy from Kepler to Ptolemy also. The spherical symmetry of the individual models was therefore preserved, even if they no longer had a common center, but various different centers.

The following examples of the epicycle-deferent technique show what a multiplicity of apparent forms of motion can be created by appropriately combining uniform circular motions [2.9]. This makes the Platonic philosophy more comprehensible in its view that behind the changes in phenomena there are the eternal and unchangeable forms. In Fig. 2.1 an elliptical orbit is produced by combining a deferent motion and an epicycle motion. Figure 2.2 shows a closed cycloid. In this way, changing distances between planets and the earth can also be represented. In principle, even angular figures can be produced. When the epicycle diameter approaches the deferent diameter, an exact straight line results. Even regular triangles and rectangles can be produced by means of appropriate combinations of an epicycle motion and a deferent motion, if one changes the speed of the east-west motion of a planet that is moving on an epicycle with a west-east motion.

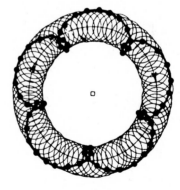

Fig. 2.1. Deferent-epicycle model of an ellipse

Fig. 2.2. Deferent-epicycle model of a cycloid

If one lets the celestial body circle on a second epicycle whose midpoint moves on the first epicycle, one can produce a multiplicity of elliptical orbits, reflection-symmetric curves, periodic curves, and also non-periodic and asymmetric curves. From a purely mathematical and kinetic standpoint, Plato's problem of "saving the phenomena" is completely solved. In principle, therefore, Plato's reduction of complexity in the sense of uniform circular motion (modified by Apollonius and Ptolemy) could influence the sciences right up until today. In any case, it cannot be disproved by phenomenological description of curved paths. In particular, from this standpoint not only the reversed roles of the earth and the sun in the so-called Copernican revolution, but also Kepler's change from circular to elliptical orbits, seem secondary, since both initiatives can be traced back to a combination of circular motions in accordance with the epicycle-deferent technique. This poses two questions: (1) How is the assertion mathematically substantiated? (2) If it is substantiated, why does it not play a role in modern scientific applications of curve theory? In order to answer the first question exactly and generally, it is necessary to go back to the modern structure of analytical geometry. But historically, Copernicus and Kepler also knew how the curves that they used (e.g., ellipses) could be reconstructed by the epicycle-deferent technique.

First of all, one must remember that points on the plane can be represented by complex numbers $z = x + iy = re^{i\theta}$ with the corresponding Cartesian coordinates (x, y) or polar coordinates (r, θ). The addition of complex numbers then corresponds to vector addition [2.10]. A uniform circular motion with center c, radius ϱ and period T can be represented by

$$z = c + \varrho e^{i((2\pi t/T)+\alpha)} = c + \varrho e^{(2\pi it/T)+i\alpha} \qquad (2.1)$$

with time t and initial phase α for the point. Now assume a point A that is moving according to the equation $z = f(t)$. Let a point B move relative to A on a circle with radius ϱ, period T, and initial phase α. The motion of B is then described by the equation

$$z = f(t) + \varrho e^{(2\pi it/T)+i\alpha} \qquad (2.2)$$

Then it is possible to describe the movement of a point B on an epicycle whose center moves around A. The addition of a new epicycle is described mathematically by the addition of a new term $\varrho e^{(2\pi it/T)+i\alpha}$ to the expression for z. Clearly, $\varrho e^{(2\pi it/T)+i\alpha} = \varrho e^{i\alpha} e^{(2\pi it/T)} = ae^{ikt}$ with a complex number $a \neq 0$ and k as a real number. In the case of a retrograde motion, T or k, respectively, is taken to be negative. A motion that results from the superposition of n epicycles is then expressed by the equation

$$z = a_1 e^{ik_1 t} + a_2 e^{ik_2 t} + \ldots + a_n e^{ik_n t} \qquad (2.3)$$

Let us proceed first from a periodic motion on the plane $z = f(t)$ (e.g., with period 2π). Mathematically, we assume f continuous with limited variation. Then for f there exists a representation with a uniformly converging series

$$f(t) = \sum_{n=-\infty}^{\infty} c_n e^{int} \qquad (2.4)$$

Therefore it can easily be proved mathematically that $f(t)$ can be approximated by means of sums

$$S_N(t) = \sum_{n=-N}^{N} c_n e^{int} \tag{2.5}$$

with any desired degree of exactitude for increasing N.

Function f is indeed uniformly convergent. Therefore for arbitrarily small $\varepsilon > 0$ one can choose an index N_0 so that for all $N \geq N_0$ and all t, it holds true that

$$|f(t) - S_N(t)| < \varepsilon \tag{2.6}$$

Astronomically, this result means that a constant-motion path (of limited variation) can be approximated to any desired degree of exactitude by means of finite superpositions of the epicycle motions.

Is it clear that so far we have used only superpositions with epicycle periods $\pm 2\pi, \pm \pi, \pm \frac{2}{3}\pi, \pm \frac{1}{2}\pi, \pm \frac{2}{5}\pi, \ldots$. In particular, therefore, only commensurable superpositions were employed, which can be expressed by means of integer number ratios in accordance with the Pythagorean tradition. But in fact non-periodic curves can also be approximated by means of epicyclical superpositions if we permit incommensurable periods. This result is mathematically supported by a proposition by Harald Bohr about almost-periodic functions (1932) [2.11]. The second question, why the epicycle-deferent technique for the explanation of the paths of motion was abandoned, cannot be answered by pointing to the observation of missing curves. Mathematically, observed curves – however exotic – could still be explained in principle (under the above, very broad mathematical frame conditions) by means of the Platonic–Apollonian use of this ancient strategy for reducing the complexity of motions.

The decisive question in this case, however, is which motions the planets "really" carry out, whether they are, in fact, combined, uniform, and unforced circular motions that seem to us on earth to be elliptical paths, or whether they are in fact compelled to follow elliptical paths by forces. This determination cannot be made geometrically and kinematically, but only dynamically, i.e., by means of a corresponding theory of forces, hence by means of physics.

Besides the epicycle-deferent-technique, Ptolemy employed imaginary balance points relative to which uniform circular motions were assumed that, relative to the earth as center, appear non-uniform. This technique proved to be useful for calculation, but constituted a violation of the central symmetry and therefore had the effect of an ad hoc assumption that was not very convincing from the standpoint of philosophy of nature, a criticism later made especially by Copernicus. The reasons that Copernicus exchanged the earth for the position of the sun were predominantly kinematic. Namely, a certain kinematic simplification of the description could be achieved in that way with a greater symmetry. Thus in the heliocentric model the retrograde planetary motions could be interpreted as effects of the annual motion of the earth, which according to Copernicus moved more slowly than the outer planets Mars, Jupiter and Saturn and faster than the inner planets Mercury and Venus. But Copernicus remained thoroughly conservative as a philosopher of nature since he considered greater simplicity in the sense of "natural" circular motion to be a sign of proximity to reality.

With Johannes Kepler, the first great mathematician of modern astronomy, the belief in simplicity was likewise unbroken. In his *Mysterium cosmographicum* of 1596, Kepler began by trying once more to base distance in the planetary system on the regular solids, alternatingly inscribed and circumscribed by spheres. The planets Saturn, Jupiter, Mars, Earth, Venus, and Mercury correspond to six spheres fitted inside each other and separated in this sequence by a cube, a tetrahedron, a dodecahedron, an icosahedra, and an octahedron. Kepler's speculations could not, of course, be extended to accommodate the discovery of Uranus, Neptune, and Pluto in later centuries.

Yet Kepler was already too much of a natural scientist to lose himself for long in Platonic speculations. His *Astronomia Nova* of 1609 is a unique document for studying the step-by-step dissolution of the old Platonic concept of simplicity under the constant pressure of the results of precise measurement. In contrast to Copernicus, Kepler supplemented his kinematic investigations with original dynamic arguments. Here the sun is no longer regarded as being physically functionless at a kinematically eccentric point, as with Copernicus, but is seen as the dynamic cause for the motion of planets. The new task was to determine these forces mathematically as well. Kepler's dynamic interpretation with magnetic fields was only a (false) initial venture. Success came later, in the Newtonian theory of gravity.

The simplicity of the celestial ("superlunar") world and the complexity of the earthly ("sublunar") are also popular themes in other cultures. Let us cast a glance at the Taoist philosophy of nature of ancient China. To be sure, it is edged with myth and less logically argued than the Greek philosophy of nature, and it also invokes more intuition and empathy; nevertheless, there are parallels between the two. Taoism describes nature as a great organism governed by cyclical motions and rhythms, such as the life cycles of the generations, dynasties, and individuals from birth to death; the food chains consisting of plant, animal, and human; the alternation of the seasons; day and night; the rising and setting of the stars; etc. Everything is related to everything else. Rhythms follow upon each other like waves in the water. What forces are the ultimate cause of this pattern in nature? As with Empedocles, in Taoism two opposite forces are distinguished, namely *yin* and *yang*, whose rhythmic increase and decrease govern the world. In the book *Kuei Ku Tzu* (4th century B.C.) it says: "Yang returns cyclically to its origin. Yin reaches its maximum and makes way for yang." [2.12] While according to Aristotle all individuals carry their natural purposes and movements in themselves, the Tao of yin and yang determines the internal rhythms of individuals, and those energies always return to their origins. The cyclical rotational model of the Tao provides explanations for making calendars in astronomy, for water cycles in meteorology, for the food chain, and for the circulatory system in physiology. It draws its great persuasiveness from the rhythms of life in nature, which people experience every day and can apply in orienting themselves to life. Nature appears as a goal-directed organism.

It is noteworthy that the Chinese philosophy of nature had no notions of atomistic particles and therefore did not develop mathematical mechanics in the sense of the occidental Renaissance. Instead, at its center there was a harmonious model of nature with rhythmic waves and fields that cause everything to be connected to everything. The preference for questions of acoustics and the early preoccupation with magnetic and electrostatic effects is understandable given this philosophy of nature. The view of the Taoists bear more resemblance to the philosophy of nature of the Stoics than to Aristotle. Here too the discussion centers on effects that spread out in a great continuum like waves on water. This continuum is the Stoics' *pneuma*, whose tensions and vibrations are said to determine the various states of nature. The multifarious forms of nature are only transitory patterns that are formed by varied tensions of the pneuma. Modern thinking leaps, of course, to the patterns of standing water waves or sound waves or the patterns of magnetic fields. Nevertheless, neither the Stoic nor the Taoist heuristic background led to the development of a physical theory of acoustic or magnetic fields comparable to Galilean mechanics with its background of an atomistic philosophy of nature. The emergence of order from complex, irregular, and chaotic states of matter was only qualitatively described, using different models for earth and for heaven.

2.2 Newton's and Einstein's Universe and the Demon of Laplace

Since antiquity, astronomers and philosophers have believed that the celestial motions are governed by simple geometric laws. Simplicity was not only understood as the demand for an economical methodology, but, still for Copernicus, as a feature of truth. Thus, the astronomical doctrine from Plato to Copernicus proclaimed: reduce the apparent complexity of the celestial system to the simple scheme of some true motions! The simple building blocks were given by the basic concepts of Euclidean geometry: circle (compass) and straight line (ruler). In contrast to the simplicity of the superlunar world, the sublunar earthly world seemed to be really complex. Thus its dynamics could not be mathematized at least in the framework of Euclidean geometry. That was the reason why Plato's mathematical atomism was soon forgotten, and Aristotle's belief in a complex qualitative dynamics of nature which cannot be mathematized in principle overshadowed scientific research until the Renaissance.

Early physicists like Galileo overcame the boundary of a superlunar ("simple") and sublunar ("complex") world. They were convinced that the dynamics of nature is governed by the same simple mathematical laws in heaven and on earth. Technically, Galileo simplified the dynamics of, e.g., free fall by selecting some observable and measurable quantities and neglecting other constraints. In short, he made a simplified mathematical model of an idealized experimental situation. Of course, even the astronomical models of antiquity only considered a few parameters, such as angular velocity and position of the planets, and neglected the complex diversity of other constraints (e.g., density, mass, friction of the celestial spheres). From a modern point of view, even the presocratic philosophers suggested qualitative "models" of a complex dynamics in nature by selecting some dominant "parameters" (e.g., water, fire, air, and earth).

In general, a system which may be physical, biological, or social, is observed in different states. The strategies for making models of observed phenomena may have changed since ancient times, but the target of the modeling activity is in some sense the same: the dynamics of the changing states in the observed systems. Obviously, the real states cannot be described by only a few observable parameters, but it is assumed that they can. In the case of early astronomy and mechanics, this was the first step of mathematical idealization and led to a geometric model for the set of idealized states which is nowadays called the state space of the model. The presocratic "models" of nature differ from modern ones not only because of their mathematization and measurability, but also because the relationship between the actual states of a real system and the points of the geometric model was believed to be ontologically necessary, while in modern systems it is a fiction maintained for the sake of theory, prediction, and so on.

The simplest scheme is the one-parameter model. Early medical experience with mammals shows that the state of health or illness can be correlated with the parameter of temperature. Many animals correlate observable features with the emotional states of other animals: the ear attitude of a dog corresponds to its state of fear, while its fang exposure is a qualitative "parameter" for its degree of rage. A combination of both attitudes represents a more adequate characterization of the dog's emotional state. The state of a planet in medieval cosmology can be defined by its angular velocity and localization. States of other systems may need more than two features (e.g., temperature, blood pressure, and pulse rate for the healthy state of a mammal).

In any case, if these parameters are numerical, then the corresponding state spaces can be represented by geometric spaces. Thus the values of two numerical parameters may be represented by a single point in a two-dimensional state space visualized by the plane of Euclidean geometry. Changes in the actual state of the system are observed and may be represented as a curve in the state space. If each point of this curve carries a label recording the time of observation, then we get a trajectory of the model. Sometimes it is useful to introduce another coordinate of time and to represent the changing parameters of states by its time series. This kind of representation is called the graph of a trajectory.

The dynamical concepts of the Middle Ages included both kinds of representation. In the 1350s, the Parisian scholastic Nicole Oresme introduced the concept of graphical representations or geometrical configurations of intensities of qualities. Oresme mainly discussed the case of a linear quality whose extension is measured by an interval or line segment of space or time ("longitude of the quality"). He proposed to measure the intensity of the quality at each point of the interval by a perpendicular ordinate ("latitude of the quality") at that point. The quantity of a linear quality is visualized by the configuration of both parameters. In the case of a uniformly accelerated motion during a time interval corresponding to the longitude AB in Fig. 2.3, the latitude at each point P of AB is an ordinate PQ whose length is the velocity at the corresponding instant [2.13]. The straight line DC of the configuration is the graph of a trajectory representing the state of velocity. The so-called Merton Rule is immediately derived with a geometrical verification of Fig. 2.3: namely, it follows from the formula for the area of the trapezoid in Fig. 2.3 that the total distance traveled is $s = \frac{1}{2}(v_0 + v_f)t$.

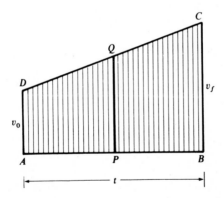

Fig. 2.3. Oresme's coordinates of a linear quality

Perhaps this interpretation was found on the basis of regarding this area as made up of very many vertical segments ("indivisibles"), each representing a velocity continued for a very short ("infinitesimal") time. The Merton Rule shows that even in the very early beginning of state space approaches a good geometric representation is not only a useful visualization, but gives new insight into the concepts of dynamics. Of course, Oresme and the Merton scholars at first only wanted to mathematize an Aristotelean-like physics of qualities. But their work was widely disseminated in Europe and led to the work of Galileo. In his famous *Discorsi* (1638), he introduced the basic concepts of modern mechanics and proceeded to the well-known distance formula $s = \frac{1}{2}at^2$ for uniformly accelerated motion from rest (free fall) with a proof and an accompanying geometric diagram that are similar to Oresme's ideas.

With Newton and Leibniz, something new was added to the theory of dynamical systems. The calculus allows one to compute the instantaneous velocity as the derivative of a velocity function and to visualize it as the tangent vector of the corresponding curve (Fig. 2.4a). The velocity vector field has become one of the basic concepts in dynamical systems theory (Fig. 2.4b). Trajectories determine velocity vectors by the differentiation procedure of the calculus. Conversely, the integration procedure of the calculus allows one to determine trajectories from velocity vectors.

The strategy of modeling a dynamical system begins with the choice of a state space in which observations can be represented by several parameters. Continuing the observations leads to many trajectories within the state space. In the sense of Newton's and Leibniz' calculus, a velocity vector may be derived at any point of these curves, in order to describe their inherent dynamical tendency at any point. A velocity vector field is defined by prescribing a velocity vector at each point in the state space. The state space filled with trajectories is called the "phase portrait" of the dynamical system (Fig. 2.4c). This basic concept of dynamical system theory was originally introduced by Henri Poincaré. The velocity vector field was derived from the phase portrait by differentiation [2.14].

Of course, the velocity vector field visualizes the dynamics of the particular system being modeled. Actually, extensive observations over a long period of time are necessary to reveal the dynamical tendencies of the system which is represented by the corresponding velocity vector field. The modeling procedure is only adequate if we assume that (a) the velocity vector of an observed trajectory is at each point exactly the same as the vector specified by the dynamical system and (b) the vector field of the model is smooth. The word "smooth" means intuitively that there are no jumps and no sharp corners. In the case of a one-dimensional state space, the vector field is specified by a graph in the plane. Thus, the graph is smooth if it is continuous and its derivative is continuous as well. Historically, condition (b) corresponds to Leibniz' famous principle of continuity playing a dominating role in the framework of classical physics.

In general, we summarize the modeling process as follows. A dynamical model is prepared for some experimental situation. We may imagine the laboratory devices of physicists like Galileo and Newton or biologists observing some organisms or even sociologists working on some social groups.

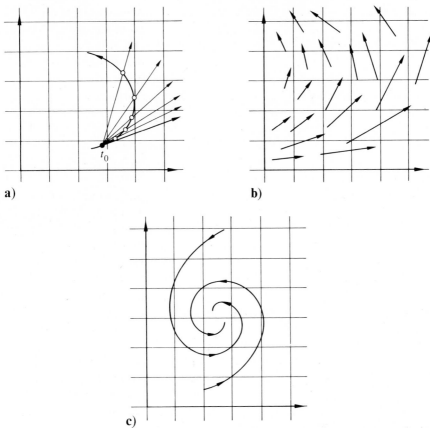

Fig. 2.4a–c. Geometric representation of a dynamical system: (**a**) Instantaneous velocity as tangent vector, (**b**) velocity vector field, (**c**) phase portrait

The dynamical model consists of the state space and a vector field. The state space is a geometrical space (e.g., the Euclidean plane or in general a topological manifold) of the experimental situation. The vector field represents the habitual tendencies of the changing states and is called the dynamics of the model. How can we find the trajectories, thus the behaviour of the system? Technically, this problem is solved by creating the phase portrait of the system. That means we have to construct the trajectories of the dynamical system. Given a state space and a ("smooth") vector field, a curve in the state space is a trajectory of the dynamical system if its velocity vector agrees with the vector field in the sense of tangent vectors (Fig. 2.5). The point corresponding to time zero is called the initial state of the trajectory. These trajectories are supposed to describe the behaviour of systems as observed over an interval of time. Moreover, physicists have been ambitious enough

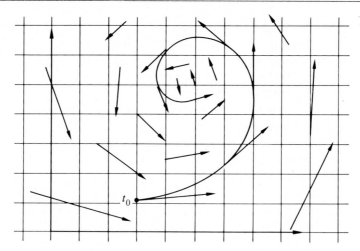

Fig. 2.5. Trajectory of a dynamical system in a vector field

to aim at making predictions indefinitily far into the future and to calculate the course of nature as if it were a huge clock.

Let us have a short glance at Newton's cosmos, which seemed to be a successful application of dynamical system theory evolving by the mathematical tools of Newton, Leibniz, Euler, etc. Newton gave three laws governing the behavior of material bodies. The first law ("lex inertiae") says that a body continues to move uniformly in a straight line if no force acts on it. If a force does act on it, then its mass times its acceleration is equal to that force (second law). The basic framework is completed by a third law: to every action there is always opposed an equal reaction. The Newtonian cosmos consists of particles moving around in a space which obeys the laws of Euclidean geometry. The accelerations of these particles are determined by the forces acting upon them. The force on each particle is obtained by adding together all the forces of other particles in the sense of the vector addition law. If the force is a gravitational one, then it acts attractively between two bodies and its strength is proportional to the product of the two masses and the inverse square of the distance between them. But, of course, there may be other types of forces.

Actually, Newton's second law was understood as a universal scheme for all forces of nature in the macrocosmos and microcosmos. With a specific law of force the Newtonian scheme translates into a precise system of dynamical equations. If the positions, velocities, and masses of the various particles are known at one time, then their positions and velocities are mathematically determined for all later times. In short, the state of a body in Newton's cosmos is specified by the two parameters of position and velocity. The Newtonian trajectories are determined by the dynamical equations of motion. If the initial states were known, then the behavior of Newton's cosmos seemed to be determined completely. This form of determinism had a great influence on the philosophy of the 18th and 19th centuries. Newton's dynamics was understood as basic science for modeling nature. But, of course, the mechanistic models are valid only in the limiting case of vanishing friction and are never fully achieved experimentally. Nature is so complex that physicists preferred to observe unnatural ("artificial") limiting cases. Later on we shall see that the physicists' belief in simple laws completely neglected the complexity of initial conditions and constraints and, thus, created an illusory model of a deterministic as well as computable nature.

According to Newton, there is only one real world of matter in one absolute framework of space-time, in which we may choose relative reference systems. This means that for any two events it is regarded as objectively decidable whether they are simultaneous and also whether they occur at the same place. Mathematically, Newton's absolute space is represented by a 3-dimensional Euclidean space the metric of which is measurable by means of rulers, while time is taken to be a 1-dimensional Euclidean space with coordinate t which is measured by standard clocks.

Because of its absolute simultaneity the Newtonian 4-dimensional space-time is stratified by maximal subsets of simultaneous events. Each stratum is a possible 3-dimensional hyperplane $t = t(e)$ of an event e which separates its causal future, with strata $t > t(e)$, from its causal past, with strata $t < t(e)$. In Fig. 2.6a the third spatial coordinate is neglected, in order to visualize each stratum as 2-dimensional plane. This causal structure includes the Newtonian assumption that there are arbitrarily fast signals by means of instantaneous action at a distance [2.15].

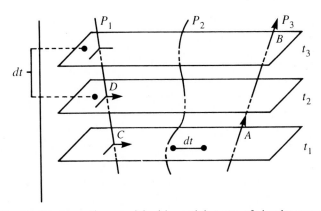

Fig. 2.6a. Newtonian space-time model with spatial strata of simultaneous events and trajectories of uniform inertial movements (*straight lines*) and accelerations (*curve*)

Newton's relative spaces are made precise by Lange as inertial systems designating reference systems for a force-free body moving in a straight line with a steady velocity. It is not stipulated which of the many possible inertial systems is used. Particular transformations (Galilean transformations) from one inertial system to another give the corresponding coordinates. Mechanical laws are preserved (invariant) with respect to these transformations. As every Galilean transformation has ten continuous parameters (one parameter for time and three times three parameters for rotation, steady velocity and translation), we can derive ten laws of conservation. Thus, e.g., the Galilean invariance of the time coordinate implicates the law of conservation of energy. Reference systems which are not inertial systems have typical effects. A disk rotating relative to the fixed stars has radial forces which cannot be eliminated by Galilean transformations. In short, in Newtonian space-time, uniform motions are considered as absolutely preferred over accelerated motions. Its structure is defined by the group of Galilean transformations.

At the beginning of this century, Einstein proved that Newton's model of space-time is restricted to mechanical motions with slow speed relative to the speed c of light. The constancy of c independently of any moving reference system is a fact of Maxwell's electrodynamics. Thus, Newton's addition law of velocities and Galilean invariance cannot hold true in electrodynamics. In his theory of special relativity (1905), Einstein assumed the constancy of the speed of light and the invariance of physical laws with respect to all inertial systems ("principle of special relativity") and derived a common framework of space-time for electrodynamics and mechanics. Einstein's special relativistic space-time was modeled by Minkowski's four-dimensional geometry. The four-dimensionality should not surprise us, because Newton's space-time has three (Cartesian) space and one time coordinate, too.

For the sake of simplicity, the units are chosen in a way that the speed of light is equal to one, and, thus, the units of length and time can be exchanged. Each point in this space-time represents an event, which means a point in space at a single moment. As a particle persists in time, it is not represented by a point, but by a line which is called the world-line of the particle. In order to visualize the Minkowskian model, we depict a space-time system with a standard time coordinate, measured in the vertical direction, and two space coordinates, measured in the horizontal direction (Fig. 2.6b) [2.16].

Uniformly moving particles are represented by straight lines, accelerated particles by curved lines. As particles of light (photons) uniformly travel with the fundamental speed c, their world-lines are straight lines at an angle of 45° to the vertical. They form a light cone centred at the common origin 0. The system of light cones at all space-time points is regarded as the Minkowskian model of relativistic space-time.

Whereas the world-line of a photon is always along the light cone at each point, the world-line of any accelerated or uniformly moved material

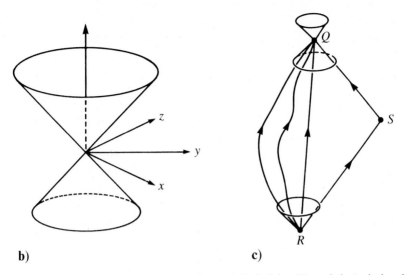

b)   **c)**

Fig. 2.6b,c. Minkowskian space-time cone in special relativity (**b**), and the twin brother paradox of special relativity: the Minkowskian distance RQ is greater than the length of RS and SQ together (**c**)

particle with a speed slower than c must always be inside the light cone at each point. As material particles or photons cannot travel faster than light, only the world-lines along and inside the light cone are determined physically. An event is called later than 0, if it is in the future cone above 0; it is called earlier than 0, if it is in the past cone below 0. Thus, the light cones determine the causal structure of relativistic space-time.

An essential difference between the Minkowskian model and ordinary Euclidian representations is the fact that the length of world-lines is interpreted as the time measured by physical clocks. Thus, time measurement becomes path-dependent, contrary to Newton's assumption of an absolute time. The so-called "twin paradox" visualizes this effect dramatically. In Fig. 2.6c, one twin brother remains on the earth R moving uniformly and very slowly, while the other makes a journey to a nearby star S at great speed, nearly that of light. Minkowskian geometry forecasts that the travelling brother is still young upon his return at Q, while the stay-at-home brother is an old man. This is not science fiction, of course, but a consequence of the time-measuring length of Minkowskian world-lines: the Minkowskian distance RQ is greater than the length of the distance RS and SQ together, contrary to the usual Euclidean interpretation. Today, these effects are experimentally well confirmed for elementary particles at high speeds near c.

In the framework of Minkowskian space-time, the invariance of physical laws with respect to particular inertial systems is realized by the Lorentz transformation. Newtonian space-time with Galilean invariance remains a limiting case for reference systems like celestial motions of planets or earthy motions of billiard balls with slow speed relative to the constant c. In this sense, Einstein's space-time is the culmination of classical physics rather than a revolutionary break with Newton.

An important concept which was first introduced into classical physics by Leibniz is energy, consisting of the kinetic energy T and the potential energy U of a system. The mechanical work done on a point mass which is displaced from a position 1 to a position 2 corresponds to the difference between the kinetic energy at position 1 and that of position 2. If this mechanical work is independent of the path followed from 1 to 2, then the corresponding force field is called conservative. Frictional forces are not conservative. In one dimension all forces must be conservative, since there is a unique path from one point to another point in a straight line, ignoring friction. The total energy $T + U$ is constant in a conservative field of force.

An important application of Newton's mechanics is the harmonic oscillator, such as the small amplitude pendulum, or the weight oscillating up and down on a spring. The harmonic oscillator appears as a model through all parts of physics and even chemistry and biology. For example, remember electromagnetic light waves, where the electric and magnetic field energies oscillate. Harmonic oscillations are also well known in technology, for example as oscillating electrical currents in a coil and a condenser, with friction corresponding to the electrical resistance. In the philosophy of the 18th and 19th centuries the pendulum was a symbol of the mechanistic universe which seemed to be completely determined and calculable by the Newtonian equations of motion.

Thus, the pendulum may be considered as a classical example of the dynamical modeling procedure. This model assumes that the rod is very light, but rigid. The hinge at the top is perfectly frictionless. The weight at the lower end is heavy, but very small. The force of gravity always pulls it straight down. In Fig. 2.7a, the pendulum is drawn in a two-dimensional Euclidean plane with the angle α of elevation, the force F of gravity, the pull $F \cos \alpha$ along the rod, and the force $F \sin \alpha$ turning it. In order to visualize the dynamical behavior of the pendulum we have to develop a dynamical model with a state space and a phase portrait. The state of the pendulum is fully determined by the angular variable α (with $\alpha = 0$ and $\alpha = 2\pi$ denoting the same angle) and the angular velocity v. Thus, we get a two-dimensional state space which can be visualized by the circular cylinder in Fig. 2.7b. The vertical circle in the center of this cylinder denotes the states of zero angular velocity $v = 0$. The straight line from front to back, at the bottom of the cylinder, is the axis of zero inclination with $\alpha = 0$, where the pendulum is lowest. At the origin with $(\alpha, v)=(0,0)$, the pendulum is at rest at its lowest position [2.17].

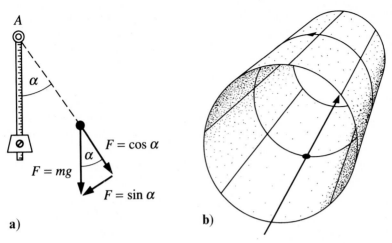

A

α

$F = \cos \alpha$

α

$F = mg$

$F = \sin \alpha$

a)   b)

Fig. 2.7a,b. Dynamical system (pendulum) (**a**) with 2-dimensional state space (circular cylinder) (**b**) [2.17]

As there is no friction and no air in the way, moving the pendulum a little to the left causes it to swing back and forth indefinitely. The full trajectory in the state space, corresponding to this oscillating motion, is a cycle, or closed loop. In the next case, the pendulum is balanced at the top, in unstable equilibrium. A tiny touch on the left causes it to fall to the right and pick up speed. The angular velocity reaches its maximum when the pendulum passes the bottom of the swing. On the way back up to the top again, the pendulum slows down. Then the pendulum balances at the top again. But when the pendulum at the beginning of its rotation shoved hard to the right, then its rate of angular velocity is rather large. Moving back up again, it slows down, but not enough to come to rest at the top. Thus, the pendulum rotates clockwise indefinitely. The corresponding trajectory in the cylindrical state space is a cycle. Unlike the slow oscillation, the fast cycle goes around the cylinder. Performing many experiments would reveal the phase portrait of this dynamical model (Fig. 2.8a). There are two equilibrium points. At the top, there is a saddle point. At the origin, there is a vortex point which is not a limit point of the nearby trajectories. The phase portrait is easier to see when the cylinder is cut open along the straight line from front to back through the saddle point at the top (Fig. 2.8b).

If the system is not closed and the effects of friction are included as in physical reality, then the equilibrium point at the origin is no longer a vortex point (Fig. 2.8c). It has become a spiraling type of point attractor. As any motion of the pendulum will come to rest because of friction, any trajectory representing a slow motion of the pendulum near the bottom approaches this limit point asymptotically.

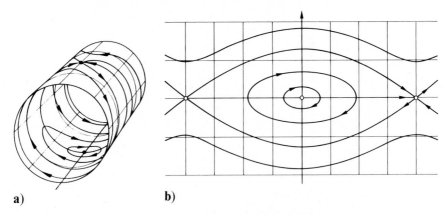

Fig. 2.8a,b. Phase portrait of the pendulum on the cylindrical state space (**a**) and cut open into a plane (**b**)

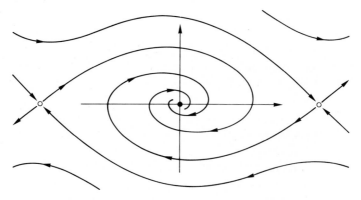

Fig. 2.8c. Phase portrait of the pendulum with friction

In two dimensions or more, other types of trajectories and limit sets may occur. For example, a cycle may be the asymptotic limit set for a trajectory (Fig. 2.9), or in a three-dimensional system a torus or even other more or less strange limit sets may occur.

Limit sets enable us to model a system's evolution to its equilibrium states. The key concepts are limit sets called "attractors" [2.18]. Mathematically, a limit set (limit point, cycle, torus, etc.) is called an attractor if the set of all trajectories approaching this limit set asymptotically is open. Roughly speaking, attractors receive most of the trajectories in the neighborhood of the limit set. Of all limit sets which represent possible dynamical equilibria of the system, the attractors are the most prominent. In the case of a limit point, an attractor represents a static equilibrium, while a limit cycle as at-

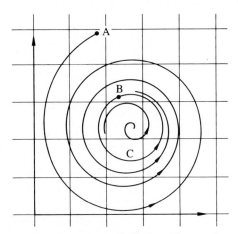

Fig. 2.9. Cycle as asymptotic limit set for a trajectory

tractor designates the periodic equilibrium of an oscillation. Vibrations on a pendulum, spring, or musical instrument are only a few of the mechanical applications. As we will see later on, periodic equilibria of oscillating dynamical systems play an important role in physics, chemistry, biology, and social sciences.

In a typical phase portrait, there will be more than one attractor. The phase portrait will be devided into their different regions of approaching trajectories. The dividing boundaries or regions are called separatrices. In Fig. 2.10, there are two point attractors with their two open sets of approaching trajectories and their separatrix.

In reality, a dynamical system cannot be considered as isolated from other dynamical systems. In order to get more adequate models, we will study two coupled systems. A simple example is provided by coupling two clocks. Historically, this particular system was observed by Christian Huygens in the 17th century. He noticed that two clocks hanging on the same wall tend to synchronize. This phenomenon is caused by nonlinear coupling through the elasticity of the wall. Indeed, any two dynamical systems can be combined into a single system by constructing the Cartesian product of the two corresponding state spaces. A small perturbation of this combined system is called a coupling of the two systems. The geometric model for the states of this combined system is formed as follows [2.19].

Each clock A and B is a kind of oscillator. For the sake of visualizing the asymptotic behaviour of both oscillators, the transient behavior is ignored and the two-dimensional state model of the Euclidean plane with a limit cycle around the origin for the two parameters of displacement and velocity is replaced by the limit cycle alone. A state of oscillator A is specified by an

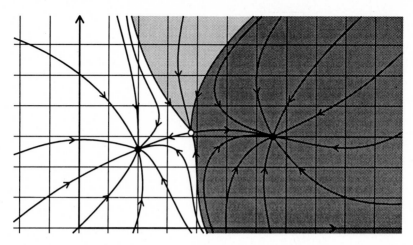

Fig. 2.10. Phase portrait with two point attractors, two open sets of approaching trajectories, and a separatrix

angle α corresponding to its phase (Fig. 2.11a), a state of oscillator B by an angle β (Fig. 2.11b).

In order to construct the state space for the combined system of both oscillators, we consider the limit cycle of clock A in a horizontal plane. Each point of this horizontal cycle represents a phase state of A. We consider such a point as the center of the limit cycle of clock B erected perpendicular to the horizontal plane of clock A (Fig. 2.11c). Each point of this vertical cycle represents a phase state of B. The pair (α, β) of phases represents the state of the combined system [2.20].

If oscillator A is stuck at phase α and oscillator B moves through a full cycle, then the combined phase point traverses the vertical cycle in Fig. 2.11c. If oscillator A also moves through a full cycle, then the vertical cycle in Fig. 2.11c is pushed around the horizontal cycle, sweeping out the torus in Fig. 2.11d. Thus, the state space for the combined system of two oscillators is the torus, which is the Cartesian product of the two cycles. Of course, the actual state model for two oscillators is four-dimensional and not only two-dimensional as in our reduced figures.

In order to get the phase portrait of the dynamical behavior for the combined system, we have to study the vector field and the trajectories on the state space of the torus. Let us first assume that each clock is totally indifferent to the state of the other. In this case, the clocks are uncoupled. The trajectory of a point on the torus corresponding to the time phase of each clock winds around the torus. If the rate of each clock is constant, then on the flat rectangular model of the torus, the trajectory is a straight line (Fig. 2.12). The slope of this line is the ratio of the rate of clock B to the rate

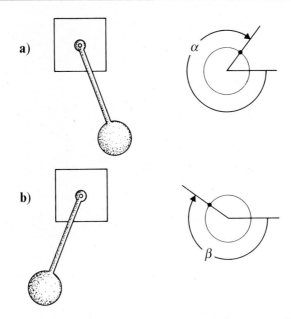

Fig. 2.11a,b. Two clocks as oscillators with two cycles as their corresponding state spaces

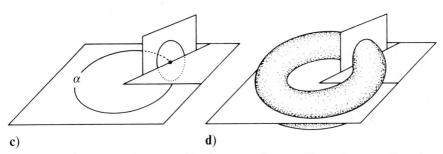

c)   d)

Fig. 2.11c,d. State space for the combined system of two oscillators (torus as Cartesian product of two cycles)

of clock A. If the two clocks run at the same rates, the ratio is one. Telling the same time means that both clocks have identical phases. Then the trajectory on the flat torus is the diagonal line in Fig. 2.12a.

A slight change in the system results in a slight change in the ratio of the rates or frequencies of the oscillators. Then the trajectory on the torus changes from a periodic trajectory to an almost periodic trajectory or to a periodic trajectory winding many times around, instead of just once (Fig. 2.12b). If two oscillators are coupled (for instance by Huygens' shared wall for the two clocks), then a small vector field must be added to the dynamical model representing the uncoupled system. It is a noteworthy theorem of geo-

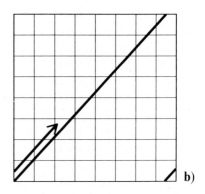

Fig. 2.12a,b. Phase portraits for the combined system of two oscillators with identical phase (**a**) and a slight change (**b**)

metric analysis that the braid of trajectories on the torus is structurally stable in the sense that a small perturbation does not result in a significant change in the phase portrait. Experimentally, this result was already confirmed by Huygens' observation of the synchronizing phenomenon of two clocks on the same wall.

Oscillators are a central dynamical paradigm for the modeling procedure of nature. They are by no means restricted to mechanical applications. In the 19th century, Hermann von Helmholtz invented an electrical vibrator and Lord Rayleigh studied coupled systems of vacuum tube oscillators used in the first radio transmitters. In this century, it was van der Pol who used the further development of radio frequency electronics in understanding coupled oscillators.

In the Newtonian universe, coupled oscillators provide examples of many-body problems. What can in general be said about the mechanics of point mass systems with several moving point masses, exerting forces upon each other? Systems with two point masses have simple and exact solutions. In a two-body problem with two point masses with isotropic central forces, the (twelve) unknowns are determined by the (ten) laws of conserved quantities and Newton's laws of motion for the two particles. The problem of two-point masses can be successfully reduced to the already solved problem of a single point mass by considering Newton's law of motion for the difference vector r and reduced mass $\mu = m_1 m_2/(m_1 + m_2)$ of both point masses m_1 and m_2. Historically, Galileo assumed that the earth moves around the sun, which is at rest. Thus he reduced the celestial motions to the simple case of a two-body problem. As we all know, the sun is actually moving around the combined centre of mass of the sun–earth system, which lies inside the surface of the sun. But this assumption is still inaccurate, of course, since many planets are simultaneously moving around the sun and all of them are exerting forces upon each other.

Another example of such a many-body-problem is given by a triple collision of three billiard balls. Provided that the balls collide only in pairs, and no triple or higher-order collisions occur, then the situation is reduced to two-body problems. The outcome depends in a continuous way on the initial state. Sufficiently tiny changes in the initial state lead only to small changes in the outcome. If three balls come together at once, the resulting behavior depends critically upon which balls come together first. Thus, the outcome depends discontinuously on the input, contrary to Leibniz' principle of continuity, which he used basically to criticize Descartes' inquiries into percussion. In the Newtonian universe, many-body problems of billiard balls and planets can be described in deterministic models in the sense that physical behavior is mathematically completely determined for all times in the future and past by the positions and velocities of the balls or planets. But the models may be non-computable in practice and in the long run. In the case of planetary theory, numerical simulations on computers for many millions of years can produce very large errors, because the initial positions and velocities are not known exactly. A very tiny change in the initial data may rapidly give rise to an enormous change in the outcome. Such instabilities in behavior are typical for many-body problems. Even in a fully deterministic world, the assumption of a Laplacean demon which can calculate the Newtonian universe in the long run will eventually be exposed as an illusory fiction.

2.3 Hamilton Systems and the Chaos of Heaven and the Quantum World

In the 18th and 19th centuries, Newtonian mechanics seemed to reveal an eternal order of nature. From a modern point of view, Newtonian systems are only a useful kind of dynamical system for modeling reality. In order to specify the initial state of a Newtonian system, the positions and the velocities of all its particles must be known. Around the middle of the 19th century, a very elegant and efficient formalism was introduced by the mathematician William Hamilton [2.21]. His fruitful idea was to characterize a conservative system by a so-called Hamiltonian function H which is the expression for the total energy (= sum of kinetic and potential energy) of the system in terms of all the position and momentum variables. While the velocity of a particle is simply the rate of change of its position with respect to time, its momentum is its velocity multiplied by its mass. Newtonian systems are described with Newton's second law of motion in terms of accelerations, which are rates of change of rates of change of position. Thus, mathematically, they are defined by second-order equations. In the Hamiltonian formulation, there are two sets of equations. One set of equations describes how the momenta of particles are changing with time, and the other describes how the positions are changing with time. Obviously, Hamiltonian equations describe the rates of change of quantities (i.e., position or momentum). Thus, we get a reduction of mathematical description with first-order equations which are, of course, deterministic. For dynamical systems of n unconstrained particles with three independent directions of space, there are $3n$ position coordinates and $3n$ momentum coordinates.

With suitable choices of the Hamiltonian function H, Hamiltonian equations can be used to characterize any classical dynamical system, not just Newtonian systems. Even in Maxwell's electrodynamics, Hamiltonian-like equations deliver the rate of change with time of the electric and magnetic fields in terms of what their values are at any given time. The only difference is that Maxwell's equations are field equations rather than particle equations, needing an infinite number of parameters to describe the state of the system, with field vectors at every single point in space, rather than the finite number of parameters with three coordinates of position and three of momentum for each particle. Hamiltonian equations also hold true for special relativity and (in somewhat modified form) for general relativity. Even, the crucial step from classical mechanics to quantum mechanics is made by Bohr's correspondence principle in the framework of the Hamiltonian formalism. These applications will be explained later on. Just now, it is sufficient to recall that Hamiltonian equations deliver a universal formalism for modeling dynamical systems in physics.

The corresponding state spaces allow us to visualize the evolution of the dynamical systems in each "phase". Thus they are called phase spaces. For systems with n particles, phase spaces have $3n + 3n = 6n$ dimensions. A single point of a phase space represents the entire state of a perhaps complex system with n particles. The Hamiltonian equations determine the trajectory of a phase point in a phase space. Globally, they describe the rates of change at every phase point, and therefore define a vector field on the phase space, determining the whole dynamics of the corresponding system.

It is a well-known fact from empirical applications that states of dynamical models cannot be measured with arbitrary exactness. The measured values of a quantity may differ by tiny intervals which are caused by the measuring apparatus, constraints of the environment, and so on. The corresponding phase points are concentrated in some small regions of a neighborhood. Now, the crucial question arises if trajectories starting with neighboring initial states are locally stable in the sense that they have neighboring final states. In Fig. 2.13a, a phase state region R_0 of initial states of time zero is dragged along by the dynamics of the vector field to a region R_t at later time t (of course, the actual large number of coordinates must be neglected in such a visualization of a phase space) [2.22].

In this case, similiar initial states lead to similiar final states. This assumption is nothing else than a classical principle of causality in the language of Hamiltonian dynamics: similar causes lead to similiar effects. Historically, philosophers and physicists from Leibniz to Maxwell believed in this causal principle, which seemed to secure the stability of the measuring process and the possibility of forecasts despite an appreciable interval of inaccuracy.

It is noteworthy that the representation in the Hamiltonian formalism allows a general statement about the causality of classical dynamical systems. Due to a famous theorem of the mathematician Liouville, the volume of any region of the phase space must remain constant under any Hamiltonian dynamics, and thus for any conservative dynamical system. Consequently, the size of the initial region R_0 in Fig. 2.13a cannot grow by any Hamiltonian dynamics if we understand "size" in the right manner as phase-space volume. But its conservation does not exclude that the shape of the initial region is distorted and stretched out to great distances in the phase space (Fig. 2.13b) [2.22].

a)   b)

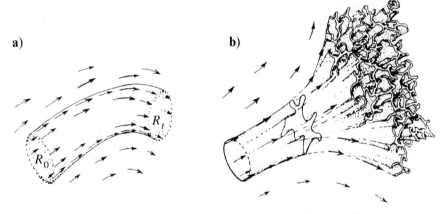

Fig. 2.13. (a) A phase state region R_0 at time 0 is dragged along by a Hamiltonian dynamics to a region R_t at later time t [2.22]. (b) According to Liouville's theorem, the volume of an initial phase state region is conserved under a Hamiltonian dynamics, although its shape may be distorted, stretched, and spread outwards [2.22]

We may imagine a drop of ink spreading through a large volume of water in a container. That possible spreading effect in phase spaces means that the local stability of trajectories is by no means secured by Liouville's theorem. A very tiny change in the initial data may still give rise to a large change in the outcome. Many-body problems of celestial mechanics and billiard balls remain non-computable in the long run, although their dynamics are deterministic. Nevertheless, Liouville's theorem implies some general consequences concerning the final regions which can be displayed by Hamiltonian dynamics, and thus by conservative dynamical systems. Remember the phase portrait Fig. 2.8c of a pendulum with friction (which is not a conservative system) with a different equilibrium point at the origin. While the non-conservative system has a spiraling type of point attractor (Fig. 2.14a), the conservative system has a vortex point (Fig. 2.14b) which is not an attractor [2.23].

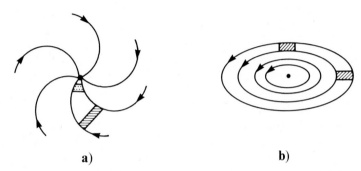

a)   b)

Fig. 2.14a,b. Point attractor of a non-conservative system without conservation (**a**), vortex point of a conservative system with conservation (**b**)

In Fig. 2.14a, trajectories are attracted to a field point, and the volume of an intial area shrinks. In Fig. 2.14b, the trajectories rotate around a vortex point, and the volume of an initial area is conserved. Thus, due to Liouville's theorem, we can generally conclude that in any conservative system attracting points must be excluded. The effect of shrinking initial areas can easily be visualized for the trajectories of limit cycles, too. So, limit cycles as attractors are also not possible in conservative systems for the same mathematical (a priori) reasons.

These results are derived a priori by a far-reaching mathematical theorem of Hamiltonian systems. We must be aware that conservative physical systems like planetary systems, pendula, free fall, etc., are only some of the empirical applications of Hamiltonian systems. Hamiltonian systems are defined by a particular kind of mathematical equation (Hamiltonian equations). Features of Hamiltonian systems are derived from the mathematical theory of the corresponding equations. Consequently, modeling reality by Hamiltonian systems means that we can epistemically forecast some a priori features, e.g., that no static equilibrium of a limit point attractor and no periodic equilibrium of a limit cycle attractor can be expected.

Philosophically, this point of view obviously conforms to Kant's epistemology in some modified sense. If we assume the mathematical framework of some dynamical systems, then, of course, we can assert some a priori statements about our empirical models, independently of their empirical applications in several sciences. But Kant's epistemology and the dynamical system approach differ in the following sense: not only is there one categorial framework (e.g., Newtonian systems), but there are many kinds of systems modeling reality with more or less success. So, it will not be physicalist or reductionist to apply conservative systems even in cognitive and economical science, later on.

A further a priori result of Hamiltonian (conservative) systems says that there are irregular and chaotic trajectories. In the 18th and 19th centuries, physicists and philosophers were convinced that nature is determined by Newtonian- or Hamiltonian-like equations of motion, and thus future and past states of the universe can be calculated at least in principle if the initial states of present events are well known. Philosophically, this belief was visualized by Laplace's demon, which like a huge computer without physical limitations can store and calculate all necessary states. Mathematically, the belief in Laplace's demon must presume that systems in classical mechanics are integrable, and, thus are solvable. In 1892, Poincaré was already aware that the non-integrable three-body problem of classical mechanics can lead to completely chaotic trajectories [2.24]. About sixty years later, Kolmogorov (1954), Arnold (1963) and Moser (1967) proved with their famous KAM theorem that motion in the phase space of classical mechanics is neither completely regular nor completely irregular, but that the type of trajectory depends sensitively on the chosen initial conditions [2.25].

As celestial mechanics is an empirically well confirmed dynamical model of a Hamiltonian system, the KAM theorem refutes some traditional opinions about the "superlunar" world. Heaven is not a world of eternal regularity, either in the sense of Aristotle's cosmos or in the sense of Laplace's demon. Obviously, it is not the seat of the Gods. Nevertheless, it is not completely chaotic. Heaven, as far as recognized by Hamiltonian systems, is more or less regular and irregular. It seems to have more similarity with our human everyday life than our forefathers believed. This may be a motivation for writers to be curious about Hamiltonian systems. But now let us see some mathematical facts.

One of the simplest examples of an integrable system is a harmonic oscillator. Practically, the equations of motion of any integrable system with n degrees of freedom are the same as those of a set of n uncoupled harmonic oscillators. The corresponding phase space has $2n$ dimensions with n position coordinates and n momentum coordinates. For a harmonic oscillator with $n = 1$ we get a circle, and for two harmonic oscillators with $n = 2$ a torus (compare Fig. 2.11d). Thus, the existence of n integrals of motion confines the trajectories in the $2n$-dimensional phase space of an integrable system to an n-dimensional manifold which has the topology of an n-torus. For an integrable system with two degrees of freedom and a four-dimensional phase-space, the trajectories can be visualized on a torus. Closed orbits of trajectories occur

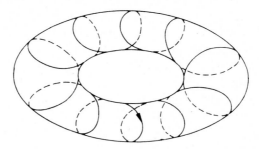

Fig. 2.15. Integrable system with two degrees of freedom on a torus and a closed orbit of a trajectory

only if the frequency ratios of the two corresponding oscillators are rational (Fig. 2.15). For irrational frequency ratios, the orbit of a trajectory never repeats itself, but approaches every point on the torus infinitesimally closely [2.26].

A nonintegrable system of celestial mechanics was studied by Hénon and Heiles in 1964. The dynamical model consists of an integrable pair of harmonic oscillators coupled by nonintegrable cubic terms of coordinates. If the initial state of the model with two position coordinates q_1, q_2 and two momentum coordinates p_1, p_2 is known, then its total energy E is determined by the corresponding Hamiltonian function H depending on these position and momentum coordinates. The trajectories of the system move in the four-dimensional phase space on a three-dimensional hyperplane which is defined by $H(q_1, q_2, p_1, p_2) = E$.

The values of E can be used to study the coexistence of regular and ir-regular motions which was forecast by the KAM theorem. For small values of E, the dynamical system has regular behavior, while for large values it becomes chaotic. In order to visualize this changing behavior, we consider the intersections of the trajectories with the two-dimensional plane of coor-dinates q_1 and q_2 (Poincaré maps). For $E = \frac{1}{24}$ (Fig. 2.16a) and $E = \frac{1}{12}$ (Fig. 2.16b), the Poincaré maps show the intersections of somewhat deformed tori which signal regular motion. Above a critical value of $E = \frac{1}{9}$, most, but not all, tori are destroyed, and spots of irregular points appear to be random. For $E = \frac{1}{8}$ (Fig. 2.16c), the Poincaré map illustrates a state of transition with the coexistence of regular and irregular motions. For $E = \frac{1}{6}$ (Fig. 2.16d), the motion seems to be almost completely irregular and chaotic [2.27].

An empirical application is given in the following three-body problem of celestial mechanics, which is nonintegrable. Consider the motion of Jupiter perturbing the motion of an asteroid around the sun (Fig. 2.17).

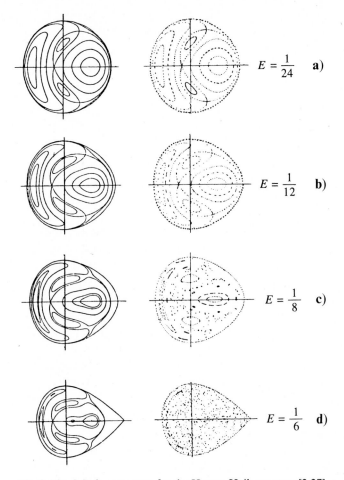

Fig. 2.16a–d. Poincare maps for the Henon-Heiles system [2.27]

Jupiter and the asteroid are interpreted as two oscillators with certain frequencies. According to the KAM theorem, stable and unstable motions of the asteroid can be distinguished, depending on the frequency ratio.

In general, we must be aware that stable as well as unstable trajectories are mathematically well defined. Consequently, even nonintegrable many-body problems describe deterministic models of the world. Metaphorically, we may say that the God of Leibniz and Newton would have no difficulty in forecasting regular and irregular trajectories *sub specie aeternitatis* and does not need to calculate their development step by step. The observed chaotic behavior is neither due to a large number of degrees of freedom (a celestial three-body problem has rather few degrees of freedom) nor to the uncertainty of human knowledge. The irregularity is caused by the nonlinearity of

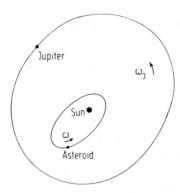

Fig. 2.17. Perturbation of an asteroid's motion by Jupiter

Hamiltonian equations which let initially close trajectories separate exponentially fast in a bounded region of phase. As their initial conditions can only be measured with finite accuracy, and errors increase exponentially fast, the long-term behavior of these systems cannot be predicted. Mathematically, initial conditions are characterized by real values which may be irrational numbers with infinite sequences of digits. Thus, computer-assisted calculations will drive the errors faster and faster with improved measurement of more and more digits.

The macrocosmos of celestial mechanics, the world of asteroids, planets, stars, and galaxies, is determined by the coexistence of regular and irregular behavior. Deterministic chaos in the heavens is not everywhere, but locally possible, and thus may cause cosmic catastrophes which cannot be excluded in principle. What about the microcosmos of quantum mechanics, the quantum world of photons, electrons, atoms, and molecules? Is there chaos in the quantum world? In order to answer this question, we first must remind the reader of some basic concepts of Hamiltonian systems and phase spaces corresponding to objects in the quantum world [2.28].

In 1900, Max Planck proposed that electromagnetic oscillations occur only in quanta, whose energy E bears the definite relation $E = h\nu$ to the frequency ν depending on the constant h ("Planck's quantum"). Besides Einstein's huge constant c of light's speed, Planck's tiny constant of quanta is the second fundamental constant of nature, according to 20th century physics. Planck's relation was experimentally supported by, e.g., the radiation of black bodies. In 1923, Louis de Broglie proposed that even the particles of matter should sometimes behave as waves. De Broglie's wave-frequency ν for a particle of mass m satisfies the Planck relation. Combined with Einstein's famous theorem $E = mc^2$ of Special Relativity ("mass is a particular state of energy and can therefore be transformed into energy by radiation"), we get a relation telling us that ν is related to m by $h\nu = mc^2$. It follows that in the quantum world, fields oscillating with some frequency can occur only in discrete units of mass, depending on Planck's and Einstein's constants. Obviously, in the quantum world, phenomena can be considered as waves as well as particles. This so-called particle-wave duality was well confirmed by many experiments which reveal features of waves or particles for quantum systems like photons or electrons, depending on the preparation of experimental conditions.

In 1913, Niels Bohr introduced his "planetary" model for the atom which could explain the observed and measured discrete stable energy levels and spectral frequencies with suprising accuracy. Bohr's rules required that the angular momentum of electrons in orbit about the nucleus can occur only in integer multiples of $\hbar = h/2\pi$. His successful, albeit somewhat ad hoc rules only delivered an approximate geometric model which must be derived from a dynamical theory of the quantum world, corresponding to Newtonian or Hamiltonian classical mechanics which can explain Kepler's planetary laws. The dynamics of the quantum world was founded by Heisenberg's and Schrödinger's quantum mechanics, which became the fundamental theory of matter in 20th century physics.

The main concepts of quantum mechanics can be introduced heuristically by analogy with corresponding concepts of Hamiltonian mechanics if some necessary modifications depending on Planck's constant are taken into account. This procedure is called Bohr's correspondence principle (Fig. 2.18). So, in quantum mechanics, classical vectors like position or momentum must be replaced by some operators satisfying a non-commutative (non-classical) relation depending on Planck's constant. If h disappears ($h \to 0$), then we get the well known classical commutative relations of, e.g., position and momentum which allow us to measure both vectors together with arbitrary accuracy. An immediate consequence of non-commutative relations in quantum mechanics is Heisenberg's uncertainty principle $\Delta p \Delta q \geq \hbar/2$. If one measures the position q with precision Δq, then one disturbs the momentum p by Δp. Thus, it is obvious that there are no trajectories or orbits in the quantum world which demand precise values of both the position and the momentum of a particle. Bohr's popular electronic orbits are only very rough geometric visualizations [2.29].

According to Bohr's correspondence principle, classical systems described by Hamiltonian functions must be replaced by quantum systems (e.g., electrons or photons) described by a Hamiltonian operator depending on operators (for position and momentum) instead of vectors. In classical physics, the states of Hamiltonian systems are determined by the points of a phase space. In quantum mechanics, the appropriate analogous concept is that of a Hilbert space. States of a quantum system are described by vectors of a Hilbert space spanned by the eigenvectors of its Hamiltonian operator.

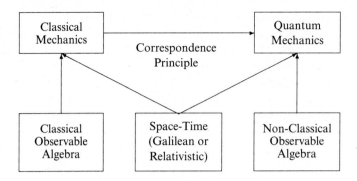

Fig. 2.18. Bohr's correspondence principle

In order to illustrate this mathematical remark a little bit more, let us imagine a single quantum particle. Classically, a particle is determined by its position in space and by its momentum. In quantum mechanics, every single position which the particle might have is a single alternative combined in a collection of all possible positions with complex-number weightings. Thus, we get a complex function of position, the so-called wave function $\psi(x)$. For each position x, the value of $\psi(x)$ denotes the amplitude for the particle to be at x. The probability of finding the particle in some small fixed-sized interval about this position is obtained by forming the squared modulus of the amplitude $|\psi(x)|^2$. The various amplitudes for the different possible momenta are also determined by the wave function. Thus, the Hilbert space is a complex state space of a quantum system.

The causal dynamics of quantum states is determined by a partial differential equation called the Schrödinger equation. While classical observables are commutative with always definite values, non-classical observables of quantum systems are non-commutative with generally no common eigenvector and consequently no definite eigenvalues. For observables in a quantum state only statistical expectation values can be calculated.

An essential property of Schrödinger's quantum formalism is the superposition principle demonstrating its linearity. For example, consider two quantum systems which once interacted (e.g., a pair of photons leaving a common source in opposite directions). Even when their physical interaction ceases at a large distance, they remain in a common superposition of states which cannot be separated or located. In such an entangled (pure) quantum state of superposition an observable of the two quantum systems can only have indefinite eigenvalues. The superposition or linearity principle of quantum mechanics delivers correlated (entangled) states of combined systems which are highly confirmed by the EPR experiments. Philosophically, the (quantum) whole is more than the sum of its parts. Non-locality is a fundamental property of the quantum world which differs from classical Hamiltonian systems [2.30]. We shall return to this question in discussing the emergence of mind–brain and artificial intelligence (Chaps. 4–5).

Bohr's correspondence principle lets the question arise of whether the existence of chaotic motion in classical Hamiltonian systems leads to irregularities in the corresponding quantum systems [2.31]. Our summary of basic quantum mechanical concepts gives some hints of changes which must be expected in passing from a classically chaotic system to its corresponding quantum mechanical version. In contrast to classical mechanics, quantum mechanics only allows statistical predictions. Although the Schrödinger equation is linear in the sense of the superposition principle and can be solved exactly, e.g., for a harmonic oscillator, and although the wave function is strictly determined by the Schrödinger equation, this does not mean that the properties of a quantum state can be calculated exactly. We can only calculate the density of probability to find a photon or electron at a space-time point.

Because of Heisenberg's uncertainty principle, there are no trajectories in the quantum world. Therefore, the determination of chaos with the exponentially fast separation of close trajectories is not possible for quantum systems. Another aspect of the uncertainty principle concerning chaos is noteworthy: remember a classical phase space with chaotical regions like in Fig. 2.16. The uncertainty principle implies that points in a $2n$-dimensional phase space within a volume h^n cannot be distinguished. The reason is that chaotic behavior smaller than h^n cannot be represented in quantum mechanics. Only the regular behavior outside these chaotic regions could be expected. In this sense, the tiny, but finite value of Planck's constant could suppress chaos.

In quantum mechanics, one distinguishes between time-independent stationary and time-dependent Hamiltonian systems. For systems with stationary Hamiltonians the Schrödinger equation can be reduced to a so-called linear eigenvalue problem which allows one to calculate the energy levels of, e.g., a hydrogen atom. As long as the levels are discrete, the wave function behaves regularly, and there is no chaos. The question arises of whether there are differences between the energy spectra of a quantum system with a regular classical limit and a quantum system whose classical version displays chaos. Time-dependent Hamiltonians are used to described the time-evolution of, e.g., elementary particles and molecules.

According to Bohr's correspondence principle, quantum chaos can be detected by starting with the investigation of some classical Hamiltonian systems. They may be integrable, almost integrable, or chaotic. Thus, the trajectories in the hyperplane of energy may be regular, almost regular, or almost chaotic. Quantizing the Hamiltonian function by replacing the vectors of position and momentum with the corresponding operators, we get the Hamiltonian operator of the corresponding quantum system. In the next step, the Schrödinger equation and eigenvalue equation can be derived. Now, we may ask if the properties of the classical system with its integrable, almost integrable, or chaotic behavior can be transferred to the corresponding quantum system. What about the spectrum, eigenfunctions, etc.? These questions are summarized under the title "quantum chaos". For instance, there are calculations which show that the energy spectrum of a free quantum particle in a stadium, for which the classical motion is chaotic, differs drastically from that of a free quantum particle in a circle, for which the classical motion is regular.

In Fig. 2.19, the distribution of distances between neighboring levels is illustrated with two examples [2.32]. In Fig. 2.19a,b, a system consisting of two coupled oscillators is shown for two different values of the coupling coefficient. While the corresponding classical dynamics of Fig. 2.19a is regular, the classical dynamics of Fig. 2.19b is almost chaotic.

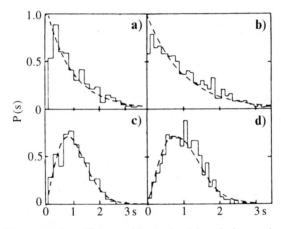

Fig. 2.19a–d. Two coupled oscillators with regular (**a**) and almost chaotic classical dynamics, (**b**) Hydrogen atom in a uniform magnetic field with corresponding regular (**c**) and almost chaotic classical dynamics (**d**) [2.32]

Figure 2.19c,d shows the example of a hydrogen atom in a uniform magnetic field. While the corresponding classical dynamics of Fig. 2.19c is regular, the classical dynamics of Fig. 2.19d is almost chaotic. The regular and chaotic cases can be distinguished by different distributions of energy levels (Poisson and Wigner distributions) which are calculated by solving the corresponding Schrödinger equation. They are confirmed by several numerical models as well as by measurements in laboratories with laser spectroscopy. In this sense, quantum chaos is no illusion, but a complex structural property of the quantum world. Hamiltonian systems are a key to discovering chaos in the macro- and microcosmos. But, of course, we must not confuse the complex mathematical structure of deterministic chaos with the popular idea of disorder.

2.4 Conservative and Dissipative Systems and the Emergence of Order

Since Poincaré's celestial mechanics (1892), it was mathematically known that some mechanical systems whose time evolution is governed by nonlinear Hamiltonian equations could display chaotic motion. But as long as scientists did not have suitable tools to deal with nonintegrable systems, deterministic chaos was considered as a mere curiosity. During the first decades of this century, many numerical procedures were developed to deal with the mathematical complexity of nonlinear differential equations at least approximately. The calculating power of modern high speed computers and refined experimental techniques have supported the recent successes of the nonlinear complex system approach in natural and social sciences. The visualizations of nonlinear models by computer-assisted techniques promote interdisciplinary applications with far-reaching consequences in many branches of science. In this scientific scenario (1963), the meteorologist Edward Lorenz, a student of the famous mathematician Birkhoff [2.33], observed that a dynamical system with three coupled first-order nonlinear differential equations can lead to completely chaotic trajectories. Mathematically, nonlinearity is a necessary, but not sufficient condition of chaos. It is necessary condition, because linear differential equations can be solved by well-known mathematical procedures (Fourier transformations) and do not lead to chaos. The system Lorenz used to model the dynamics of weather differs from Hamiltonian systems à la Poincaré mainly by its dissipativity. Roughly speaking, a dissipative system is not conservative but "open", with an external control parameter that can be tuned to critical values causing the transitions to chaos.

More precisely, conservative as well as dissipative systems are characterized by nonlinear differential equations $\dot{x} = F(x, \lambda)$ with a nonlinear function F of the vector $x = (x_1, \dots, x_d)$ depending on an external control parameter λ. While for conservative systems, according to Liouville's theorem, the volume elements in the corresponding phase space change their shape but retain their volume in the course of time, the volume elements of dissipative systems shrink as time increases (compare Figs. 2.13, 2.14) [2.34].

Lorenz's discovery of a deterministic model of turbulence occurred during simulation of global weather patterns. The earth, warmed by the sun, heats the atmosphere from below. Outer space, which is always cold, absorbs heat from the outer shell of the atmosphere. The lower layer of air tries to rise, while the upper layer tries to drop. This traffic of layers was modeled in several experiments by Bénard. The air currents in the atmosphere can be visualized as cross-sections of the layers. The traffic of the competing warm and cold air masses is represented by circulation vortices, called Bénard cells. In three dimensions, a vortex may have warm air rising in a ring, and cold air descending in the center. Thus, the atmosphere consists of a sea of three-dimensional Bénard-cells, closely packed as a hexagonal lattice. A footprint of such a sea of atmospheric vortices can be observed in the regular patterns of hills and valleys in deserts, snowfields, or icebergs.

In a typical Bénard experiment, a fluid layer is heated from below in a gravitational field (Fig. 2.20a). The heated fluid at the bottom tries to rise, while the cold liquid at the top tries to fall. These motions are opposed by viscous forces. For small temperature differences ΔT, viscosity wins, the liquid remains at rest, and heat is transported by uniform heat conduction. The external control parameter of the system is the so-called Rayleigh number Ra of velocity, which is proportional to ΔT. At a critical value of Ra, the state of the fluid becomes unstable, and a pattern of stationary convection rolls develops (Fig. 2.20b) [2.35].

Fig. 2.20a,b. Bénard experiment: a heated fluid layer

Beyond a greater critical value of Ra, a transition to chaotic motion is observed. The complicated differential equations describing the Bénard experiment were simplified by Lorenz to obtain the three nonlinear differential equations of his famous model. Each differential equation describes the rate of change for a variable X proportional to the circulatory fluid flow velocity, a variable Y characterizing the temperature difference between ascending and descending fluid elements, and a variable Z proportional to the deviation of the vertical temperature profile from its equilibrium value. From these equations, it can be derived that an arbitrary volume element of some surface in the corresponding phase space contracts exponentially in time. Thus, the Lorenz model is dissipative.

This can be visualized by computer-assisted calculations of the trajectories generated by the three equations of the Lorenz model. Under certain conditions, a particular region in the three-dimensional phase space is attracted by the trajectories, making one loop to the right, then a few loops to the left, then to the right again, etc. (Fig. 2.21) [2.36].

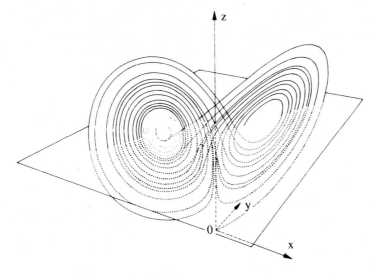

Fig. 2.21. Lorenz attractor

The paths of these trajectories depend very sensitively on the initial conditions. Tiny deviations of their values may lead to paths which soon deviate from the old one with different numbers of loops. Because of its strange image, which looks like the two eyes of an owl, the attracting region of the Lorenz phase was called a "strange attractor". Obviously, the strange attractor is chaotic. But which topological structure do the trajectories achieve by winding more and more densely without intersecting each other? An example illustrates the definition of so-called fractal dimensions [2.37]:

Let M be the subset of the attractor in the n-dimensional phase space. Now, the phase space is covered by cubes with edge length ε. Let $N(\varepsilon)$ be the number of cubes which contain a piece of the attractor M. If ε contracts to zero ($\varepsilon \rightarrow 0$), then the negative limit of the ratio of the logarithm of $N(\varepsilon)$ and the logarithm of ε, i.e., $D = -\lim \ln N(\varepsilon)/\ln \varepsilon$, is called the fractal dimension.

If the attractor is a point (Fig. 2.14a), the fractal dimension is zero. For a stable limit circle (Fig. 2.9) the fractal dimension is one. But for chaotic systems the fractal dimension is not an integer. In general, the fractal dimension can be calculated only numerically. For the Lorenz model, the strange attractor has the fractal dimension $D \approx 2.06 \pm 0.01$.

Another dissipative system in which chaotic motion has been studied experimentally is the Belousov–Zhabotinsky reaction. In this chemical process an organic molecule is oxidized by bromate ions, the oxidation being catalyzed by a redox system. The rates of change for the concentrations of the reactants in a system of chemical reactions are again described by a system of nonlinear differential equations with a nonlinear function. The variable which signals chaotic behavior in the Belousov–Zhabotinsky reaction is the concentration of the ions in the redox system. Experimentally, irregular oscillations of these concentrations are observed with a suitable combination of the reactants. The oscillations are indicated by separated colored rings. This separation is a fine visualization of nonlinearity. Linear evolutions would satisfy the superposition principle. In this case the oscillating rings would penetrate each other in superposition.

The corresponding differential equations are autonomous, i.e., they do not depend on time explicitly. For computer-assisted visualization it is often convenient to study the flow in a dynamical system described by differential equations of motion via discrete equations which construct the intersecting points of the trajectories with the $(d-1)$-dimensional Poincaré map in the corresponding d-dimensional phase space (compare Fig. 2.16). The constructed points are denoted by $x(1), x(2), \ldots, x(n), x(n+1), \ldots$ with increasing time points n. The corresponding equation has the form $x(n+1) = G(x(n), \lambda)$ for the successor point $x(n+1)$ of $x(n) = (x_1(n), \ldots, x_{d-1}(n))$. The classification of conservative and dissipative systems can be generalized from flows to Poincaré maps. A discrete map equation is called dissipative if it leads to a contraction of volume in phase space.

A famous example of a discrete map is the so-called logistic map with many applications in the natural sciences as well as the social sciences. The basic concepts of complex dynamical systems from nonlinearity to chaos can be illustrated by this map with rather simple computer-assisted methods. Thus, let us have a short glance at this example. Mathematically, the logistic map is defined by a quadratic (nonlinear) recursive map $x_{n+1} = \alpha x_n(1 - x_n)$ of the interval $0 \leq x \leq 1$ onto itself with control parameter α varying between $0 \leq \alpha \leq 4$. The function values of the sequence $x_1, x_2, x_3, \ldots$ can be calculated by a simple pocket computer. For $\alpha < 3$ the sequence converges towards a fixed point (Fig. 2.22a). If α is increased beyond a critical value α_1, then the values of the sequence jump periodically between two values after a certain time of transition (Fig. 2.22b). If α is increased further beyond a critical value α_2, the period length doubles. If α is increased further and further, then the period doubles each time with a sequence of critical values $\alpha_1, \alpha_2, \ldots$. But beyond a critical value α_c, the development becomes more and more irregular and chaotic (Fig. 2.22c) [2.38].

The sequence of period doubling bifurcations which is illustrated in Fig. 2.23a is governed by a law of constancy which was found by Grossmann and Thomae for the logistic map and recognized by Feigenbaum as a universal property for a whole class of functions (the Feigenbaum-constant) [2.39]. The chaotic regime beyond α_c is shown in Fig. 2.23b.

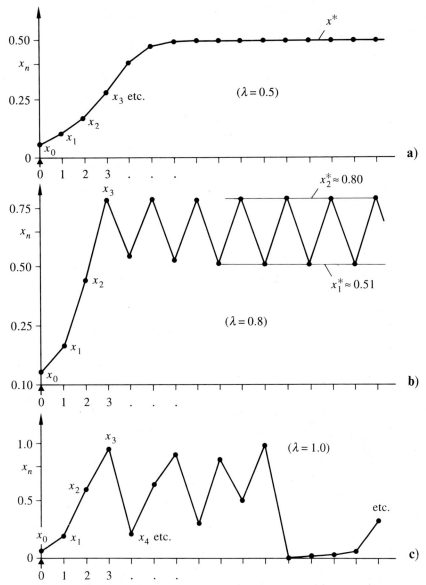

Fig. 2.22a–c. Logistic curve as nonlinear recursive map with control parameter $\alpha = 4\lambda = 2$ (**a**), $\alpha = 4\lambda = 3,2$ (**b**), and $\alpha = 4\lambda = 4$ (**c**)

In Fig. 2.24a–c the mappings of x_n onto x_{n+1} are illustrated for different control parameters, in order to construct the corresponding attractors of a fixed point, periodic oscillation between two points, and complete irregularity without any point attractor or periodicity.

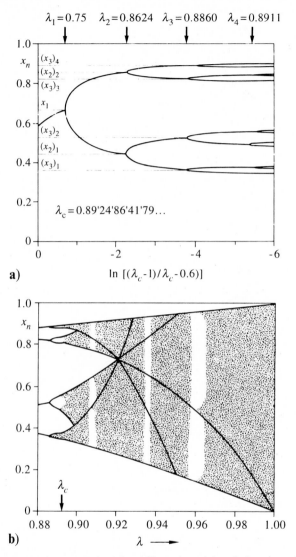

Fig. 2.23a,b. Sequence of period doubling bifurcations (**a**) and chaotic regime of the logistic map beyond $\alpha_c = 4\lambda_c$ (**b**)

It is rather astonishing that a simple mathematical law like the logistic map produces a complexity of bifurcations and chaos for possible developments as shown in Fig. 2.23a,b. A necessary, but not sufficient reason is the nonlinearity of the equation. In this context, the degrees of increasing complexity are defined by the increasing bifurcations which lead to chaos as the most complex and fractal scenario. Each bifurcation illustrates a possible

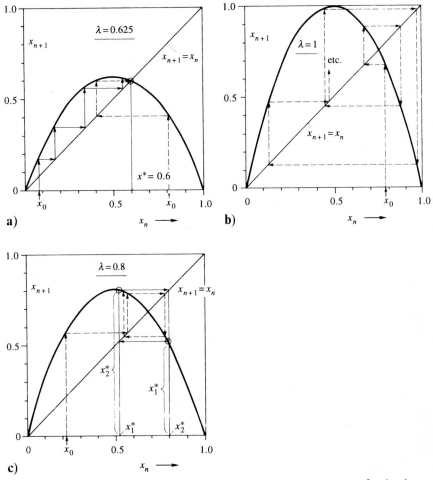

Fig. 2.24a–c. Attractors of logistic map with differing control parameter: fixed point attractor (**a**), periodic oscillation (**c**), and chaos (**b**)

branch of solution for the nonlinear equation. Physically, they denote phase transitions from a state of equilibrium to new possible states of equilibria. If equilibrium is understood as a state of symmetry, then phase transition means symmetry breaking which is caused by fluctuational forces.

Mathematically, symmetry is defined by the invariance of certain laws with respect to several transformations between the corresponding reference systems of an observer. In this sense the symmetry of Kepler's laws is defined by its Galileantransformations (compare Fig. 2.6a). The hydrodynamical laws describing a fluid layer heated from below (Fig. 2.20a) are invariant with respect to all horizontal translations. The equations of chemical reactions (in an infinitely extended medium) are invariant with respect to all translations, rotations, and reflections of a reference system used by an observer [2.40].

Nevertheless, these highly symmetric laws allow phase transitions to states with less symmetry. For example, in the case of a Bénard experiment, the heated fluid layer becomes unstable, and the state of stationary convection rolls develops (Fig. 2.20b). This phase transition means symmetry breaking, because tiny fluctuations cause the rolls to prefer one of two possible directions. Our examples show that phase transition and symmetry breaking is caused by a change of external parameters and leads eventually to a new macroscopic spatio-temporal pattern of the system and emergence of order.

Obviously, thermal fluctuations bear in themselves an uncertainty, or more precisely speaking, probabilities. A particle which is randomly pushed back or forth (Brownian motion) can be described by a stochastic equation governing the change of the probability distribution as a function of time. One of the most important means to determine the probability distribution of a process is the so-called master equation. To visualize the process we may think of a particle moving in three dimensions on a lattice.

The probability of finding the system at point x at time t increases due to transitions from other points x' to the point under consideration ("rate in"). It decreases due to transitions leaving this point ("rate out"). As the "rate in" consists of all transitions from initial points x' to x, it is composed of the sum over these initial points. Each term of the sum is given by the probability of finding the particle at point x', multiplied by the transition probability (per unit time) for passing from x' to x. In an analogous way the "rate out" can be found for the outgoing transitions. Thus, the rate of change for the probability distribution of a process is determined by a stochastic differential equation which is defined by the difference between "rate in" and "rate out".

Fluctuations are caused by a huge number of randomly moving particles. An example is a fluid with its molecules. So a bifurcation of a stochastic process can only be determined by the change of probabilistic distribution. In Fig. 2.25 the probabilistic function changes from a sharp centration at a single attractor (Fig. 2.25a) to a flat distribution (Fig. 2.25b) and finally to a distribution with two maxima at two attractors (Fig. 2.25c), when the

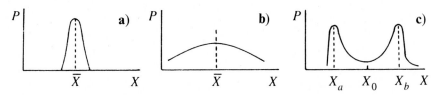

Fig. 2.25a–c. Probabilistic function with single attractor (**a**), flat distribution (**b**), and two attractors as stochastic symmetry breaking (**c**)

control parameter increases beyond corresponding critical values. Figure 2.25c illustrates stochastic symmetry breaking [2.41].

In this context, complexity means that a system has a huge number of degrees of freedom. When we manipulate a system from the outside we can change its degrees of freedom. For example, at elevated temperature the molecules of water vapor move freely without mutual correlation. When the temperature is lowered, a liquid drop is formed. This macroscopic phenomenon is formed when the molecules keep a mean distance between each other with correlated motion. At the freezing point water is transformed into ice crystals with a fixed molecular order. Since the early days of mankind people have been familiar with these phase transitions. The different aggregate states may have been a reason for philosophical ideas that water is a basic element of matter (compare Sect. 2.1).

Another example is taken from the material sciences. When a ferromagnet is heated, it loses its magnetization beyond a critical value. But the magnet regains its magnetization when the temperature is lowered. Magnetization is a macroscopic feature which can be explained by changing the degrees of freedom at the microscopic level. The ferromagnet consists of many atomic magnets. At elevated temperature, the elementary magnets point in random directions. If the corresponding magnetic moments are added up, they cancel each other. Then, on the macroscopic level, no magnetization can be observed. Below a critical temperature, the atomic magnets are lined up in a macroscopic order, giving rise to the macroscopic feature of magnetization (Fig. 4.9a). In both examples, the emergence of macroscopic order was caused by lowering the temperature. The structure is formed without loss of energy at low temperature. Thus, it is a kind of conservative (reversible) self-organization. Physically, it can be explained by Boltzmann's law of distribution demanding that structures with less energy are mainly realized at low temperatures.

On the other hand, there are systems whose order and functioning are not achieved by lowering temperature, but by maintaining a flux of energy and matter through them. Familiar examples are living systems like plants and animals which are fed by biochemical energy. The processing of this energy may result in the formation of macroscopic patterns like the growth of plants, locomotion of animals, and so on. But this emergence of order is by no means reserved to living systems (compare Chap. 3). It is a kind of dissipative (irreversible) self-organization far from thermal equilibrium which can be found in physics and chemistry as well as in biology.

As is well-known from the second law of thermodynamics, closed systems without any exchange of energy and matter with their environment develop to disordered states near thermal equilibrium. The degree of disorder is measured by a quantity called "entropy". The second law says that in closed systems the entropy always increases to its maximal value. For instance, when a cold body is brought into contact with a hot body, then heat is exchanged so that both bodies acquire the same temperature, i.e., a disordered and homogeneous order of molecules. When a drop of milk is put into coffee, the milk spreads out to a finally disordered and homogeneous mixture of milky coffee. The reverse processes are never observed. In this sense, processes according to the second law of thermodynamics are irreversible with a unique direction [2.42].

An example from hydrodynamics is the Bénard instability, which was already described in the beginning of Sect. 2.4. When the heated fluid layer (Fig. 2.20a) reaches a critical value, it starts a macroscopic motion (Fig. 2.20b). Thus a dynamic well-ordered spatial pattern emerges out of a disordered and homogeneous state as long as a certain flux of energy is maintained through the system.

Another example from fluid dynamics is the flow of fluid round a cylinder. The external control parameter is the Reynolds number *Re* of fluid velocity. At low speed the flow happens in a homogeneous manner (Fig. 2.26a). At higher speeds, a new macroscopic pattern with two vortices appears (Fig. 2.26b). With yet higher speeds the vortices start to oscillate (Fig. 2.26c–d). At a certain critical value, the irregular and chaotic pattern of a turbulent flow arises behind the cylinder (Fig. 2.26e). Figure 2.26a–e presents a survey of possible attractors with one and more fixed points, bifurcations, oscillating, and quasi-oscillating attractors, and finally fractal chaos [2.43].

A famous example from modern physics and technology is the laser. A solid state laser consists of a rod of material in which specific atoms are embedded. Each atom may be excited by energy from outside leading it to the emission of light pulses. Mirrors at the end faces of the rod serve to select these pulses. If the pulses run in the axial direction, then they are reflected several times and stay longer in the laser, while pulses in different directions leave it. At small pump power the laser operates like a lamp, because the atoms emit independently of each other light pulses (Fig. 2.27a). At a certain pump power, the atoms oscillate in phase, and a single ordered pulse of gigantic length emerges (Fig. 2.27b) [2.44].

The laser beam is an exampel of macroscopic order emerging by a dissipative (irreversible) self-organization far from thermal equilibrium. With its exchange and processing of energy, the laser is obviously a dissipative system far from thermal equilibrium.

In former days of history, scientists would have postulated certain demons or mystic forces leading the elements of these systems to new patterns of order. But, as in the case of conservative self-organization, we can explain dissipative self-organization by a general scheme which is made precise by well-known mathematical procedures. We start with an old structure, for instance a homogeneous fluid or randomly emitting laser. The instability of the old structure is caused by a change of external parameters, leading eventu-

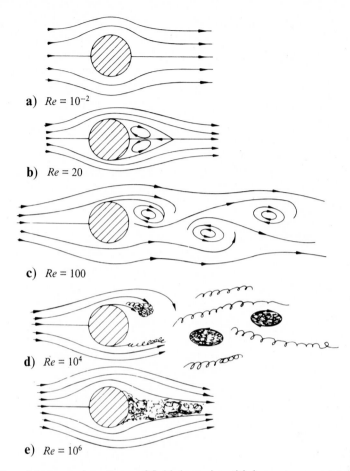

a) $Re = 10^{-2}$

b) $Re = 20$

c) $Re = 100$

d) $Re = 10^4$

e) $Re = 10^6$

Fig. 2.26a–e. Macroscopic patterns of fluid dynamics with homogeneous state (**a**), two vortices (**b**), oscillations (**c**), quasi-oscillations (**d**), and chaos (**e**) behind a cylinder depending on increasing fluid velocity as control parameter

ally to a new macroscopic spatio-temporal structure. Close to the instability point we may distinguish between stable and unstable collective motions or waves (modes). The unstable modes start to influence and determine the stable modes which therefore can be eliminated. Hermann Haken calls this process very suggestively a "slaving principle". Actually, the stable modes are "enslaved" by the unstable modes at a certain threshold.

Mathematically, this procedure is well known as the so-called "adiabatic elimination" of fast relaxing variables, for instance, from the master equation describing the change of probabilistic distribution in the corresponding system. Obviously, this elimination procedure enables an enormous reduction of the degrees of freedom. The emergence of a new structure results

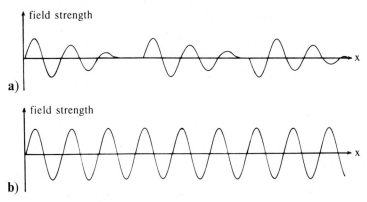

Fig. 2.27a,b. Wave patterns emitted from a lamp (**a**) and from a laser (**b**)

from the fact that the remaining unstable modes serve as order parameters determining the macroscopic behavior of the system. The evolution of the macroscopic parameters is described by differential equations. In contrast to properties of the elements of a system at the microscopic level (for instance, atoms, molecules, etc.), the order parameters denote macroscopic features of the whole system. In the case of the laser, some slowly varying ("undamped") amplitudes of modes may serve as the order parameters, because they start to enslave the atomic system. In the language of biology, the order parameter equations describe a process of "competition" and "selection" between modes. But, of course, these are only metaphoric formulations which can be made precise by the mathematical procedure mentioned above [2.45].

In general, to summarize, a dissipative structure may become unstable at a certain threshold and break down, enabling the emergence of a new structure. As the introduction of corresponding order parameters results from the elimination of a huge number of degrees of freedom, the emergence of dissipative order is combined with a drastic reduction of complexity. Dissipative structures are a fundamental concept of complex systems which are used in this book to model processes in natural and social sciences. The irreversibility of dissipative structures may remind us of Heraclitus' famous quotation that nobody can enter a stream in the same state. Obviously, irreversibility violates the time-invariance symmetry which characterizes the classical (Hamiltonian) world of Newton and Einstein. But the classical view will turn out to be a special case in a steadily changing world. On the other hand, Heraclitus believed in an ordering law harmonizing irregular interactions and creating new order states of matter. We have to see wether the mathematical scheme of a dissipative system will satisfy the universal features of such a law.

A general framework for the evolution of matter would be based on a unified theory of all physical forces (Fig. 2.28). The standard models of cos-

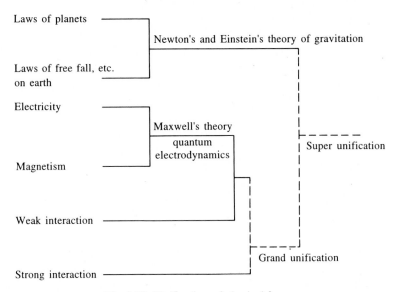

Fig. 2.28. Unification of physical forces

mic evolution which are derived from Einstein's general theory of relativity must be explained by the principles of quantum theory. Until today there are only several more or less satisfying mathematical models of cosmic evolution which can only partially be tested and confirmed by experiments. Nevertheless, it is the general idea of these models that the emergence of structures with increasing complexity (elementary particles, atoms, molecules, planets, stars, galaxies, etc.) can be explained by cosmic phase transitions or symmetry breaking [2.46].

In cosmic evolution an initial state is assumed to be nearly homogeneous and symmetric in the sense that in general no elementary particles can be distinguished, but they can be transformed into one another. During cosmic evolution, critical values have been realized step by step at which symmetries break down by deviations and fluctuations and new particles and forces emerge: "C'est la dissymétrie, qui crée le phénomène," said Pierre Curie [2.47]. But we must be aware that the cosmic processes of symmetry breaking and phase transitions are mathematical extrapolations from experiments and theories in high energy physics.

Nowadays, physics distinguishes four fundamental forces, the electromagnetic, strong, weak, and gravitational forces which are mathematically described by so-called gauge fields. Elementary particle physics aims to unify the four physical forces in one fundamental force corresponding to the initial state of the universe. Electromagnetic and weak forces have already been unified at very high energies in an accelerator ring at CERN (Fig. 2.28).

Unification means that at a state of very high energy the particles that "feel" the weak force (electrons, neutrinos, etc.) and those that "feel" the electromagnetic force cannot be distinguished. They can be described by the same symmetry group (U(1) × SU(2)), i.e., they are invariant with respect to transformations of this group. At a particular critical value of lower energy the symmetry breaks down into partial symmetries (U(1) and SU(2)) corresponding to the electromagnetic and weak forces.

Physically, this kind of symmetry breaking means a phase transition which is connected with the emergence of two new physical forces and their elementary particles. The process of spontaneous symmetry breaking is well known. For instance, our breakfast egg is not stable in its symmetric position on its top. Any tiny fluctuation causes it to fall spontaneously down to an asymmetric, but energetically stable position. The phase transition of a ferromagnet from a non-magnetic to a magnetic state is caused by cooling down the temperature to a critical point. The elementary dipoles spontaneously take one of the two possible magnetic orientations, break the spin-rotation symmetry, and cause the emergence of a new macroscopic property (magnetization).

The complex variety of baryons (protons, neutrons, etc.) and mesons interacting via the strong force are constructed from the so-called quarks with three degrees of freedom, i.e., the so-called "colors" red, green, and blue. A baryon, for instance, is built up from three quarks which are distinguishable by three different colors. These three colors are complementary in the sense that a hadron is neutral (without color) to its environment. The mathematical symmetry group (SU(3)) characterizing the color transformation of quarks is well known.

After the successful unification of the electromagnetic and weak interactions physicists try to realize the "grand unification" of electroweak and strong forces, and in a last step the "superunification" of all four forces (Fig. 2.28). There are several research programs for superunification, such as supergravity and superstring theory. Mathematically, they are described by extensions to more general structures of symmetry ("gauge groups") including the partial symmetries of the four basic forces. Technically, the unification steps should be realized with growing values of very high energy. But the "grand unification" demands states of energy which cannot be realized in laboratories. Thus, the high energy physics of grand unification could only be confirmed by certain consequences which could be tested in a laboratory or observed in the universe (e.g., the decay of protons). The superunification of all forces would demand infinitely increasing states of energy whose physical principles are still unknown.

The theory of the so-called "inflationary universe" assumes an early state of the universe with small size, but very high energy ("quantum vacuum") which expands very rapidly to macroscopic dimensions driven by a repulsive force of the quantum vacuum state ("anti-gravity"). This cosmic phase transition allows one to explain some well-known properties of the observed universe such as the relatively homogeneous distribution of stars and matter. During the inflationary period, some tiny deviations from symmetry and uniformity would have been amplified until they were big enough to account for the observed structures of the universe. In the expanding universe the density of matter varied slightly from place to place. Thus, gravity would have caused the denser regions to slow down their expansion and start contracting. These local events led to the formation of stars and galaxies [2.48].

In general, the emergence of the structural variety in the universe from the elementary particles to stars and living organisms is explained by phase transitions, corresponding to symmetry breaking of equilibrium states (Figs. 2.29, 2.30). In this sense the cosmic evolution of matter is understood as a self-organizing process with the emergence of conservative and dissipative structures. But we must be aware that cosmic self-organization is today only a "regulative idea of research", as Kant had said: we have more or less plausible dynamical models which are more or less empirically confirmed. The very beginning of cosmic evolution is still unknown.

If we only assume the classical principles of Einstein's general relativity, then, as Roger Penrose and Stephen Hawking have mathematically proved, the standard models of cosmic evolution have an initial singularity which may be interpreted as the Big Bang, i.e., the emergence of the universe from a mathematical point. But if we assume a unification of the general theory of relativity (i.e., Einstein's relativistic theory of gravitation) and quantum mechanics with imaginary (instead of real) time, then, as Hawking has mathematically proved, a "smooth" cosmic model is possible without any beginning, which simply exists, according to the mathematical principles of a unified relativistic quantum physics [2.49].

Philosophically, this model reminds us of Parmenides' world of unchangeable being. But the uncertainty principle of quantum mechanics implies that the early universe cannot have been completely uniform because there must have been some uncertainties or fluctuations in the positions and velocities of the particles. Thus, the universe would have undergone a period of rapid expansion which is described by the inflationary model, leading to our complex universe in the long run. The equilibrium of the Parmenidean world broke down and changed to the evolutionary and complex world of Heraclitus, caused by a basic principle of quantum physics under the hypothesis of a "smooth" time without singularities.

A cosmological model of an "eternal" universe without beginning and without end was already introduced by Hermann Bondi, Thomas Gold, and Fred Hoyle in 1948. These authors did not only assume spatial homogeneity and isotropy of the universe at every time ("Cosmological Principle" of the standard models with Big Bang), but also temporal homogeneity and

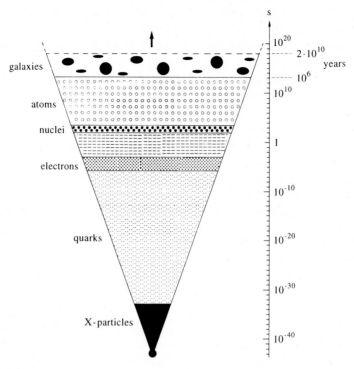

Fig. 2.29. The emergence of structural variety in the universe from elementary particles to galaxies [2.48]

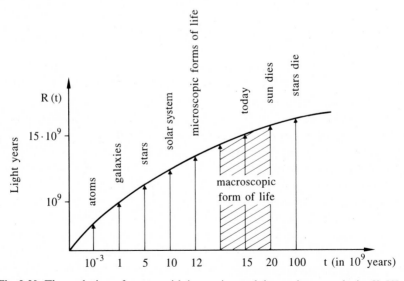

Fig. 2.30. The evolution of matter with increasing and decreasing complexity [2.50]

isotropy: Their "perfect Cosmological Principle" suggests that the universe globally looks the same not only at all points and in all directions, but at all times, leading to a steady state model. According to Hubble, there is a correlation between the red shift and the increasing distances of expanding galaxies. So if the average number of galaxies per unit proper volume is to remain constant, new galaxies must appear to fill up the holes in the widening comoving coordinate mesh. An ad-hoc hypothesis of steady state cosmology was the necessity of continuous creation of matter.

In recent quasi-steady state cosmologies, the strange assumption of a contingent and nonlocal creation of matter is explained by the local birth of new galaxies everywhere and at every time in the universe. The conditions of local big bangs are assumed to be realizable in the supermassive centers of old galaxies. The red shifts seems also to indicate the age of a galaxy. The uniform evolution with the sequential emergence of elementary particles, atoms, molecules, galaxies, stars, etc. after the global Big Bang (Fig. 2.29) is replaced by an autocatalytically self-reproducing universe without global beginning and without end, but with local births, growths, and deaths of galaxies. In this case, old dying galaxies create the matter of new galaxies like plants and organisms bearing the seed of new life. The universal dynamics would be a gigantic never ending nonlinear recycling process of matter [2.49].

From a theological point of view, these models do not need any creator, because their worlds simply have been and will be self-contained and self-organizing without beginning and without end. From a mathematical point of view, these models may be very elegant. But from a methodological point of view, we must conclude that we do not yet have a complete and consistent theory combing quantum mechanics and relativistic gravity which could explain the evolution of matter with its increasing complexity. Thus we are only certain of some of the properties that such a unified theory could have.

2.5 Complex Systems and the Self-Construction of Materials

In the evolution of matter, self-organizing processes can be observed from the level of elementary particles to the cosmic structures of galaxies. They are not only interesting from an epistemic point of view, but for applications in materials and life science, too. At the boundary between materials science and life science, supramolecular systems play a tremendous role. In this case molecular self-organization means the spontaneous association of molecules under equilibrium conditions into stable and structurally well-defined aggregates with dimensions of $1-10^2$ nanometers (1 nm = 10^{-9} m = 10 Å).

Nanostructures may be considered as small, familiar, or large, depending on the view point of the disciplines concerned. To chemists, nanostructures are molecular assemblies of atoms numbering from 10^3 to 10^9 and molecular weights of 10^4 to 10^{10} daltons. Thus, they are chemically large supramolecules. To molecular biologists, nanostructures have the size of familiar objects from proteins to viruses and cellular organelles. But to materials scientists and electrical engineers, nanostructures are at the current limit of microfabrication and thus they are rather small [2.51].

Nanostructures are complex systems which evidently lie at the interface between solid-state physics, supramolecular chemistry, and molecular biology. It follows that the exploration of nanostructures may deliver hints about the emergence of life and about the fabrication of new materials. But engineering of nanostructures cannot be mastered in the traditional way of mechanical construction. There are no man-made tools or machines for putting together their building blocks like the elements of a clock, motor, or computer chip. We must thus understand the principles of self-organization which are used by nanostructures in nature. Then, we only need to arrange the appropriate constraints under which the atomic elements of nanostructures associate themselves in a spontaneous self-construction: The elements adjust their own positions to reach a thermodynamic minimum without any manipulation by a human engineer.

Historically, the idea of supramolecular interactions dates back to a famous metaphor of Emil Fischer (1894), who described a selective interaction of molecules as the lock and key principle. Today, supramolecular chemistry has far surpassed its original focus. Molecular self-assemblies combine several features of covalent and noncovalent synthesis to make large and structurally well-defined assemblies of atoms. The strengths of individual van der Waals interactions and hydrogen bonds are weak relative to typical covalent bonds and comparable to thermal energies. Therefore, many of these weak noncovalent interactions are necessary in order to achieve molecular stability in self-assembled aggregates. In biology, there are many complex systems of nanoscale structures such as proteins and viruses which are formed by self-assembly. Living systems sum up many weak interactions between chemical entities to make large ones. How can one make structures of the size and complexity of biological structures, but without using biological catalysts or the informational devices coded in genes?

Many nonbiological systems also display self-organizing behavior and furthermore provide examples of useful interactions. Molecular crystals are self-organizing structures. Liquid crystals are self-organized phases intermediate in order between crystals and lipids. Micelles, emulsions, and lipids display a broad variety of self-organizing behavior. An example is the generation of cascade polymers yielding molecular bifurcational superstructures of fractal order [2.52]. Their synthesis is based on the architectural design of trees. Thus, these supramolecules are called dendrimers (from the Greek word *dendron* for tree and polymer). The generation of dendrimers has followed two basic procedures for monomer addition. A divergent construction begins at the core and builds outward via an increasing number of repeating bifurcations. A convergent construction begins at the periphery and builds inward via a constant number of transformations. The divergent construction transforms the chemical reaction centers from the center into the periphery, generating a network of bifurcating branches around the center. The bifurcations increase exponentially up to a critical state of maximal size. They yield fractal structures such as molecular sponges which can contain smaller molecules, which can then be dispersed in a controlled way for medical applications.

Examples of cave-like supramolecules are the Buckminsterfullerenes, forming great balls of carbon [2.53]. The stability of these complex clusters is supported by their high geometric symmetry. The Buckminsterfullerenes are named after the geodesic networks of ball-like halls which were constructed by the American architect Richard Buckminster Fuller (1895–1983). The cluster C_{60} of 60 carbon atoms has a highly Platonic symmetry of atomic pentagons forming a completely closed spheroid.

Cave-like supramolecules can be arranged using chemical templates and matrices to produce complex molecular structures. Several giant clusters comparable in size to small proteins have been obtained by self-assembly. Figure 2.31 shows a ball-and-stick model of the largest discrete cluster (700 heavy atoms) ever characterized by X-ray structure analysis. This cluster containing 154 molybdenum, 532 oxygen, and 14 nitrogen atoms has a relative molecular mass of about 24 000. The highly symmetric "big wheel" was synthesized by Achim Müller and coworkers [2.54]. Giant clusters may have exceptional novel structural and electronic properties: There are planes of different magnetization which are typical for special solid-state structures and of great significance for materials science. A remarkable structural property is the presence of a nanometer-sized cavity inside the giant cluster. The use here of templates and the selection of appropriate molecular arrangements may well remind us of Fischer's lock and key principle.

Molecular cavities can be used as containers for other chemicals or even for medicaments which need to be transported within the human organism. An iron-storage protein that occurs in many higher organisms is ferritin. It is an unusual host-guest system consisting of an organic host (an aprotein) and a variable inorganic guest (an iron core). Depending on the external demand, iron can either be removed from this system or incorporated into it. Complex chemical aggregates like polyoxometalates are frequently discovered to be based upon regular convex polyhedra, such as Platonic solids. But their collective electronic and/or magnetic properties cannot be deduced from the known properties of these building blocks. According to the catchphrase "from molecules to materials" supramolecular chemistry applies the "blue-prints" of conservative self-organization to build up complex materials on the nanometer scale with novel catalytic, electronic, electrochemical,

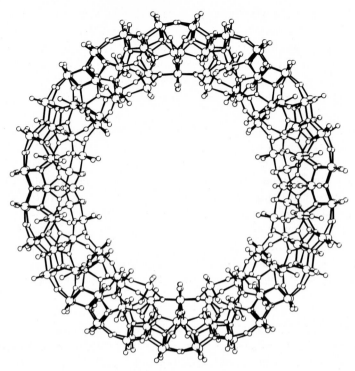

Fig. 2.31. Giant supramolecular cluster ("big wheel") in a ball-and-stick representation: An example of a complex near-equilibrium system [2.54]

optical, magnetic, and photochemical properties. Multi-property materials are extremely interesting.

Supramolecular transistors are an example that may stimulate a revolutionary new step in the development of chemical computers. Actually, there is a strong trend towards nanostructures in electronic systems which may realize small, fast devices and high-density information storage. But one can also imagine nonelectronic applications of nanostructures. They could be used as components in microsensors or as catalysts and recognition elements in analogy to enzymes and receptors in living systems. In natural evolution very large complex molecular systems are also produced by stepwise gene-directed processes. The conservative self-organization processes of nanomolecular chemistry are non-gene-controlled reactions. Only a clever combination of conservative and non-conservative self-organization could have initiated prebiotic evolution before genes emerged. But even during the evolution of complex organisms, conservative self-organization must have occurred. Open ("dissipative") physical and chemical systems lose their structure when the input of energy and matter is stopped or changed (e.g., laser, BZ-reaction). Organismic systems (like cells) are able to conserve much of their structure at least for a relatively long time. On the other hand, they need energy and matter within a certain interval of time to keep their structure more or less far from thermal equilibrium. In the technical evolution of mankind, the principles of conservative and dissipative self-organization have once more been discovered and open new avenues of technical applications.

The complex systems approach enables engineers to endow materials with more and more of the attributes of living organisms. Self-regulation and self-adaption to a changing environment are well-known capabilities of living systems. They can be considered as specific forms of self-organizing open systems in a changing environment. Analogously, engineers aim to create complex materials systems that can sense their own state, the state of their environment and respond to it. Dramatic examples are materials for bridges that could detect and counter corrosion before a pylon gives way, buildings that could brace themselves against seismic waves, or skins of aeroplanes that could spontaneously react against dangerous material fatigue.

Actuators are materials which can change their features according to changing states of the system [2.55]. Examples are piezoelectric ceramics and polymers acting either as pressure sensors or as mechanical actuators. The electrical polarity of their crystal or molecular structures allows a transformation of mechanical forces exerted on them into electrical current or, conversely, a transformation of electrical stimuli into vibrations. Piezoelectric polymers could be embedded in the skin of a robotic hand in order to get a high degree of sensitivity (e.g., to decipher braille).

Other examples are alloys with a so-called shape memory that can be used as actuators. Below a certain control value of transition temperature, a shape-memory wire will take any shape it is bent into. When the wire is heated beyond the transition, it returns to its original shape. Engineers propose the incorporation of a shape-memory metal into a material system in its low- temperature shape. It exerts a force whenever it is heated. The force-generating transition takes place as the atoms in the alloy's crystal grains toggle between different geometric arrangements. Damage-resisting bridges or airplane wings would be possible applications of these control structures.

There are even actuator materials that can reversibly transform their mechanical properties from a liquid to a solid state. They consist of fine polarizable particles of ceramic or polymer suspended in a liquid such as silicone oil. When subjected to strong electric fields, such fluids organize themselves into filaments and networks which stiffen the material into a gel-like solid. When the electric field is removed, the organization dissipates, and the material becomes fluid again. Other applications are optical fibers acting as sensor materials. The properties of these hair-thin fibers are affected by changes in temperature, pressure or other physical or chemical conditions within the materials. They can be considered as "glass nerves" providing optical signals of the material's internal "health".

Sometimes, modern materials scientists call their systems smart or even "intelligent" materials. The goal of their research is sometimes described as "animation of the inanimate world" [2.56]. From a philosophical point of view, this slogan seems to hark back to alchemistic traditions. Some philosophers of science may perhaps criticize the vocabulary of materials scientists as non-scientific animism. But, from the view point of complex systems, there is a hard scientific core. Properties of self-organization are not necessarily combined with conscious behavior on the basis of nervous systems. They even do not necessarily depend on biological catalysts or the informational devices coded in genes. Thus, there is no break between the so-called inanimate and animate world. In the evolution of matter, we observe systems with more or less high degrees of organization. It is clear that we have only made the very first steps in understanding their full potential.

3. Complex Systems and the Evolution of Life

How can one explain the emergence of order in the Darwinian evolution of life? In the history of philosophy and biology, life was explained teleologically by non-causal ("vital") forces aiming at some goals in nature. In a famous quotation Kant said that the "Newton for explaining a blade of grass" could never be found (Sect. 3.1). Boltzmann could show that living organisms are open dissipative systems which do not violate the second law of thermodynamics: Maxwell's demons are not necessary to explain the arising order of life in spite of the increasing entropy and disorder in closed systems according to the second law. Nevertheless, in the statistical interpretation from Boltzmann to Monod the emergence of life is only a contingent event, a local cosmic fluctuation at the boundary of the universe (Sect. 3.2). In the framework of complex systems the emergence of life is not contingent, but necessary and lawful in the sense of dissipative self-organization. The growth of organisms and species is modeled as the emergence of macroscopic patterns caused by nonlinear (microscopic) interactions of molecules, cells, etc., in phase transitions far from thermal equilibrium (Sect. 3.3). Even ecological populations are understood as complex dissipative systems of plants and animals with mutual nonlinear interactions and metabolism with their environment (Sect. 3.4). Spencer's idea that life is determined by a structural evolution with increasing complexity seems to be mathematized by complex dynamical systems. Is the "Newton of life" found? The theory of complex dynamical systems does not explain what life is, but it can model how forms of life can arise under certain conditions. Thus, the existence of our life is still a wonder for us as well as for our ancestors, even if we shall eventually model the complex dynamics of life.

3.1 From Thales to Darwin

Before we discuss complex systems and the evolution of life, let us a have a glance at early philosophies of life [3.1]. It is a surprising fact that many aspects of modern ecology remind us of early ideas of self-organisation. In mystic interpretations, life was understood as a cyclic movement of growth and decay, birth and death. Animals and humans only survived in adapting to the great cycles of nature such as high tide and low tide, the change of seasons, changing constellations of stars, fertile and sterile periods of nature, and so on. Nature itself seemed to be a great organism, and humans were considered as partly involved in its natural development. Mythologies of natural religions and their rituals were used to conjure the forces of nature and to live in harmony with the natural order.

Mythology was given up in favour of natural philosophy, when people asked for the basic principles of life, and when they no longer accepted demons and gods as personified forces of nature. In the 6th century B.C. the presocratic philosopher Thales of Miletus declared water to be the fundamental source of life. Anaximander seems to have some early ideas of evolution:

> It is said that in the wet element the first living beings come to be, in a husk of prickly rinds; with increasing age they climbed onto the dry element, the rind tore of on all sides and so, for a short time they took on a different live form. [3.2]

Regarding the derivation of humans, Anaximander expressed an utterly modern conception. Observing the long period that human children need for care and protection, he drew the conclusion that if they had always required that, humans would have not been able to survive. So, earlier they must have been different. Empedocles explained the processes of life with certain mixtures and transformations of the familiar elements water, air, fire, and earth.

While these organic explanations of life seemed to be intuitively convincing for former contemporaries, Democritus' atomism with his reduction of life to the interaction of invisible atoms was considered as rather abstract. Even consciousness and the soul of man was explained by microscopic interactions of tiny material elements. So, Democritus and his school were not only attacked as materialists, but as atheists too. Plato tried to model the first elements of matter and their combination by geometric figures and constructions.

From a modern scientific point of view, Democritus' atomism and Plato's mathematical models were early reductionist programs for life. They tried to reduce physiological and biological processes to the interactions of physical elements. But the idea of explaining the changing and pulsating processes of life on the basis of the rigid and dead figures of geometry or material atoms must have seemed thoroughly unnatural, speculative, and far-fetched to the contemporaries of that time. In short, "real" life seemed to be hopelessly "complex", and Euclid's mathematics too "simple". So, Euclid's mathematics was reserved for the "superlunar" world of stars, but not applied to the "sublunar" world of earthly life.

This is where the Aristotelian philosophy of life begins. While Plato, in the Pythagorean tradition, drew his concept from geometry, Aristotle formulated his concept of processes in nature mainly on the basis of the ways in which living organisms such as plants and animals function. The processes and courses of life are known to us from everyday experience. What is more obvious than to compare and explain the rest of the world, which is unknown and strange, with the familiar? According to Aristotle, the task of physics is to explain the principles and functions of nature's complexity and changes. In modern terms, Aristotle rejected atomic reductionism as well as the mathematization of life as speculative and unrealistic.

Life was defined by the feature of self-movement, in contrast to a dead stone which must be pushed from outside in order to move. In this Aristotelean sense, life meant "having a soul", which was understood as an organizing force (entelechy) of matter (vitalism). In modern terms, the self-organisation of life was interpreted by Aristotle as a functionally governed process aiming at certain "attractors" of purposes (teleology). For instance, a tree grows out of a seed with the purpose of reaching its final form. In modern terms, the change of forms characterizing the growth of an organism is something like the (qualitative) evolution of an order parameter which Aristotle called the "potentiality" of that organism. But, of course, the main difference compared to modern concepts of order parameters is the fact that Aristotle criticized any reduction of macroscopic forms to atomic or microscopic interactions.

It is noteworthy that Aristotle proposed a continuous scale of more or less animated states of nature (*scala naturae*) and denied an absolute contrast of "alive" and "dead". He was always seeking for the intermediate or connecting links between organisms with different complexity. For instance, for a Greek like Aristotle, living by the Mediterranean Sea with its plentiful flora and fauna, it was easy to observe organisms like water-lilies "which may be doubted to be animals or plants, because they grow to the floor like plants, but eat fishes like animals" [3.3]. On the background of continuity, Aristotle suggested a kind of biogenetic law: "In the beginning the fetus of an animal seems to have a kind of life like a plant; during its later development, we may speek of a sensitive and thinking soul" [3.4].

Aristotle was not only a theorist, but one of the first observing botanists, zoologists, and physiologists. He designed a taxonomy of plants and animals according to different features, and tried to describe the physiological processes of life. His leading paradigm of life was the idea of a self-organizing organism, rejecting any atomic, molecular, or anorganic reductionism. Aristotle's philosophy of life has overshadowed the development of biology until today.

In the Roman period, even medicine was influenced by Aristotelean tradition. Galen, the physician of the Roman emperor Marcus Aurelius, taught that organs had to be adapted completely to their functions in our body. Following Aristotelean teleology, he described the digestive organs selecting the "purposeful" parts of food for the life processes and separating the "useless" ones. In the Middle Ages, Albertus Magnus combined Aristotelean philosophy of life with Christianity. On the background of Aristotle's teleology, Albertus developed an early ecology demanding that humans have to live in harmony with their natural environment. Organisms and their environments are connected with each other by numerous exchanges of air, food, excreta, etc. which are in a natural balance ("equilibrium") governed by divine ordinances. Albertus thought that even the health of the human soul depends on a healthy environment with healthy air, climate, plants, and animals. Soul and body are not separated, but an organic whole.

The decisive condition of modern physics was the connection of mathematics, observation, experiment, and engineering which was realized by Galilei in the Renaissance. Newton founded a new mathematical and experimental philosophy of nature which he called *Philosophiae naturalis principia mathematica* (1687). Geometry and mechanics became the new paradigm of natural sciences. In the history of science this period is called the mechanization of nature, which was imagined to be nothing else than a huge mechanical clock. The mathematician and philosopher René Descartes and the physicist Christian Huygens taught that every system in nature consists of separated elements like the cog wheels of a clock. Every effect of nature was believed to be reducible to linear causal chains like sequences of cog wheels of a clock. Obviously, Cartesian mechanism is contrary to Aristotelian holism.

Even the physiology of life processes should be explained mechanically. The heart, for instance, was considered as a pumping machine. In general, Descartes believed that the motions of an animal and human body can be derived from the mechanism of organs "and that with the same necessity as the mechanism of a clock from the position and form of its weights and wheels" [3.5]. The anatomy of human bodies which was practised by dissection since the Renaissance was an application of the analytical method of Descartes. According to Descartes, each system can be separated into its basic building blocks, in order to explain its functions by the laws of geometry and mechanics.

The Italian physicist and physiologist Borelli (1608–1679) founded the so-called iatrophysics as an early kind of biophysics. He transferred a famous quotation of Galileo from physics to biology, and declared in his book *De motu animalium* (About motions of animals) emphatically:

As the scientific recognition of all these things is founded on geometry, it will be correct that God applied geometry by creating animal organisms, and that we need geometry for understanding them; therefore it is the only and suitable science, if one wants to read and understand the divine script of the animal world. [3.6]

While Descartes still believed in an immortal soul of man, Lamettrie reduced man to an automaton without soul, according to his motto *L'homme machine* (1747). Human and animal bodies were only distinguished by their level of complexity and organization. After physics, teleology in the tradition of Aristotle should be eliminated in physiology and medicine, too. During the Enlightenment, the mechanism of life was understood as materialistic and atheistic philosophy. The following story by Voltaire about Lamettrie is rather amusing: when Lamettrie suddenly fell ill during a plentiful banquet, and died because of indigestion a few days later, the God-fearing contemporaries were said to be thankful about the fact that a materialist had to die because of his own insatiability.

Nevertheless, some Aristotelean concepts were discussed during the age of mechanization. For instance, Leibniz assumed a hierarchical order of nature with a continuous scale of animation from the smallest building blocks ("monads") to the complex organisms. Leibniz tried to combine Aristotelean ideas with physical mechanics, and became one of the early pioneers for a theory of complex dynamical systems. Concerning the status of man in nature, Leibniz declared:

Thus, every organic body of a living being is a kind of divine machine or natural automaton surpassing all artificial automata infinitely. [3.7]

Inspired by Leibniz, the zoologist Bonnet (1720–1793) proposed a hierarchy of nature ("Echelles des êtres naturelles") with a measure of complexity which seems to be rather modern. Bonnet underlined "organization" as the most important feature of matter. An organization realizing the most effects with a given number of different parts is defined as the most perfect one [3.8].

At the end of the 18th century, Immanuel Kant criticized the application of Newtonian mechanics to biology: "The Newton explaining a blade of grass cannot be found." The main reason for Kant's critique is that in the 18th century the concept of a machine was only made precise in the framework of Newtonian mechanics. Thus, in his famous *Kritik der Urteilskraft*, Kant wrote that an organism "cannot only be a machine, because a machine has only moving force; but an organism has an organizing force ... which cannot be explained by mechanical motion alone" [3.9]. Kant also criticized Aristotelean teleology and the assumption of "aims" and "purposes" in nature as a metaphorical anthropomorphism. An organism must be described by the model of a "self-organizing being".

Like Kant, Goethe rejected the materialistic-mechanical explanation of life which was defended by, e.g., the French encyclopedist Holbach in his

Système de la Nature. For Goethe, the mechanist model of nature is "grey, ... like death ... like a ghost and without sun" [3.10]. He believed that life develops organically and harmonically like the metamorphosis of a plant or the mental maturity of man.

On the background of Goethe's age and Kant's critique of mechanistic rationalism, a romantic philosophy of nature arose in Germany in the beginning of the 19th century. It was a renaissance of the organic paradigm against mechanism. Friedrich Schelling (1775–1854) designed a "science of the living" assuming that organization and reproduction are main features of the living [3.11]. Oken (1779–1851), physician and philosopher of nature, described a "planetary process", in which living organisms were explained by a synthesis of magnetism, chemism, and galvanism. From a modern point of view, "self-organization" and "self-reproduction" were far-reaching concepts of the romantic philosophy of nature. But in those days they were only speculations or inspired intuitions, because the experimental and mathematical base was still missing.

The peaceful picture of an organic and harmonic metamorphosis was soon pushed aside by biology. Charles Darwin's theory of evolution does not need teleological forces to explain life. The "survival of the fittest" (Herbert Spencer) depends on the greater advantage of selection with respect to certain conditions of the environment (for instance nourishment, climate) [3.12]. Darwin was inspired by some ideas of Lamarck (1744–1829), for instance the heredity of acquired properties. Darwin's evolution is governed by (genetic) variability of species ("mutation") and natural selection driving development in one direction. Spencer taught that life is driving to more complexity, controlled by selection. Many contemporaries considered Darwinism not only as a theory of natural sciences. Darwin's theory seemed to present a scenario of life with a strong analogy to the society of the 19th century. The "selection of the fittest" became a slogan of the "social Darwinism" as a political attitude.

In the second half of the 19th century, Haeckel generalized the evolution of life from monocellular organisms to humans. But in those days the theory of evolution could not be compared with the highly confirmed physical and chemical theories. Darwin could only deliver some comparative studies of morphology. He described the variability of species and natural selection, but he could not explain it by mathematized and testable laws like physics. Mendel's laws of heredity (1865) were still unknown to Darwin as well as to many contemporaries. Nevertheless, one of the great physicists of the 19th century, Ludwig Boltzmann, declared, casting a retrospective glance at his century:

When you ask me for my deepest conviction if our century will sometimes be called the iron century or the century of steam or the century of electricity, then I answer without hesitation, it will be called the century of Darwin. [3.13]

3.2 Boltzmann's Thermodynamics and the Evolution of Life

In the 19th century, the dominant topics of natural science, social science, and philosophy became "evolution" and "history". While the biological sources of these ideas date back to Darwin's theory of evolution, physical examples of irreversible processes were at first discussed in thermodynamics. The initial principles of thermodynamics were developed by Carnot (1824). His principles were discovered in analyzing mechanical forces produced by steam engines. Roughly speaking, the first law of thermodynamics says that energy cannot be created or destroyed. Despite mechanical work, electrical energy, or chemical transformations that energy is constantly undergoing in nature, the total energy within a closed system remains unchanged. In accordance with Einstein's equivalence of mass and energy (compare Sect. 2.2), the first law has been enlarged to a conservation principle of mass and energy in this century.

The basic importance of the second law in the context of physical evolution was recognized by Clausius (1865) who borrowed the term "entropy" from the Greek word for evolution or transformation [3.14]. Mathematically, the entropy change of a system is defined by the reversible heat addition to the system divided by its absolute temperature. According to Ilya Prigogine, one must refer to the fact that every system has surroundings [3.15]. Thus, the variation of entropy during a time moment is more generally the sum of the rate at which entropy is supplied to the system by its surroundings and the rate at which entropy is produced inside the system. The second law of thermodynamics demands that the rate at which entropy is produced inside the system is greater than or equal to zero. For closed and isolated systems without an entropy supply from (or sink to) the surroundings we get the classical statement of Clausius that entropy increases or remains constant when thermodynamic equilibrium has been reached. In other words, there is no process in nature involving physical, chemical, biological, or (as we shall see) informational transformations occurring spontaneously without some energetic cost in terms of entropy.

Entropy is a macroscopic property of systems like volume and size. Therefore, thermodynamics was at first only a phenomenological theory describing possible heat distributions of macroscopic systems. Boltzmann was not happy with this positivistic attitude and tried to deliver a statistical-mechanical explanation reducing such macroscopic states of systems as, e.g., heat to the mechanics of microscopic molecules. Inspired by the microstatemacrostate distinction which has become crucial for the theory of evolution, Boltzmann gave thermodynamics its first statistical interpretation [3.16]. Irreversibility in statistical thermodynamics is based on this distinction.

In general, statistical mechanics explains a macrostate like density, temperature, etc., by microstates. In this sense, an observable macrostate is said to be realized by a large number W of microstates. In order to define the number W, a large number of independent mechanisms of the same kind like atoms, molecules, crystals, etc., are considered. They develop their microstates according to their equations of motion with different initial phase states. If a macrostate is realized by W microstates of this kind, then Boltzmann's entropy quantity H of the corresponding macrostate is assumed to be proportional to the logarithm of W, i.e., $H = k \ln W$ with the Boltzmann constant k. In a continous phase space, the Boltzmann expression can be generalized by an integral of a velocity distribution function. For Boltzmann, H is a measure for the probability of molecular arrangements corresponding to the observable macrostates of the system.

Boltzmann's reductionism met historically with violent objections from physicists, mathematicians, and philosophers. Positivistic physicists and philosophers like Ernst Mach criticized Boltmann's hypothesis of molecules and atoms which were not empirically confirmed in those days. But after their successful discovery this critique is only of historical interest.

One of the most important objections is Loschmidt's reversibility paradox. Since the laws of mechanics are invariant (symmetric) with respect to the inversion of time, to each process there belongs a corresponding time-reversed process. This seems to be in contradiction with the existence of irreversible processes. Boltzmann answered that the second law of thermodynamics in the form of his so-called H-theorem cannot be derived only from the (reversible) mechanical laws, but requires the additional assumption of extremely improbable initial conditions, too. The second law is assumed to hold true with very high probability, but not with security. Irreversible processes are only frequent or probabilistic, reversible ones seldom and improbable. Thus, the second law allows local deviations or fluctuations (for instance Brownian motion) [3.17].

Another objection, by Henri Poincaré and Ernst Zermelo, underlined that each state of a mechanical system with finitely many degrees of freedom must recur at least approximately after a certain time [3.18]. Thus, an arrow of time connected with an increase of entropy cannot exist. Boltzmann answered that the times of return become extremely long with increasing number of degrees. Cosmologically, there are two possible points of view in the sense of Boltzmann: (1) the universe started with extremely improbable initial conditions, or (2) when the universe is large enough, there may be deviations from the equal distribution in some places. Figure 3.1 illustrates Boltzmann's hypothesis of fluctuations. He assumed that the whole universe is in thermal equilibrium, i.e., in maximum disorder. Boltzmann believed that both directions of time are completely symmetric. So, the curve of local entropy increases similarly in both directions of time, becoming flat with maximum entropy [3.19].

Life as a developing system of order is only possible in regions with strongly changing entropy, i.e., on the two slopes of the entropy curve in Fig. 3.1. The two arrows denote Boltzmann's local worlds in which life may occur. So, in the sense of Boltzmann, there cannot be an objectively unique arrow of time, but only one of the two possible directions of increasing entropy which people subjectively experience, living in one of the two possible local worlds on the slopes in Fig. 3.1. Before we criticize Boltzmann's view in detail, let us have a glance at his theory of life against the background of his thermodynamics of thermal equilibrium, which overshadowed science in this century until recently.

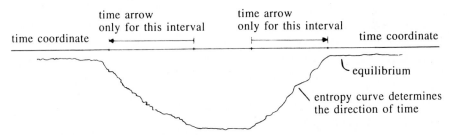

Fig. 3.1. Entropy curve in Boltzmann's universe in thermal equilibrium with symmetric directions of time

Ludwig Boltzmann (1844–1906) was the first scientist who tried to reduce the biological theory of evolution to the thermodynamics and chemistry of the 19th century. For scientists at the end of the last century, it was a great problem that the second law of thermodynamics seemed to forecast the final disorder, death, and decay of nature, while Darwinian evolution seemed to develop living systems of order with increasing complexity. Of course, the second law is reserved to closed systems, and living systems must be open, in a permanent exchange of energy, matter, and information with their environment. Nevertheless, how is the local increase of complexity possible in a sea of disorder and thermal equilibrium?

Boltzmann suggested some explanations which already remind us of modern biochemical concepts of molecular autocatalysis and metabolism. The origin of first primitive living beings like cells was reduced to a selection of inanimate molecular building blocks which Boltzmann imagined as a process like Brownian motion. Plants as cellular aggregates are complex systems of order. Thus, in the sense of the second law of thermodynamics they are improbable structures which must fight against the spontaneous tendency of increasing entropy in their body with sunlight. Because of the high temparature of the sun, the earth gets energy with relatively low entropy which can be used to compensate the spontaneous increase of entropy in the plants. This process is realized by photosynthesis, which was physically explained by Boltzmann in 1886:

> Thus, the general struggle of life is neither a fight for basic material ... nor for energy ... but for entropy becoming available by the transition from the hot sun to the cold earth. [3.20]

Boltzmann extended his physically founded theory of evolution to the history of the nervous system and the emergence of memory and consciousness. He argued that the sensitivity of primitive organisms to outer impressions has led to the development of special nerves and organs of seeing, hearing, feeling, moving, etc.:

> The brain is considered as the apparatus or the organ to develop models of the world. Because of the great utility of these models for the survival of the race, the human brain has been developed according to Darwin's theory with the same perfection as the giraffe's neck or the stork's bill. [3.21]

Even the ability to develop concepts and theories was explained by evolution. Boltzmann tried to justify human categories of space, time, and causality as tools developed by the brain for the racial survival of the fittest. He did not hesitate to extend biological evolution even to the socio-cultural development and history of mankind. In 1894, the Viennese physician S. Exner wrote about the subject "Morality as weapon for the struggle of life" in the sense of Boltzmann. In 1905, Boltzmann himself gave a lecture with the amazing title "Explanation of the entropy law and love by the principles of the probability calculus". Obviously, Boltzmann's Darwinism had reached its limits.

At the beginning of this century, life still could not be explained by physical and chemical foundations. Classical mechanics, the foundation of natural sciences in the 17th and 18th centuries, assumed deterministic and time-reversible laws of nature delivering no explanation of irreversible processes of life. A frictionless pendulum clock moves time-reversibly as an oscillating mechanical system, in principle without limit. Humans are born, grow, and die – why? The thermodynamics of the 19th century deals with irreversible processes of closed systems being driven to a state of maximum entropy or disorder. But how can the development of complex living systems be explained? In the sense of Boltzmann's statistical interpretation, the emergence of order and biological complexity can only be an improbable event, a local cosmic fluctuation "at the edge of the universe" (as Jacques Monod said later), which will disappear without significance for the whole universe in thermal equilibrium [3.22]. Following Monod, we have only the philosophical choice of an existentialism à la Camus, to perish in a finally senseless biological and cultural evolution with human dignity. The tragic death of the genius Ludwig Boltzmann, who committed suicide in 1909, seems to be a symbol of this attitude. But Boltzmann's thermodynamics did not definitely explain the origin of life. He only proved that his statistical interpretation of the second law is not contrary to Darwinian evolution.

After classical mechanics in the 17th and 18th centuries and thermodynamics in the 19th century, quantum mechanics has become the fundamental theory of physics. In spite of Heisenberg's principle of uncertainty, the laws of quantum mechanics and classical mechanics are characterized by the reversibility of time. Concerning the reductionist program for treating complexity, it was a great success that the quantum chemistry of molecules could be explained by the laws of quantum mechanics. In 1927, Heitler and London succeeded in modifying Schrödinger's equation of atomic and subatomic systems for molecules. There are no particular chemical forces in chemistry besides the well-known physical forces. After physics, teleology seemed to be eliminated in chemistry.

But is chemistry completely reducible to physics [3.23]? Definitely only in a restricted sense! The structural models of molecular orbits can only be introduced by abstraction from quantum mechanical correlations. While, for instance, the electrons of an atom cannot be distinguished in the sense of the Pauli principle, they are used by chemists as quasi-classical objects moving around the atomic nucleus in well-distinguished orbits. There are well-known chemical procedures of abstraction (Born–Oppenheimer and Hartree–Fock procedure) for introducing electronic orbits in approximately quasi-classical models of the non-classical quantum world. Further on, we have to consider practical limitations of computability with Schrödinger's equations for complex molecules in spite of all the fantastic successes of numerical quantum chemistry. This weak reduction of chemistry to physics seemed to prove that scientists should continue on the path of reductionism, in order to reduce elementary particles, atoms, molecules, cells, and finally organisms to physics and chemistry.

In the 1920s and 1930s, the struggle between physical reductionism and neovitalism could not actually be decided. For example, the physicist Heitler, the biologist Driesch, and the philosophers Bergson and Whitehead supported explicit neovitalistic opinions in the Aristotelean tradition [3.24]. They argued that particular biological laws may sometimes invalidate the laws of physics and chemistry. From Aristotle to Goethe and Schelling, teleological self-organization and the spontaneity of life from living cells to consciously acting humans have been mentioned to demonstrate that physical reductionism is impossible. Wholeness is a primary feature of an organism which cannot be reduced to the sum of its building blocks. Inspired by Niels

Bohr's so-called Copenhagen interpretation of quantum mechanics, some physicsts tried to mediate between physicalism and vitalism with Bohr's concept of complementarity. Bohr used complementarity to justify such excluding concepts of quantum mechanics as, e.g., particle–wave dualism. Thus, complementarity is assumed for the two classes of physical-chemical and biological laws, which are believed to be incommensurable. We must remember that complementarity is not a physical law, but a philosophical interpretation of quantum mechanics which was not supported by Erwin Schrödinger. He knew that in the 1930s and 1940s the struggle of physicalism and vitalism could not be decided, and complementarity was only a concept to describe the status quo. In his book *What is Life?* Schrödinger wrote:

> After all we have heard about the structure of living matter we must be ready for the fact that it works in a way which cannot be reduced to the usual physical laws. The reason is not that a "new force" or something similiar governs the behavior of the single atoms in a living organism, but because its structure differs from all that we have ever studied in our laboratories. [3.25]

Schrödinger recalls the image of an engineer who is familiar with a steam engine and wants to explore an electromotor. As the two motors work in quite different ways, he will hit upon the idea that the electromotor is driven by a ghost. In the tradition of Leibniz, Schrödinger expects to understand a living organism as "the finest master piece which was ever achieved according to the leading principles of God's quantum mechanics" [3.26].

The problem for Schrödinger was that he as well as Monod tried to describe the emergence of order and life in the framework of Boltzmann's thermodynamics. He was right with his critique of teleological forces or ordering demons, which were even postulated by physicists at the end of the last century. The fiction of a demon which can reverse the increase of entropy in a closed system according to the second law without outer influence, and therefore cause it to act as a *perpetuum mobile* of the second kind, dates back to James Clerk Maxwell. In 1879 William Thompson (the later Lord Kelvin) introduced "Maxwell's sorting demon", which was able to separate the gas molecules in a closed container in static equilibrium and with a homogeneous distribution of molecular velocities spontaneously into two parts with faster and slower molecules [3.27].

Obviously, "sorting demons" are an ad hoc hypothesis which cannot be explained in the physical framework of the 19th century. Boltzmann's thermodynamics as well as Newton's mechanics are insufficient to model the emergence of complex order, and thus the origin and growth of living systems. The first and second laws of thermodynamics are subjected to an important condition that is generally not true of all nature. These laws assume that all energy exchanges take place in a closed and isolated system. As energy and material fluxes through most regions of the universe, natural systems are rarely closed. As solar energy bombards the earth, it cannot be considered as a closed and isolated system.

So the first and second laws of thermodynamics are not false, but they are empirically restricted to approximately isolated microscopic subsystems, cosmic systems, or prepared conditions in laboratories. The situation can be compared with Newton's classical mechanics. After Einstein's special theory of relativity, it has not become false, but it is no longer the universal scheme of physics, and is restricted to motions that are slow relative to the speed of light. Most of nature must be modeled by dynamical systems which do not live in Boltzmann's general condition of equilibrium, because they are subjected to energy and material fluxes.

Historically, fundamental contributions such as those of Maxwell or Gibbs deal uniquely with situations corresponding to equilibrium or to situations infinitesimally near to equilibrium. Pioneering work on non-equilibrium thermodynamics was started by, e.g., Pierre Duhem at the beginning of this century. But his work remained unnoticed until Onsager (1931), and later the Prigogine school, the Haken school, and others began to study the behavior of complex systems far from thermal equilibrium. From a historical point of view, the situation can be compared with the development of chaos theory and complex Hamiltonian systems (compare Sect. 2.3). Chaotic phenomena were already discovered and well known to Poincaré, Maxwell, and others. But the mathematical problems connected with nonlinear complex systems deterred most scientists from searching for corresponding models.

3.3 Complex Systems and the Evolution of Organisms

Open systems not only have internal sources of entropy production but also an external source of entropy production associated with energy or mass transformations to or from their surroundings. These systems maintain their structure by dissipation and consumption of energy and are called "dissipative structures" by Ilya Prigogine. We have already become acquainted with nonliving dissipative systems like fluids, lasers, and clouds which are dependent on outside energy fluxes to maintain their structure and organization. Nonequilibrium systems exchange energy and matter with their environment, maintaining themselves for some period of time in a state far from thermal equilibrium and at a locally reduced entropy state. Small instabilities and fluctuations lead to irreversible bifurcations and thus to an increasing complexity of possible behavior.

A mathematical theory of dissipative structures with nonlinear evolution equations seems to offer the framework for modeling Aristotle's "sublunar" world of growing and dying nature. It is amazing to recognize that Aristotle's idea of a cyclic nature corresponds to periodic attractors or limit cycle solutions of corresponding differential equations. The cyclic nature of these systems allows them not only to develop stability but also to develop a hierarchy of complex structures within themselves. A cycle of living systems as it was already described in antiquity becomes autocatalytic by virtue of an evolutionary feedback.

The main idea was already expressed by Spencer and Boltzmann when they assumed that a pre-biological system may evolve through a whole succession of transitions leading to a hierarchy of more and more complex states. But, contrary to Boltzmann, these transitions can only arise in nonlinear systems far from thermal equilibrium. Beyond a critical threshold the steady-state regime becomes unstable and the system evolves to a new configuration. Evolving through successive instabilities, a living system must develop a procedure to increase the nonlinearity and the distance from equilibrium. In other words, each transition must enable the system to increase its entropy production. The evolutionary feedback of Ilya Prigogine, Manfred Eigen, and others means that changing the control parameter of the system beyond a certain threshold leads to an instability through fluctuations, which causes increased dissipation and thus influences the threshold again.

It follows that life did not originate in a single extraordinarily unlikely event and that the evolution of life did not proceed against the laws of physics. As we have learnt, Boltzmann's and Monod's idea of a gigantic fluctuation which would unfold over the time of biological evolution stems from equilibrium thermodynamics. While the probability of a dissipative structure (for instance, a periodic temporal process like a Bénard problem) is tiny in equilibrium statistical mechanics, it occurs with probability one in conditions far from equilibrium. Thus, Prigogine can argue:

> Far from being the work of some army of Maxwell demons, life appears as following the laws of physics appropriate to specific kinetic schemes and too far from equilibrium conditions. [3.28]

In the mathematical framework of nonlinear complex systems, many models have already been suggested to simulate the molecular origin of life. Complexity on the molecular scale is characterized by a large potential number of states which could be populated given realistic limits of time and space.

> For instance, a typical small protein molecule contains a polypeptide chain which is made of about 10^2 amino-acid residues. Given the 20 classes of natural amino acids, there are 20^{100} or 10^{130} alternative sequences of this length. The DNA molecule that comprises the total genome of a single bacterial cell represents one or few choices out of more than $10^{1000000}$ alternative sequences. Obviously, only a minute fraction of all such alternatives could have been tested by nature. Mathematically, a sequence containing ν residues of λ

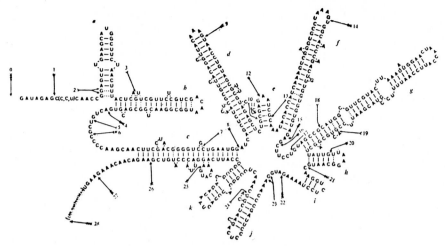

Fig. 3.2. Gene coding for a sequence of 129 amino acids

classes allows for $\binom{\nu}{k}(\lambda-1)^k$ alternative copies having substitutions at k positions. Figure 3.2 shows a gene which codes for a sequence of 129 amino acids [3.29].

Certain microstates may strongly influence macroscopic behavior. Such fluctuations may amplify and cause a breakdown of formerly stable states. Nonlinearity comes in through processes far from the thermal equilibrium.

Classical and only necessary conditions for life demand: (1) self-reproduction (in order to preserve a species, despite steady destruction), (2) variability and selection (in order to enlarge and perfect the possibility of a species, biased by certain value criteria), and (3) metabolism (in order to compensate for the steady production of entropy) [3.30].

Manfred Eigen has suggested a realization of these criteria by a mathematical optimization process. In this model, the nucleation of a self-reproducing and further evolving system occurs with a finite expectation value among any distribution of random sequences of macromolecules such as proteins and nucleic acids. The initial copy choice for self-reproduction is accidental, but the subsequent evolutionary optimization to a level of unique efficiency is guided by physical principles. In this model, life should be found wherever the physical and chemical conditions are favorable, although some molecular structures should show only slight similarity with the systems known to us.

The final outcome will be a unique structure, e.g., an optimized molecular sequence. Darwin's principle of the survival of the fittest is mathematized by an optimization principle for possible microstates of molecular sequences. It is assumed that in simple cases biomolecules multiply by autocatalysis.

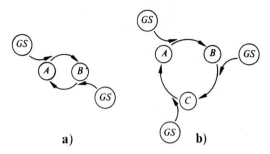

Fig. 3.3a,b. Autocatalysis with two kinds of biomolecules (**a**) and cyclic catalysis (hyper-cycles) with more kinds (**b**)

For instance, two kinds of biomolecules A and B from ground substances GS are multiplied by autocatalysis, but in addition the multiplication of one kind is assisted by that of the other kind and vice versa (Fig. 3.3a). In more complicated cases with more kinds of biomolecules, the latter are assumed to multiply by cyclic catalysis (Eigen's "hypercycles") (Fig. 3.3b). This mechanism combined with mutations is able to realize an evolutionary process.

Eigen suggests the following simplified model of an evolutionary opti-mization [3.31]: the machinery of a biological cell is codified in a sequence of four chemical substances A, T, G, C which consititute the genes. Each gene represents a functional unit, which is optimally adapted to the special purpose of its environment. The length of a gene in nature is seldom more than 1000 sequential positions. Thus for 4 symbols, there are 4^{1000} alternative genes ("mutations") of length 1000. In scientific notation that means about 10^{600} possibilities. In order to get an impression of this huge number, we should recall that the content of matter in the whole universe corresponds to 10^{74} genes and that the age of the universe is less than 10^{18} seconds.

Thus if all the matter of the universe since its very beginning (the "Big Bang") were used to alter and to produce new genes of length 1000 in each second, then by today only 10^{93} mutations could have been tested. Eigen concludes that genes which represent optimal functional unities cannot be produced by random processes, but must be developed by a self-optimizing process.

Mathematically, the process of adaptation can be imagined as a succes-sive replacement of positions which aims at a final ("optimal") sequence. This is a typical interpretation of the problem-solving approach of com-puter science. In order to solve a problem successfully, we must find an ade-quate spatial representation of the self-optimizing strategies. Because of the huge numbers, a 3-dimensional space is obviously not suitable. The length of strategies, i.e., the distance from a gene to its optimal variant, is too long. One would go astray. Further, sequences with a high degree of similarity cannot be represented adequately by a neighborhood in a 3-dimensional space. Thus it is proposed to alter the dimensions in the following way.

A sequence with n positions is defined as a point in n-dimensional space. For two symbols 0 and 1 there are 2^n alternative sequences, which are the points of the space. Each point has n nearest neighbors, which represent the mutations differing only in one position ("1-mistake mutation"). Between the two extremal points with only 0 or only 1, there are $n!$ possible connections. In Fig. 3.4a–d there are some examples of n-dimensional sequential spaces for the dual case. The great advantages of these spaces are the very short distances and the dense network of possible connections. As an example, the longest distance in a 1000-dimensional space is only 1000 units, in a 23-dimensional space with 10^{14} points only 23 units.

The 23-dimensional space suffices to represent all the points on the earth's surface in units of one meter. In this space optimal strategies can be given to find the highest mountain in some region of the earth. For this purpose we introduce a valuation function which associates each point with a numerical "height". Imagine a tour in the Alps. You have no fixed aim (for instance a special mountain), but a rough idea of orientation: you want to wander "uphill" without losing too much height. Mathematically, the gradient of your path is known and determines your decision as to direction. In the real Alps you encounter 1-dimensional edges and passes in the mountains, and your chances of reaching optimal points are restricted. In a 23-dimensional space you can go in 23 directions and distinguish pathways with different gradients, i.e., k in direction "uphill" and $23 - k$ "downhill" ($k \leq 23$). The chances of reaching optimal points in your neighborhood are very high.

In the n-dimensional sequential space of genes, the valuation of points is given by "selection values". Mutations do not arise completely irregularly or chaotically, but depend on which predecessors occur in the distribution most frequently. The question of which predecessors occur in the distribution most frequently depends again on their selection value relative to the optimal variant in the distribution. The selection values are not distributed irregularly, but in connected regions. For instance, on the earth a high mountain like Mount Everest is not situated in flat countryside, but among the Himalayas.

For the replication and self-reproduction of life, Eigen presupposes a self-optimizing machinery of high efficiency. Freeman Dyson proposed a mathematical model, according to which primitive life systems first occur without the correct machinery for replication and selection and fulfill only metabolic functions in relations to their environment [3.32]. The essential characteristic of such molecular systems is their homeostasis, i.e., the ability to maintain a stable and more or less constant equilibrium in a chang-

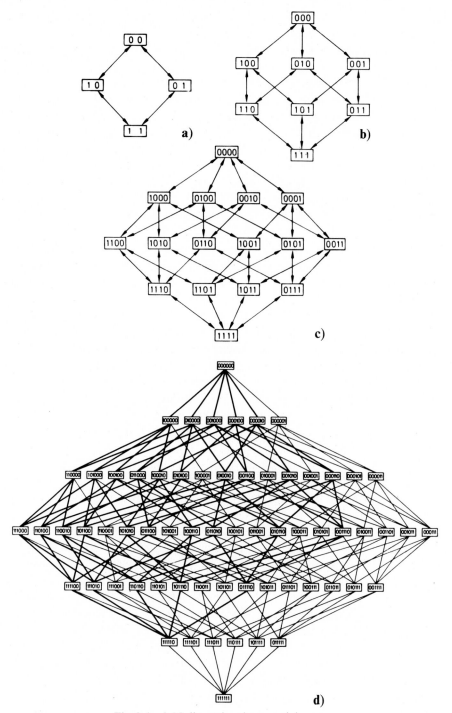

Fig. 3.4a–d. N-dimensional sequential spaces

ing environment. According to Dyson, the configuration of self- replication mechanisms occurs only in a second step. The nucleic acids required for that purpose are explained by Dyson as "non-assimilated" byproducts of the early metabolic life process, which first assumed a parasitic status in the whole system, and then ultimately developed through symbiotic intermediate states to fully-integrated functional mechanisms for reproduction and evolution.

The dual function of metabolism and replication in Dyson's model has much in common with the complex system of prebiotic evolution that was suggested by Stuart Kauffman [3.33]. He rejects the idea that life sprung from an RNA world. His hypercycle-like system is a complex autocatalytic network of reactions self-organizing themselves in such a way that metabolism, as a macroscopic order state of the system, becomes possible. Metabolism draws materials and energy from the environment around to increase and maintain internal order. Thus, it is an open dissipative system.

In general, an evolutionary process is expected to produce new kinds of species [3.34]. A species may be considered as a population of biomolecules, bacteria, plants, or animals. These populations are characterized by genes which undergo mutations producing new features. Although mutations occur at random, they may be influenced by external factors in the environment, such as changing temperature or chemical agents. At a certain critical mutation pressure new kinds of individuals of a population come into existence. The rate of change of these individuals is described by an evolution equation. As these individuals have new features, their growth and death factors differ. A change (mutation) is only possible when fluctuations occur in the population and the environment. Thus, the evolution equation determines the rate of change as the sum of fluctuations and the difference of growth and death factors.

A selection pressure can be modeled when different subspecies compete for the same living conditions (e.g., the same food supply). If the mutation rate for a special mutant is small, only that mutant survives which has the highest gain factor and the smallest loss factor and is thus the fittest. The competition procedure can be simulated by the slaving principle: unstable mutants begin to determine the stable ones. It is noteworthy that the occurrence of a new species due to mutation and selection can be compared with a nonequilibrium phase transition of a laser [3.35].

A living cell is an open system with a flow of energy passing through it. As already shown by Erwin Schrödinger, the energy flow creates conditions that allow strong deviations from thermodynamic equilibrium. According to Prigogine et al., this results in models of dissipative self-organization and pattern formation, the parameters of which are set up by genetic as well as epigenetic constraints. However, it would be misleading to expect that the process of self-organization in living cells simply represents a reduced copy of the pattern-formation phenomena in macroscopic reaction–diffusion systems. The laws of physics, when applied at a different scale typical for intracellular processes, can influence the mechanisms involved and produce a wealth of new properties. This also makes spatial pattern formation based on such reactions and diffusion impossible at very small length scales. The temporal self-organization of chemical processes, expressed in the generation of different periodicities and interactions between them, plays a fundamental role in living cells [3.36]. Thus, from a methodological point of view, it is not sufficient to know the general scheme of dissipative self-organization. But we must inquire experimentally into its cellular application under particular temporal, spatial, and chemical constraints.

Nevertheless, the link between physical-chemical systems and biological structures can be modeled by dissipative structures which may be involved in living systems. An important example is provided by the immune system, the disturbance of which causes many very dangerous illnesses, such as AIDS. Concerning the antibody-antigen kinetics, new types of antibodies can be generated successively, where some antibodies act as antigens. This process leads to a very complex dynamics of the total system [3.37].

As we have learnt, among the most striking features of dissipative systems are oscillatory phenomena. At the subcellular level, there are series of oscillating enzymatic reactions. Glycolysis is a process of great importance in living cells. The regulatory enzyme gives rise to oscillations with periods ranging from two to ninety minutes. Experimental oscillations can be identified with limit cycle type oscillations arising when the uniform state is unstable.

Another metabolic oscillation is that of the periodic synthesis of cyclic AMP in cellular slime molds. This species exhibits a transition between two different states of organization. First, the amebas are independent and separate cells. The transition to an aggregation and finally to a multicellular fruiting body takes place after starvation. Single cells, deprived of nutrient, aggregate in concentric waves around centers as a response to cyclic AMP being emitted from the centers. The synthesis of cyclic AMP realizes the limit cycle type. The aggregation process itself represents a self-organization occurring beyond instability.

Modeling this process in the framework of complex systems, we first consider a population of separated and homogeneous cells. A control parameter denotes the supply of nutrition which can be tuned to a critical value of starvation. Then the cyclic AMP is emitted and overcomes the random diffusive motion of amebas, and the uniform state becomes unstable. On the macroscopic level, the cells start to differentiate into several functions and to cooperate. On the macroscopic level, intermediate states of aggregation can be observed which finally lead to the new form of the mature multicellar body. Producing isolated spores, the life cycle of the slime mold is repeated in the described states of phase transition (Fig. 3.5) [3.38].

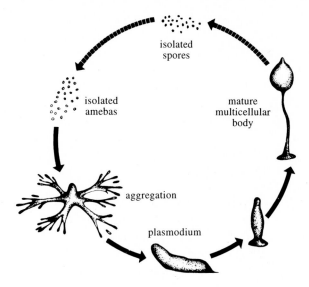

isolated
spores

isolated
amebas

mature
multicellular
body

aggregation

plasmodium

Fig. 3.5. Dynamical model of morphogenesis with states of cell formation (life cycle of the slime mold)

The spontaneous emergence of organic forms has seemed to be a miracle of life. Thus, in the history of science morphogenesis was a prominent counterexample against physical reductionism in biology. Today, morphogenesis is a prominent example for modeling biological growth by complex dynamical systems. What would Goethe have said about a mathematical model of his beloved metamorphosis? In this context, pattern formation is understood as a complex process wherein identical cells become differentiated and give rise to a well-defined spatial structure. The first dynamic models of morphogenesis were suggested by Rashevsky, Turing, and others. Let us regard Rashevsky's model for the morphogenesis of plant growth ("phyllotaxis") [3.39].

Figure 3.6a shows an idealized vine stalk sprouting one branchlet at a time with a symmetrically rotating direction for the three branchlets. At the tip of the growing stalk is a growth bud containing a mass of undifferentiated and totipotent cells. The problem of phyllotaxis was the emergence of the growth pattern with differentiated cells as the leaf bud cells, the branch cells, and others leading to the leaf buds and branchlets. The Rashevsky model is based on a ring of growth cells around the circumference of the stalk, near the growth bud at the top.

A cell is considered as a bag of fluid with homogeneous chemical composition. One of the chemical constituents is a growth hormone called morphogen. The concentration x of this morphogen is the observed parameter

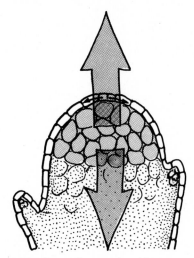

Fig. 3.6a. Dynamical model of morphogenesis with states of cell differentiation (Rashevsky model of phyllotaxis) [3.39]

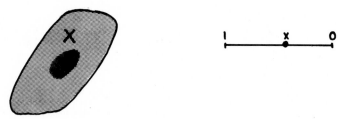

Fig. 3.6b. A cell with concentration x of morphogen and the corresponding state space with states x on a line segment

of the model. As the parameter varies between 0 and 1, the state space of the model is a line segment (Fig. 3.6b). If the concentration of this morphogen exceeds a certain critical value, the growth function of the cell is turned on, the cell devides, and a branchlet comes into existence.

In the next step, two cells are considered as an open system in which one morphogen can be exchanged between the two-celled system and its environment. If the concentration of the morphogen in the second cell is denoted by y, then the state of the whole system corresponds to a point (x, y) in the unit square which is interpreted as the state space of the system. In Fig. 3.6c, the state space is divided into four regions corresponding to the situations "cell 1 off and cell 2 growing" (A), "both cells' growth turned off" (B), "cell 1 growing, cell 2 off" (C), "both cells growing" (D).

In the final step three cells form a ring with a uniform concentration of the morphogen in each. The point (x, y, z) in the unit cube represents

c)

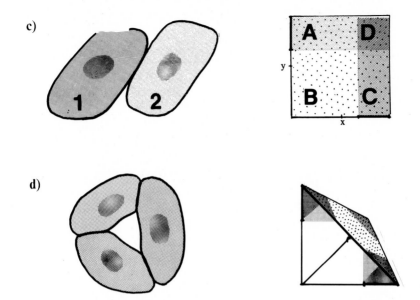

d)

Fig. 3.6c,d. A two-celled system with morphogens x and y and the corresponding state space with states (x, y) in a unit square (**c**) and a three-celled system with a uniform concentration of morphogens x, y, and z and the corresponding state space with states (x, y, z) in the unit cube (**d**) [3.39]

a state of the system. In a three-dimensional space, the state space for the closed system of three cells with one morphogen is realized by a triangle with $x + y + z = 1$, i.e., the sum of the concentrations is constant (Fig. 3.6d).

In Fig. 3.6e, a dynamical system with a periodic attractor is added to the state space. One after another of the three cells is turned on, then off, periodically. In Fig. 3.6f, the stalk is modeled as a stack of rings of cells, each ring represented as an identical copy of the triangular model of Fig. 3.6d. Growth of the stalk upwards in time is represented by associating time with the upward direction. The periodic attractor of Fig. 3.6e is transformed to a periodic time series spiraling upwards in time.

In this simplified dynamical model of morphogenesis, a central problem remains open. How do the originally undifferentiated cells know where and in which way to differentiate? Experiments indicate that this information is not originally given individually to the cells but that a cell within a cellular system receives information on its position from its surroundings. A famous example is the hydra, which is a tiny animal, consisting of about 100 000 cells of about 15 different types. Along its length it is subdivided into different regions, e.g., its head at one end. If a part of a hydra is transplanted to a region close to its old head, a new head grows by an activation of the cells.

e)

f)

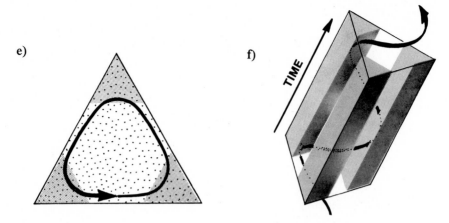

Fig. 3.6e,f. A three-celled system with a periodic attractor (**e**) and the growth of a stalk modeled as a stack of three-celled systems (Fig. 3.6d) with periodic attractor (Fig. 3.6e) transformed to a spiraling time series (**f**) [3.39]

There is some experimental evidence that activator and inhibitor molecules actually exist.

In a mathematical model due to Gierer and Meinhardt, two evolution equations were suggested, describing the rate of change of activator and inhibitor concentrations, which depend on the space-time coordinates. The change of rates is due to a production rate, decay rate, and diffusion term. Obviously, inhibitor and activator must be able to diffuse in some regions, in order to influence the neighboring cells of some transplant. Further on, the effect of hindering autocatalysis by the inhibitor must be modeled. In Fig. 3.7 the interplay between activator and inhibitor leads to a growing periodic structure which is calculated and plotted by computer-assisted methods [3.40].

To derive such patterns, it is essential that the inhibitor diffuses more easily than the activator. Long range inhibition and short range activation are required for a non-oscillating pattern. By methods of mathematical analysis, the evolving patterns described by the evolution equations of Gierer and Meinhardt can be determined. A control parameter allows one to distinguish stable and unstable configurations ("modes").

At a critical value, the unstable modes start to influence and dominate the stable ones according to the slaving principle. Mathematically, the stable modes can be eliminated, and the unstable ones deliver order parameters determining the actual pattern. Thus, actual patterns come into existence by competition and selection of some unstable solution. Selection according to

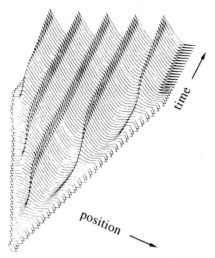

Fig. 3.7. Computer-assisted model of morphogenesis leading to a periodic structure [3.40]

the slaving principle means reduction of the complexity which stems from the huge number of degrees of freedom in a complex system.

Biochemically, this kind of modeling of morphogenesis is based on the idea that a morphogenetic field is formed by diffusion and reaction of certain chemicals. This field switches genes on to cause cell differentiations. Independently of the particular biochemical mechanism, morphogenesis seemes to be governed by a general scheme of pattern formation in physics and biology. We start with a population of totipotent cells corresponding to a system with full symmetry. Then, cell differentation is effected by changing a control parameter which corresponds to symmetry breaking. The consequence is an irreversible phase transition far from thermal equilibrium. In Fig. 3.8, the phase transition of activator and inhibitor concentration is illustrated in a computer simulation.

Independently of the common scheme of symmetry breaking, there is an important difference between physico-chemical and biological pattern formation. Physical and chemical systems lose their structure when the input of energy and materials is stopped (compare, e.g., the laser or the Zhabotinsky reaction). Biological systems are able to conserve much of their structure at least for a relatively long time. Thus they seem to combine approximately conservative and dissipative structures.

Since antiquity, living systems were assumed to serve certain purposes and tasks. Organs of animals and humans are typical examples of functional structures which are explored by physiology and anatomy. How can the functional structures of medicine be understood in the framework of complex systems [3.41]?

Fig. 3.8. Two computer-assisted models of morphogenesis simulating the phase transitions of activator and inhibitor concentrations

The complex bifurcations of vessel networks are examples of fractal structures. The form of trees, ferns, corals, and other growing systems are well described by fractals. In Chap. 5 *Artificial Intelligence*, we shall discuss recursive and computer assisted procedures to simulate the fractal growth of trees. The vascular tree in the heart reminds us of a complex network of branches and roots. This is quite natural when one appreciates that vascular growth occurs by the budding of capillaries into regions of cell division and differentiation.

Trees branching into open space have room to expand. But hearts, lungs, and other organs occupy a limited space. The networks of nerves or vessels that penetrate them are servants to the principal occupants of the space. The structure of the microvascular network is virtually completely defined by the cells of the organ. In skeletal and cardiac muscles the capillaries are arrayed parallel to the muscle cells, with some cross-branches. The system is guided in its growth by the need for the nerve or the vascular system to follow the lines of least resistance.

This leads to medically quite interesting questions of whether the fractal growth and form of the vascular network can give rise to observed heterogeneities of flow in the heart. A simple algorithm for a branching network as shown in Fig. 3.9 leads to an appropriate probability density function of regional flow. The fractal system of an organ has become a functional structure [3.42].

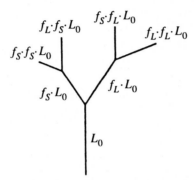

Fig. 3.9. A branching vascular network of the heart with fractal recursions for branch lengths starting with length L_0 of a main stem vessel and decreasing lengths of following daughter vessels by factors f_L and f_S (subscripts L and S indicating longer and shorter vessels)

Fractal illustrations of a bronchial network are an inspiration for physicians to apply these approaches to the lung. Physical systems, from galactic clusters to diffusing molecules, often show fractal behavior. Obviously, living systems might often be well described by fractal algorithms. The vascular network and the processes of diffusion and transmembrane transport might be fractal features of the heart. These fractal features provide a basis which enables physicians to understand more global behavior such as atrial or ventricular fibrillation and perfusion heterogeneity.

As we have seen in Sect. 2.4, nonlinear dynamics allows us to describe the emergence of turbulence, which is a great medical problem for blood flow in arteries. Turbulence can be the basis of limit cycling, as can be shown with water flowing through a cylindrical pipe. A variety of control systems produce oscillations. It might also be expected that some oscillating control systems show chaotic behavior.

Atrial and ventricular fibrillation are the classic phenomena that appear chaotic. The clinical statement on the heart rate in atrial fibrillation is that it is irregularly irregular. The observations are that the surface of the atrium is pulsing in an apparently chaotic fashion. However, the studies of reentry phenomena and of ventricular fibrillation show that there are patterns of excitation, again illustrating that this is organized ("mathematical") chaos.

Fig. 3.10. Regular and chaotic heart rate

Fractal and chaotic algorithms for this have been described. The two curves of Fig. 3.10 show a regular and chaotic heart rate [3.43].

Nevertheless, chaotic states cannot generally be identified with illness, while regular states do not always represent health. There are limited chaotic oscillations protecting the organism from a dangerous inflexibility. Organs must be able to react in flexible ways, when circumstances change rapidly and unexpectedly. The rates of heart beat and respiration are by no means fixed like the mechanical model of an idealized pendulum.

Single organs and the whole organism of the human body must each be understood as a system of nonlinear complex dynamical systems of high sensibility. Tuning their control parameters to critical values may cause phase transitions of irreversible developments representing more or less dangerous scenarios of human health. Dissipative complex structures are open systems which cannot be separated from their surrounding environment. Thus, on the background of the complex dynamical system approach, the classical "mechanical" view of medicine separating the human body into particular parts for highly specialized experts must be heavily criticized. The whole body is more than the sum of its parts. It is amazing to recognize that from the modern view of complex dynamics an old demand of traditional physicians since antiquity is supported again, namely that medicine is not only an analytical science, but an art of healing which has to consider the wholeness of health and illness.

3.4 Complex Systems and the Ecology of Populations

Ecosystems are the results of physical, chemical, and biotic components of nature acting together in a structurally and functionally organized system. Ecology is the science of how these living and nonliving components function together in nature. Obviously, in the framework of the complex system approach, ecology has to deal with dissipative and conservative structures of very high complexity depending on the complexity of the individual physical, chemical, and biotic systems involved in them, and the complexity of their interactions [3.44].

In 1860, one of the first empirical case studies on ecology was provided by Henri Thoreau in a lecture about "the succession of forest trees". He observed that nature displayed a process of plant development resulting in a sequential change of species that seemed to be observable and predictable. If the ecosystem is left undisturbed, the progression from bare field to grassland to grass-shrub to pine forest and finally to an oak-hickory forest is a predictable, 150-year process, at least in Massachusetts in the 19th century [3.45].

In nearly the same year, Charles Darwin published his famous theory of evolution based on the mechanisms of variation and selection. Darwin saw progressive changes by organisms resulting from competition and adaption to fit optimally into ecological niches. It is the flux of energy from the sun and chemical reactions that sets the process of life in motion and maintains it. Boltzmann already recognized that the biosphere extracts a high energy-entropy cost for the organization of living things. These processes are not only based on the biotic components of an ecosystem but affect the nonbiotic components as well.

James Lovelock has proposed that living systems drive the major geochemical cycles of the earth. He proposed that the global atmospheric composition was not only developed by living systems but also controlled by the global ecosystem. The "balance of nature" has become the popular title for a complex network of equilibria characterizing the human ecosystem on earth [3.46].

The mathematical theory of complex systems allows one to model some simplified ecological case studies. The phenomena to be explained are, mainly, the abundance and distribution of species. They may show typical features of dissipative structures like temporal oscillations. At the beginning of the 20th century, fishermen in the Adriatic Sea observed a periodic change of numbers in fish populations. These oscillations are caused by the interaction between predator and prey fish. If the predators eat too many prey fish, the number of prey fish and then the number of predators decreases. The result is that the number of prey fish increases, which then leads to an increase in the number of predators. Thus, a cyclic change of both populations occurs.

In 1925, Lotka and Volterra suggested a nonlinear dynamical model. Each state of the model is determined by the number of prey fish and the number of predator fish. So the state space of the model is represented by a two-dimensional Euclidean plane with a coordinate for prey fish and a coordinate for predator fish. The observations, over time, of the two populations describe a dotted line in the plane. Births and deaths change the coordinates by integers, a few at a time. To apply continuous dynamics, the dotted lines must be idealized into continuous curves.

The vector field on the state space can be roughly described in terms of four regions (Fig. 3.11a). In region A, both populations are relatively low. When both populations are low, predator fish decrease for lack of prey fish while prey fish increase because of less predation. The interpretation of this habitual tendency as a bound velocity vector is drawn as an arrow. In region B, there are many prey fish, but relatively few predators. But when there are many prey fish and few predator fish, both populations increase. This is interpreted by the vector in region B. In region C, both populations are relatively large. The predator fish are well fed and multiply, while the prey fish population declines. This tendency is shown by the vector in region C. In region D, there are few prey fish but many predator fish. Both populations decline. This tendency is shown by the vector in region D. The phase portrait of this system can be visualized by a closed trajectory, because the flow tends to circulate.

In Fig. 3.11b, the phase portrait is a nest of closed trajectories, around a central equilibrium point. As dynamical systems theory tells what to expect in the long run, the phase portrait enables the ecologist to know what happens to the two populations in the long run. Each initial population of predator fish and prey fish will recur periodically [3.47].

If some kind of ecological friction were added to the model, the center would become a point attractor. This would be a model for an ecological system in static equilibrium (Fig. 3.11c). A different but perhaps more realistic modification of the model results in a phase portrait like Fig. 3.11d, with only one periodic trajectory.

From an analytical point of view, the evolution of a population is governed by an equation for the rate of change of its size [3.48]. Obviously, the number of individuals in the population changes according to its growth rate minus its death rate. A further parameter which has to be considered refers to the limited food supply or depletion of the food resources. There are several living conditions of populations which must be modeled. If different species live on different kinds of food and do not interact with each other, they can coexist.

If different species live in similiar conditions, then the overlapping food supplies must be considered in the evolution equations of the populations. An enormous reduction of complexity is realized if the temporal change of the food supply is neglected. The resulting evolution equations allows several scenarios of coexistence, when stable configurations are realized.

Biologically, stable states correspond to ecological niches which are important for the survival of a species. The predator-prey relation of two populations is realized by the Lotka–Volterra equations characterizing the phase

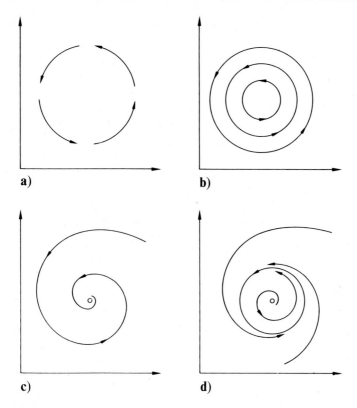

a)   b)

c)   d)

Fig. 3.11a–d. Phase portraits of an ecological system with a prey and a predator population (Lotka-Volterra): (**a**) a closed trajectory, (**b**) a nest of closed trajectories, (**c**) a point
attractor, (**d**) a periodic trajectory [3.47]

portraits in Fig. 3.11. A particular form of cooperation in nature is the symbiosis of two species. Modeling a symbiosis by evolution equations, it must
be considered that the multiplication rate of one species depends on the
presence of the other.

Animal populations can be characterized on a scale of greater or lesser
complexity of social behavior. There are populations of insects with a complex social structure which is rather interesting for sociobiology. Nicolis and
others have tried to model the social organization of termites by a complex dynamical system. The interactions between individuals are physically
realized by sound, vision, touch, and the transmission of chemical signals.

The complex order of the system is determined by functional structures
like the regulation of the castes, nest construction, formation of paths, the
transport of materials or prey, etc. Ants synthesize chemical substances which
regulate their behavior. They have a tendency to follow the same direction at
the place where the density of the chemical molecules reaches a maximum.

Collective and macroscopic movements of the animals are regulated by these chemical concentrations.

In order to model the collective movements, two equations are suggested, considering the rate of change for the concentrations of insects and chemical substances. There is a critical value of an order parameter ("chemicotactic coefficient") for which a stationary homogeneous solution becomes unstable. The system then evolves to an inhomogeneous stationary state. Accordingly, different branching structures will appear, as observed in different ant societies. Figure 3.12 shows the collective movement of ants with two types of structure characteristic of two different species [3.49].

The social complexity of insects can also be characterized by such coordinated behavior as nest construction. This activity has been well observed and explored by experimental studies. A typical observation is that the existence of a deposit of building material at a specific point stimulates the insects to accumulate more building material there. This is an autocatalytic

Fig. 3.12. Branching networks of collective movements in two different species of ants [3.49]

reaction which, together with the random displacement of insects, can be modeled by three differential equations. These equations refer to the observation that the termites, in manipulating the construction material, give it the scent of particular chemical substance, which diffuses in the atmosphere and attracts the insects to the points of highest density, where deposits of building material have already been made.

Thus, the first equation describes the rate of change of the concentration of building material, which is proportional to the concentration of insects. A second evolution equation refers to the rate of change of the scent with a certain diffusion coefficient. A third evolution equation describes the rate of change of the concentration of insects including the flow of insects, diffusion, and motion directed toward the sources of the scent.

The complex social activity of nest construction correponds to the solutions of these equations. Thus, an uncoordinated phase of activity in the beginning corresponds to the homogeneous solution of these equations. If a sufficiently large fluctuation with a larger deposit of building material is realized somewhere, then a pillar or wall can appear. The emergence of macroscopic order, visualized in the insect's architecture of nests, has been caused by fluctuations of microscopic interactions.

Models of the above types are now often used in ecology. It must be mentioned that they are still on a rather simplified level. In reality, one has to take into account many additional effects, such as time-lag, seasons, different death rates, and different reaction behavior. In general, there is not only the interaction of one or two complex populations with their (simplified) environments, but a huge number of different interacting populations. The phase portraits of their nonlinear dynamics at least allow global forecasts in the long run [3.50].

In the traditional Darwinian view, there are two important forces at work in biological evolution which must be modeled in complex dynamical systems: mutation pressure and selection. In biological populations in which the behavior of individuals is governed uniquely by their genes, the amplification of a new type individual corresponds to the Darwinian evolution by natural selection of mutants which appear spontaneously in the system. In the case of higher animals, there is the possibility of behavioral change ("innovation") and its adaption by information. In ecological evolution, new ecological niches have arisen which are occupied by specialized species. Obviously, there is no simple scheme of evolution, but a complex hierarchy of changing and stabilizing strategies which have been layered from prebiological evolution to ecological and finally cultural evolution with human learning strategies (compare Chap. 6).

The complex system approach shows a great variety of possible evolutions with unexpected directions, caused by stochastic fluctuations. There is no global optimizer, no global utility function, no global selection function, no other simplified strategy of evolution but successive instabilities near bifurcation points. In short, Darwin's view is only a particular aspect of evolution. For many contemporaries, he seemed to replace a personal deity called "God" by an impersonal deity called "evolution" governing the world by simple laws. This secularized religious attitude of the 19th century was later on continued by political thinkers like Karl Marx, who believed in an impersonal deity called "history" governing human fate by simplified laws of society.

In the 18th century, Kant had already complained that the term "nature" seems to denote an impersonal deity. But "nature" is, as Kant argued, nothing more than a "regulative idea" of man. From a modern point of view we can actually only recognize dynamical models with more or less high degrees of complexity which may fit observational data with more or less accuracy. The abandonment of some mighty supervisor in nature and human history may leave us feeling alone with perhaps dangerously chaotic fluctuations. But, on the other hand, these fluctuations may enable real innovations, real choices, and real freedoms.

4. Complex Systems and the Evolution of Mind–Brain

How can one explain the emergence of brain and mind? The chapter starts with a short history of the mind-body problem. Besides religious traditions, the concepts of mind and body held by our ancestors were often influenced by the most advanced standards in science and technology (Sect. 4.1). In the framework of complex systems the brain is modeled as a complex cellular system with nonlinear dynamics. The emergence of mental states (for instance pattern recognition, feeling, thoughts) is explained by the evolution of (macroscopic) order parameters of cerebral assemblies which are caused by nonlinear (microscopic) interactions of neural cells in learning strategies far from thermal equilibrium. Pattern recognition, for instance, is interpreted as a kind of phase transition by analogy with the evolution equations which determine pattern emergence in physics, chemistry, and biology (Sect. 4.2). In recent studies in neurobiology and cognitive psychology, scientists even speculate that the emergence of consciousness and self-consciousness depends on the production rate of "meta-cell-assemblies" as neural realizations of self-reflection. The Freudian unconscious is interpreted as a (partial) switching off of order parameters referring to certain states of attention. Even our dreams and emotions seem to be governed by nonlinear dynamics (Sect. 4.3).

Is the "Newton of the human brain and mind" found? Of course not. The complex system approach cannot explain what mind is. But we can model the dynamics of some mental states under certain conditions. Even the modeling of intentional behavior cannot be excluded in principle. Complex systems do not need a central processor like the fiction of "a little man" in the brain. Thus, Virchow's cynical observation that he did not find any soul in human bodies even after hundreds of operations is obsolete. A mental disposition is understood as a global state of a complex system which is caused by the local nonlinear interactions of its parts, but which cannot be reduced to its parts (Sect. 4.4). The wonder of our feeling, imagination, and creativity which has been celebrated by poets and artists since the beginning of human culture is not touched by the complex system approach, although we shall sometimes model some aspects of their nonlinear dynamics.

4.1 From Plato's Soul to Lamettrie's "L'Homme machine"

One of the most complex organs in nature is the human brain. Nowadays we know that it enables the emergence of human mind, consciousness, and personality, which has been considered as one of the greatest miracles of mankind since the early beginning of human thinking. The complex system approach allows us to model the emergence of human perception and thinking with respect to nonlinear interactions of complex neural networks. Thus, models of complex systems help us to understand *how* the mind-brain process may work, and *how* it may have emerged in natural evolution under certain conditions. From this point of view, it will be, in the long run, no wonder *how* consciousness and mind emerge by well-known laws in the evolution of nature. But it still remains a wonder *that* it has arisen.

Before we explore complex systems and the evolution of mind-brain, let us glance at the early philosophy of mind and the history of neural physiology. On the historical background we can decide which questions of the traditional mind-body problem have been solved by the complex system approach and which questions are still unsolved.

In the previous chapters, we have already remarked that early myths and religious beliefs were attempts to explain the human living world and to conjure the forces of nature. Obviously, human desire, fear, anger, and imagination govern the human living world like the might of nature. Consciousness or spirit or mind or soul are experienced in life, and they seem to "leave" the body of dead people. Humans have tried to model these unknown processes by familiar experiences of interacting physical things. Mental or conscious states are hypostatized as a particular substance called "soul", or something like that, which is responsible for the intentional behavior of humans. With the hypostatization of mental states, the problem of the position of the soul in the body has been raised, and is usually answered by the idea that it pervades the body, or is centered in some organs, such as the heart and the lungs. Although the effects of this miraculous "thing" are obviously real, it cannot be seen or seized like a God or ghost. Thus, it has been generally believed to have divine origin. Criticizing traditional myths and religious beliefs, the presocratic philosophers searched for natural causes and principles. Some thinkers regarded the "soul" as material stuff like "air" or "fire" because they were believed to be the finest and lightest forms of matter. For Anaxagoras mind is the principle of motion and order, and therefore the principle of life. For Heraclitus soul is like a flame of fire ruled by the law (*logos*) of the universe. The soul, like fire, is killed by water: "It is death for souls to become water" [4.1]. These approaches are nothing else than modeling the unknown by the familiar and known.

It is noteworthy that one of the early medical thinkers, the Pythagorean Alcmaeon of Croton, seems to have been the first Greek thinker to locate sensation and thought in the brain [4.2]. Like the early Greek astronomical model of a heliocentric universe, this genial idea was soon overshadowed by the authority of Aristotle who taught that the heart is the seat of consciousness, while the brain is only a mechanism for cooling by means of air. Although Aristotle was greatly influenced by the early Greek medical thinkers, he disagreed with Hippocrates' great insight that the brain "is the messenger to consciousness (*sunesis*) and tells it what is happening".

An early reductionistic philosophy of mind was defended by Democritus, who tried to reduce the mental states to interactions of the smallest atoms [4.3]. The problem with his reductionism is, of course, that "soul" is only identified with particular material (but unobservable) atoms. In contrast to material identifications or analogies, the Pythagorean philosophers explained that the human soul or mind must be an immaterial essence because it is able to think immaterial ideas like mathematical numbers and relations. In other words, the soul is modeled as a mathematical proportional system whose harmonies or disharmonies represent mental states like musical melodies.

The Pythagorean concept influenced Plato's philosophy of the human soul, which is related to his theory of forms or ideas. In his dialogue *Menon* Plato demonstrates that an untrained slave can solve mathematical problems. The reason is that in Plato's view every man has an eternal knowledge before (*a priori*) any empirical experience. Man has this kind of a priori knowledge, e.g., in mathematics, by participation in the eternal forms and ideas which are *ante rem*, which means independent of the fuzzy and transitory appearance of being [4.4].

Aristotle criticized Plato's hypothesis of an idealistic world behind reality. Ideas are human abstractions of forms which are acting in nature (*in re*). The soul is described as the form ("essence") of the living body, the "first entelechy", which is a teleological force. But it is not separated from matter. The soul is involved as potentiality in a human body. According to Aristotle, the human organism is understood as a wholeness.

Nevertheless, in the Aristotelean and Stoic tradition the anatomy of the nervous system had been discovered. Galen believed that the nerves transport a psychic pneuma to the muscles which are caused to produce movements. The psychic pneuma was not only a material stuff like breath or air, but a kind of vital spirit [4.5]. In the Middle Ages, the Aristotelean and Stoic philosophy of nature had a great influence on medical thinkers in the Islamic tradition like Avicenna, who founded a medical school with impressive activities in the fields of surgery, pharmacology, and practical healing and helping ill people [4.6]. Later on, these medical standards of the Persian and Arabic world were only realized by a few thinkers of the Christian Middle Ages like Albertus Magnus. Concerning the philosophy of mind and brain, the scientific discussions on, for example, the problem of human consciousness were always overshadowed by religious ideologies, and for a long time it was rather dangerous to dissect dead bodies in anatomy.

In contrast to holistic philosophy in the tradition of Aristotle and Avicenna, Descartes' rationalism taught a dualistic ontology strictly separating mind and matter, soul and body. The human body (*res extensa*) is a material machine constructed by the laws of mechanics and geometry. It is directed and controlled by innate ideas (*ideae innatae*) which are incorporated in the human mind (*res cogitans*). In his *Meditations*, Descartes gets to his primal intuition of the human mind by methodically doubting everything. Methodical doubt is designed to uncover what, if anything, is indubitable. Whereas Descartes can doubt the results of all sciences, of common sense, of perception, he cannot doubt that he exists as a kind of thing which is capable of engaging in cognitive processes such as doubting:

> But what then am I? A thing which thinks. What is a thing which thinks? It is a thing which doubts, understands, affirms, denies, wills, refuses, which also imagines and feels. [4.7]

The difficulty with Descartes' theory is, of course, the interaction of mind and body. He assumes that the human organism with its various organs is directed by the mind with its seat in the brain. The nerves are message cables to and from the brain. They work as causal chains between the ordering mind and the executing muscles. According to the clockwork paradigm of his mechanics, Descartes believed in tiny material particles called "animal spirits" moving and pushing each other very quickly in the cables of the nerves, in order to transport some input from the brain to the muscles.

In contrast to all mechanical effects in nature, the human mind is able to decide the direction of a movement voluntarily. Thus, the action of the mind on animal spirits is to deflect the direction of their motion. This can be done without violating Descartes' laws of physics as long as the "amount of motion" (the later so-called conservation law of momentum) was conserved. In Fig. 4.1, Descartes' mechanical model of perception is illustrated: the minute particles of light beams bomb the human eyes, exciting the brain by the transmission of particular nerves and their "animal spirits". The movements of arm and hand are coordinated with perception by the mind in the brain [4.8].

In his book *Passions of the soul* Descartes even tried to analyze all emotional states like fear and love as the passive physical outcomes of the way various "animal spirits" are induced to flow by external events. If Descartes' mechanical model of animal spirits is replaced by biochemical substances

Fig. 4.1. Descartes' geometric model of perception and arm movement

and electrophysical effects like hormones and neurotransmitters, then his concept of nervous activity seems to be rather modern.

His main difficulty is the immediate interaction of an immaterial, and therefore unextended and indivisible mind (*res cogitans*) with a material, and therefore extended and divisible body (*res extensa*). Descartes locates the mind within a very small organ of the brain, the pineal gland, steering the movements of the animal spirits. But how can the unextended mind exert a push on an extended particle like an animal spirit? In the framework of mechanics, this problem of interaction was unsolvable in principle and gave rise to several developments in the philosophy of mind.

For occasionalistic philosophers like Malebranche all causation is miraculous. God must intervene on the occasion of every particular case of causal action. Thus, the mind-body problem is explained by a theological and ad-hoc hypothesis. Spinoza reduced Descartes' dualism of mind and matter to a monism of one unique substance. God alone is the only substance of everything. All appearances of nature, mind, and body are only attributes ("states") of the universal substance. There are no miracles at any occasion. But God and human mind are naturalized, and nature has become divine in a universal pantheism [4.9].

According to the complex system approach, it was Leibniz who delivered a most remarkable philosophy of mind. Concerning his philosophy of nature, we recall of Leibniz' universe that, contrary to Descartes and Spinoza, consists of infinitely many substances ("monads"), corresponding to different points of view in space and with a more or less clear perspective on the whole. Thus, the monads are considered as soul-like substances endowed with perception and memory, differing in the degree of clarity of their consciousness. There are substances with a rather great perspective and rather high degree of consciousness like humans, when they are compared with animals, plants, and stones with decreasing degrees of consciousness. Even God may be embedded in Leibniz' monadology as the central monad with the highest degree of consciousness and best perspective on the whole, but still as an individual and different entity [4.10].

Obviously, Leibniz did not have the metaphysical problems of Descartes' interactionism. He actually tried to combine classical mechanics with the traditional Aristotelean teleology of nature, because he was aware of the mechanistic lack of an adequate philosophy of mind. From a modern point of view, Leibniz' notion of more or less animated soul-like substances with perception and memory seems to be rather strange. But it was no problem for him to model his monadology in the framework of automata with more or less complexity. Leibniz suggested that each substance can be modeled by an automaton with different states corresponding to the monad's perceptions. Its degree of consciousness is measured by its degree of complexity, characterizing the size of the monad's state space and its information processing capacity. The states of Leibniz' more or less complex automata are correlated with each other in networks, according to his famous quotation that the monads have "no windows" and do not interact, but reflect each other like the mirrors in a baroque palace. Leibniz' complex networks of monads will be discussed in Chap. 5, *Complex systems and the evolution of artificial intelligence*, in more detail. In short, Leibniz assumed that mind is not reserved to humans, but is a feature of systems which emerges with different intensity according to the system's degree of complexity.

The English empiricist philosophers like Locke and Hume criticized the Cartesian-Platonic belief that mental states can be analyzed by introspection and pure thinking without sensory experience. For empiricist philosophers mind is nothing more than a *tabula rasa*, an empty store for receiving sensory data, in order to form concepts by association and abstraction. Images are merely less vivid copies of sense-impressions which can be imaginatively combined, like for instance the notion of an unicorn.

According to the complex system approach, Hume developed a remarkable psychological theory of association. He proclaimed that there is neither a causal mechanism in nature nor a causal law in our mind, but only an unconscious reflex of associating those sense-impressions which occur in a correlated way on several occasions like flashes of lightning and thunder. We may say that the brain has an unconscious capacity to build up patterns of sense-impressions. Notions are nothing more than terms designating more or less complex patterns of sense-impressions. Apart from mathematics, there are no sharp and definite concepts founded on perception, but only more or less fuzzy patterns allowing more or less probable assertions about events. In *A treatise of human nature*, Hume wrote:

> The table before me is alone sufficient by its view to give me the idea of extension. This idea, then, is borrow'd from, and represents some impression, which this moment appears to the senses. But my senses convey to me only the impressions of colour'd points, dispos'd in a certain manner. If the eye is sensible of any thing farther, I desire it may be pointed out to me. But if it be impossible to shew any thing farther, we may conclude with certainty, that the idea of extension is nothing but a copy of these colour'd points, and of the manner of their appearance. [4.11]

According to Descartes' rationalism, the human mind rules the body's mechanics like a monarch governing the state in his century of absolutism. For Hume, there is no separate substance of the human mind, but only a self-organizing field of permanently emerging and disappearing patterns excited by associations of more or less intense sense-impressions. Hume's spontaneously associating and separating sense-impressions can be compared with the free citizens of a democratic society who may associate in groups and parties without the orders and prohibitions of a sovereign.

Kant tried to synthesize rationalism and empiricism. According to empiricism, cognition starts with experience and sensory data. But rationalism is right, because we need mental structures, cognitive schemes, and categories in order to organize experience and cognition. Kant tried to introduce the philosophical categories which found the axioms of Newtonian mechanics. It is the main feature of his epistemology that recognition does not arise by passive impressions of the external world on the tabula rasa of our brain. Recognition in the Kantian sense is an active process producing models of the world by *a priori categories*. The spatial and temporal order of physical events is reduced to geometrical forms of intuition. Perception in the Kantian sense is active information processing regulated by a priori anticipations. The causal connections of events is made possible philosophically by an a priori category of causality.

Hume was right that a causal relation cannot be perceived. But, in order to forecast and calculate the path of a pushed billiard ball exactly, it is not sufficient to repeat several pushes of billiard balls and to associate several sense-impressions à la Hume. We must anticipate that causes and effects can be connected by some deterministic relation. That is done by Kant's general scheme of (deterministic) causality in epistemology. But the question is which particular causal function is adequate to decide and test by physical experience. Cognitive schemes are already applied in everyday life. They are even modeled by data schemes of programming languages in computer science (compare Sect. 5.2). Thus, Kant's epistemology may be interpreted as an important forerunner of modern cognitive sciences, where cognitive schemes are presumed to order the mass of experienced data. But, contrary to Kant, they may be changed in the development of history, as was shown by the change from Euclidean space to non-Euclidean spaces, e.g., in the theory of relativity [4.12].

While Spinoza suggested a spiritual monism as a way out from Descartes' dualism, Lamettrie supported a kind of materialistic monism. Descartes' assumption of a separate soul-like substance (*res cogitans*) was criticized as superfluous, because all mental states should be reduced to mechanical processes in the human body: "L'Homme machine". Lamettrie claimed there were no fundamental differences between humans and animals. Intelligent and reflex behavior should be explained by "irritation" of the nerves and not by a "ghost in the machine". But, in the mechanistic framework of the 18th century, Lamettrie's revolutionary ideas could only be an inspiring program of physiological research [4.13].

The famous mathematical physicist and physiologist Hermann von Helmholtz (1821–1894) was a post-Kantian philosopher [4.14]. He supported a kind of naturalized framework of cognitive categories which must be presumed before any particular perception of the world can be constructed. Of course, the categories had changed since Kant. Nevertheless, there are some general schemes like the concept of space, numbers, measurement, and causality characterizing the physical theories of the 19th century. Helmholtz was aware, for instance, of the mathematical possibility of non-Euclidian spaces. Thus, he thought that the correct physical geometry must be decided by physical measurement.

Concerning his theories of physiology, Helmholtz started as a student of Johannes Müller (1801–1858), who is sometimes called the father of modern physiology [4.15]. Müller defended a law of specific nerve energies demanding that each nerve has its own particular energy or quality. He found that sensations could be elicited by mechanical or chemical influences, heat, electricity, etc. A Kantian aspect of perception is now naturalized, because it became evident that the brain has to reconstruct the world from its effects on nerves. Nevertheless, Müller defended an immaterial conception of animal spirit. He believed that animal spirits cannot be measured, because their speed is too high.

Helmholtz had explored the mathematical conservation law of energy. As energy could be transformed but neither created nor destroyed, an immaterial energy of life beyond the conservation law seemed to be senseless. Helmholtz preferred the theory that so far as the question of energy was concerned, the body could be viewed as a mechanical device for transforming energy from one form to another without special forces and spirits. Chemical reactions were capable of producing all the physical activity and heat generated by the organism. Muscular activity was realized by chemical and physical changes in the muscles. Furthermore, Helmholtz measured the velocity of nerve conduction and demonstrated that it was slower even than the speed of sound [4.16]. Philosophically, these results were interpreted as a refutation of Müller's vitalism.

Emil du Bois-Reymond (1818–1896), who was another student of Müller, showed that nervous affect was actually a wave of electrical activity. In those days, histologists began to discover separate cell bodies and fibers through the microscope. According to these results, nervous activity and the brain seemed to be a complex system of nerve cells ("neurons") with a complicated network of connections. The communication structure of neurons transmitting a signal from one neuron to another one was first described at the beginning of this century. But observations of synaptic junctions were not possible before the use of the electron microscope around the middle of this century.

How can the emergence of perceptions, thoughts, and feelings be explained by these descriptions of neuroanatomy and neurophysiology? One of the first thinkers who explained mental states by cell assemblies of neural networks was the American philosopher and psychologist William James. In his brief course "Psychology" (1890), James defended the Darwinian and evolutionary position that the brain is not constructed to think abstractly, but is constructed to ensure survival in evolution. In a pragmatic way, he assumed that the brain has many of the features of a good engineering solution applied to mental operations:

Mental facts cannot properly be studied apart from the physical environment of which they take cognizance . . . . Mind and world in short have evolved together, and in consequence are something of a mutual fit. [4.17]

Brain organization seems to be very poor at doing arithmetic and formal logic. But the ability to form concepts and associations, to make good guesses and to assume hypotheses is a characteristic feature of the brain. James presents a mechanistic model of association that stems from Hume's pioneering work and reminds us of the later associative neural networks. In a more qualitative way, he formulated some principles which are partially incorporated in the modern mathematical models of complex neural networks:

1. James believed that association was mechanistic and a function of the cerebral cortex.

2. James' principle of association:

When two elementary brain processes are active together or in immediate succession, one of them, on reoccurring tends to propagate its excitement into the other.

3. James' summing rule for brain activity:

The amount of activity at any given point in the brain cortex is the sum of the tendencies of all other points to discharge into it, such tendencies being proportionate (1) to the number of times the excitement each other point may have accompanied that of the point in question; (2) to the intensities of such excitements; and (3) to the absence of any rival point functionally disconnected with the first point, into which the discharges might be diverted. [4.18]

If in the second principle the term "brain process" is replaced by "neuron", then we get a description of a synapse which was later introduced by Hebb (compare Sect. 4.2). If in the third rule the term "point in the brain cortex" is replaced by "neuron", we get a linear summation rule of synaptic inputs which is very close to some network models of the Hebbian type. James also discussed the ability of networks of partial associations to reconstruct the missing pieces through some particular procedure of cell connecting. Although James was, of course, not familiar with computer-assisted modeling, he had the essential insight of the complex system approach that complex events are made up of numerous subassociations which are interconnected by elementary mechanisms like synapses.

In his chapter on "Association", James considered someone thinking of a certain dinner-party. The only thing which all the components of the dinner-party could combine to recall would be the first concrete occurrence which ensued upon it. All the details of this occurrence could in turn only combine to awaken the next following occurrence, and so on. In relation to Fig. 4.2, James described this process schematically:

If a, b, c, d, e, for instance, be the elementary nervetracts excited by the last act of the dinner-party, call this act A, and l, m, n, o, p be those of walking home through the frosty night, which we may call B, because a, b, c, d, e will each and all discharge into l through the paths by which their original discharge took place. Similarly they will discharge into m, n, o, and p; and these latter tracts will also each reinforce the other's action because, in the experience B, they have already vibrated in unison. The lines in Fig. 4.2 [4.19, Fig. 57] symbolize the summation of discharges into each of the components of B, and the consequent strength of the combination of influences by which B in its totality is awakened. [4.19]

James is convinced that "the order of presentation of the mind's materials is due to the cerebral physiology alone." In the modern complex system approach, order parameters are used to describe mental states which are caused by macroscopic neural cell assemblies. In the following sections we will see that many basic insights into the operations of the mind from the

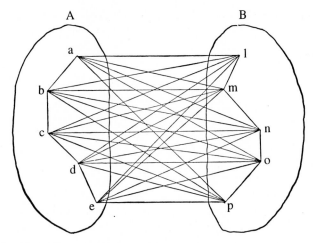

Fig. 4.2. William James' geometric model of an association network

presocratic philosophers to Kant and James have not been fundamentally altered even today.

4.2 Complex Systems and Neural Networks

In the 19th century, the physiologists discovered that macro-effects like perception, vision, muscular motion, etc., displayed by the nervous system depend on individual cells. These cells are able to receive and transmit signals by causing and responding to electric current. Obviously, the nervous system and the brain have turned out to be one of the most complex systems in the evolution of nature. There are at least ten billion nerve cells (neurons) in the human brain. Each neuron receives inputs from other cells, integrates the inputs, generates an output, and sends it to other neurons. The inputs are received by specialized synapses, while outputs are sent by specialized output lines called axons.

A neuron itself is a complex electrochemical device containing a continuous internal membrane potential. If the membrane potential exceeds a threshold, the neuron propagates a digital action potential to other neurons. The nerve impulses originate in the cell body, and are propagated along the axon with one or more branches. Neurologists usually distinguish excitatory and inhibitory synapses, which make it more or less likely that the neuron fires action potentials. The dendrites surrounding the neuron might receive incoming signals from tens or thousands of other neurons. The activity of a neuron is measured by its firing frequency. Biological neurons are not binary, because outputs are continuous. However, many models of neural networks are simplified and use binary computing elements [4.20].

Brains are complex systems of such cells. But while an individual neuron does not see or reason or remember, brains are able to do so. Vision, reasoning, and remembrance are understood as higher-level functions. Scientists who prefer a bottom-up strategy recommend that higher-level functions of the brain can be neither addressed nor understood until each particular property of each neuron and synapse is explored and explained.

An important insight of the complex system approach discloses that emergent effects of the whole system are system effects which cannot be reduced to the single elements. Philosophically, the whole is more than the sum of its parts. Thus, a purely bottom-up-strategy of exploring the brain functions must fail. On the other hand, the advocates of a purely top-down strategy proclaiming that cognition is completely independent of the nervous system are caught in the old Cartesian dilemma "How does the ghost drive the machine?".

Traditional positions in the philosophy of mind (compare Sect. 4.1) have more or less defended one of these strategies of research. In the 18th century, Leibniz and later on the zoologist Bonnet already suggested that there is a scale of complexity in nature with more or less highly developed levels of organization. In Fig. 4.3 the levels of organization in the nervous system are illustrated [4.21]. The hierarchy of anatomical organizations varies over different scales of magnitude, from molecular dimensions to that of the entire central nervous system (CNS).

The scales consider molecules, membranes, synapses, neurons, nuclei, circuits, networks, layers, maps, systems, and the entire nervous system. On the right side of the figure, a chemical synapse is shown at the bottom, in the middle a network model of how ganglion cells could be connected to simple cells in visual cortex, at the top a subset of visual areas in visual cortex, and on the left the entire CNS.

The research perspectives on these hierarchical levels may concern questions, for example, of how signals are integrated in dendrites, how neurons interact in a network, how networks interact in a system like vision, how systems interact in the CNS, or how the CNS interact with its environment. Each stratum may be characterized by some order parameters determining its particular structure, which is caused by complex interactions of subelements with respect to the particular level of hierarchy. Beginning at the bottom we may, for instance, distinguish the orders of ion movement, channel configurations, action potentials, potential waves, locomotion, perception, behavior, feeling, and reasoning.

It is quite obvious that an important function of the nervous system is to monitor and control the living conditions of the organism relative to its environment. An example of an elementary controllable state is, for instance,

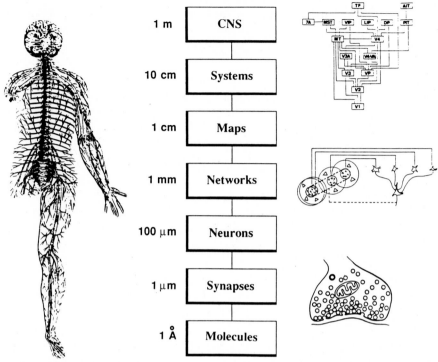

Fig. 4.3. Degrees of complex systems in the central nervous system (CNS): a chemical synapse, a network model of cellular connections in the visual cortex, and subsystems of the visual cortex [4.21]

the temperature of an organism. At the highest level the change of states in the environment needs anticipatory planning and social interactions, which have led to the human skills of verbal communication, creating art, solving mathematical problems, etc., during a complex cultural evolution.

From a Darwinian point of view, the evolution of the nervous system with its levels of increasing complexity seems to be driven by the fundamental purpose in nature to survive as the fittest. Some scientists of the brain even defend the strong opinion that the emergence of mental phenomena like abstract thinking is only some kind of "epiphenomenon" which was not originally intended by nature. But the belief in intentions and purposes of nature is, of course, only a human metaphor presuming some secularized divinity called "nature" governing evolution. According to the complex system approach, each level of the CNS has its own functional features which cannot be reduced to the functional features of lower levels. Thus, abstract thinking can only be regarded as an "epiphenomenon" from the perspective of a level like, say, the control system of the body's temperature.

In order to model the brain and its complex abilities, it is quite adequate to distinguish the following categories. In neuronal-level models, studies are concentrated on the dynamic and adaptive properties of each neuron, in order to describe the neuron as a unit. In network-level models, identical neurons are interconnected to exhibit emergent system functions. In nervous-system-level models, several networks are combined to demonstrate more complex functions of sensory perception, motor functions, stability control, etc. In mental-operation-level models, the basic processes of cognition, thinking, problem-solving, etc. are described. Their simulation is closely related to the framework of artificial intelligence (compare Chap. 5).

From a methodological point of view, we must be aware that models can never be complete and isomorphic mappings of reality. In physics, for instance, models of the pendulum neglected friction. In chemistry, models of molecules treated electrons in orbitals like planets in the solar system, contrary to Heisenberg's uncertainty principle. Nevertheless, these models are useful with respect to certain conditions of application. The conditions of brain models are given by the levels of brain organization. If a function of a certain level of brain organization is modeled, the model should take into account the conditions from the levels below and above. Higher-level properties are often not relevant. In general, the methodology of modeling is determined by a calculation of methodological costs and benefits. A model of the human brain which is intended to be realistic in every respect needs over-expensive analysis and construction. It would never satisfy the desired purpose, and thus it is impractical. Scientists will be more successful if they try to model each level of brain organization with simplifications concerning the levels below. On the other side, the models must be rich enough to reveal the essential complex features of brain organization.

According to the complex system approach, brain functions should be modeled by an appropriate state space and a phase portrait of its dynamical trajectories describing the brain's activities. René Descartes, the French mathematician and philosopher, already described the coordination of perception, arm moving, and brain in the framework of (Euclidian) geometry (Fig. 4.1).

Today, neural networks are geometrically characterized by vector spaces and neural matrices. The electrochemical input of neurons are connected with the outputs by weights. In a schematic section of the cerbellum (Fig. 4.4) the weights w_{ij} from a neural matrix allow the network to calculate the output vector from the input vector by matrix multiplication [4.22].

The example of Fig. 4.4 concerns a 3×4-neuron matrix. Neural physiology demands great flexibility of modeling, because the neural network may be rather complex. But the connectivity matrices can effect transformations on state spaces of high dimensionality into others with different dimensionality. Mathematically, these transformations of high dimensionalities may

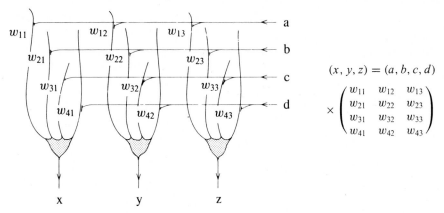

Fig. 4.4. Schematic section of the cerebellum modeled by neural matrix multiplication [4.22]

provoke geometrical problems which cannot be solved by the elementary formalism of analytical geometry. In this case, a generalized tensor network theory is necessary, in order to manage complex coordination tasks. From a historical point of view, it is amazing that the change from Euclidian to more general topological and metric spaces can be stated not only for the physics of the outer world in general relativity but also for the intrinsic features of the nervous system.

With respect to Descartes' early approach, let us regard an elementary sensorimotor coordination which is represented by vector or tensor transformations. How can an animal seize an object which it perceives by its sensory organs (Fig. 4.5a)? In a simplified model, the position of the two eyes is at first codified in a 2-dimensional space of sensory data. The state space can be visualized by a 2-dimensional topographic map. An impulse is sent from a point of the sensory state space to a corresponding point of the motor state space which is also represented by a 2-dimensional topographic map. A point of the motor state space codifies the corresponding position of the arm (Fig. 4.5b) [4.23].

Another example of sensorimotor coordination is given by the vestibulo-ocular reflex. This is the neural arrangement whereby a creature can stabilize an image on the retina by short-latency movements of the eyes in the opposite direction to head movement. There are two neural structures involved in this neural system, which can be represented by different coordinate systems intrinsic to the CNS. First we have to analyze the semicircular canals of the vestibular apparatus in the ear, three canals on each side, which can be represented by a 3-dimensional coordinate system. Second, each eyeball has six extraocular muscles corresponding to a 6-dimensional coordinate

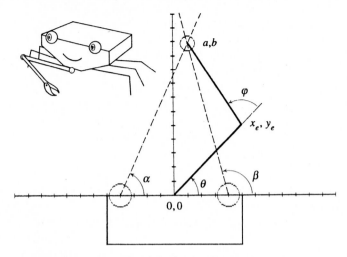

Fig. 4.5a. Sensorimotor coordination of perception and arm movement [4.23]

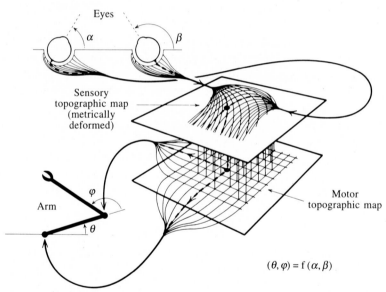

Fig. 4.5b. Geometric model of sensorimotoric coordination by sensory and motor topographic maps [4.23]

system. Thus, the sensorimotor coordination of the vestibulo-ocular reflex is geometrically described by a tensor transformation of a 3-dimensional (covariant) vector. The mathematical scheme can be used for calculations of any eye-muscle activation emerging from a given vestibular input.

On the level of neurons and networks, nets of artificial units are used to simulate and explore the brain organization [4.24]. These units are assumed to vary between 0 and 1. Each unit receives input signals from other units via synaptic connections of various weights. The incoming and outgoing representations are ordered sets of values, and the output units are activated appropriately. Mathematically, the procedure can be interpreted as a mapping of some inputs as arguments onto corresponding outputs as function values. The function rule is determined by the arrangement of the weights, which depends on the topology of the neural network.

In the brain, neurons sometimes constitute a population as input layer (Fig. 4.6). The axons of these cells are sent to a second layer of neurons. Axons from cells in this second layer can then project to a third population of cells, and so on. The assembled set of simultaneous activation levels in all the input units is the network's representation of the input stimulus, as input vector. This input vector with its activation levels is propagated upward to the middle layer. The result is a set of activation levels determined by the input vector of the input layer and the several connection weights at the ends of the terminal branches of the input units to the neurons of the middle layer. This activation vector of the middle layer is propagated upward to the topmost layer of units, where in the case of a three-layer network an output vector is produced. Again, this ouput vector is determined by the activation vector at the middle layer and the connection weights at the ends of the terminal branches of the middle units to the output units.

A two-layer network with only an input and an output layer corresponds to a simple stimulus-response scheme with observable and measurable inputs and outputs. In the case of a three-layer network, the units of the middle layer with their weights are sometimes not directly measurable or only hypothesized as a hidden mechanism in a black box. Thus, they are called hidden units.

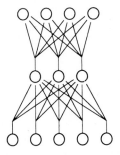

Fig. 4.6. Three-layer network with hidden units

Real nervous systems, of course, display many more units and layers. In humans, for instance, the structure of the cerebral cortex contains at least six distinct layers of neurons. By the way, the mapping of inputs onto outputs by a neural matrix in the cerebellum in Fig. 4.4 can equivalently be described by a two-layer network with input and output layer. A three-layered network is equivalent to a pair of neural matrices connected in series. But this kind of a many-layered network cannot be generalized for the whole brain and nervous system, because cell populations in the real brain often show extensive cell-to-cell connections within a given layer, which will be considered in some different models.

According to the complex system approach, the neurons of a particular layer can be interpreted as the axes of a state space representing the possible activity states of the layer. The development of states, their dynamics, is illustrated by trajectories which may be caused by some learning process of the particular network.

For instance, a perception can be explained by vector-processing in a neural network. At first, there is a sample of stimuli on the input neurons from the outer world (e.g., electromagnetic light signals, colors in the eye, or sound waves in the ear), which is processed in the neural network to produce an output vector which represents, e.g., a visual or auditory picture of the outer world. But the neural networks must learn to distinguish and recognize the correct forms, colors, sounds, etc., in a huge mass of input data.

The learning procedure is nothing else than an adjustment of the many weights so that the desired output vector (e.g., a perception) is achieved. The learning procedures can be simulated by mathematical algorithms which are an important topic of research in artificial intelligence (compare Sect. 5.3). They produce weight configurations at each neural layer which can be represented in terms of vectors, too. At any given time, the complete set of synaptic values defines a weight space with points on each axis specifying the size of a particular weight. In general, learning means minimizing the errors or differences between a most adequate solution (perception, idea, etc.) and a less adequate one. Thus, a learning process can be visualized by a trajectory in the weight space, starting from the initial randomly set position to the final minimal-error position (Fig. 4.7a). The key to this kind of modeling means that weights in a network can be set by an algorithmic procedure to embody a function. It is assumed that any representable world can be represented in a network, via configurations of the weights.

Figure 4.7a shows a trajectory in the synaptic weight space during a learning process. This space, simplified for three weights, represents all possible weight combinations from the synapses in the three-layer network (Fig. 4.6). Figure 4.7b shows a corresponding activation-vector space whose axes are the hidden units of the three-layer network (Fig. 4.6) [4.25].

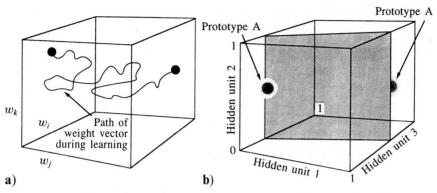

Fig. 4.7a,b. Synaptic weight space (**a**) and activation vector space (**b**) of the three-layer network in Fig. 4.6 [4.25]

The weight space and the activation space are similarity spaces, because similar vectors representing similar things are reflected by proximity of position. Weight configurations cluster similar things, taking into account that weight configurations may be sensitive to very tiny differences between things. Thus, in the activation space, we can distinguish prototype vectors representing similar things with tiny differences measured by their distances to the prototype vector. On the macroscopic level of observation and behavior, these prototype vectors may represent particular categories of trees, plants, fruits, persons, etc., which are more or less similar. In the framework of complex system dynamics, prototype vectors can be interpreted as point attractors dividing the state space into several regions.

Similar motor behavior (like seizing, walking, etc.) is represented by similar trajectories in a motor state space. Learning, as we saw, means reconfiguration of weights according to some algorithmic procedure. The crucial question arises: how do thousands of cells and synapses know when they should change their states without the guiding hand of a demon?

In his famous book *The Organization of Behavior* (1943), Donald Hebb suggested that learning must be understood as a kind of self-organization in a complex brain model. As in the evolution of living organisms, the belief in organizing "demons" could be dropped and replaced by the self-organizing procedures of the complex system approach. Historically, it was the first explicit statement of the physiological learning rule for synaptic modification. Hebb used the word "connectionism" in the context of a complex brain model. He introduced the concept of a synapse, which was later called a "Hebb-synapse". The connection between two neurons should be strengthened if both neurons fired at the same time:

> When an axon of cell A is near enough to excite a cell B and repeatedly or persistently takes part in firing it, some growth process or metabolic change takes place in one or both cells such that A's efficiency, as one of the cells firing B, is increased. [4.26]

In 1949, the "Hebb-synapse" could only be a hypothetical entity. Today, its neurophysiological existence is empirically confirmed. Hebb's rule is by no means a mathematically exact statement. Later on, we shall see that many Hebb-like learning rules of connectionism are possible. A simple mathematical version of the Hebb rule demands that the change Δw_{BA} of a weight w_{BA} between a neuron A projecting to neuron B is proportional to the average firing rate v_A of A and v_B of B, i.e., $\Delta w_{BA} = \varepsilon v_B v_A$ with the constant ε.

Hebb-like rules suggest schemes tending to sharpen up a neuron's predisposition "without a teacher" from outside. In this sense, it is a self-organizating method for a neuron's firing to become better and better correlated with a cluster of stimulus patterns. Hebb was aware that the brain uses global patterns of connected neurons to represent something. He explicitly used the term "cell assemblies", which has become a key to modern neuroscience. Active cell assemblies could correspond to complex perceptions or thoughts. Philosophically, Hebb's idea of cell assemblies reminds us of Hume's concept of association, which was only psychological without the physiological basis of the brain.

How are Hebb's physiological ideas incorporated in modern complex systems of neural networks? The basic concept of an associative network demands that an input vector is "associated" with an output vector by some transformation. Mathematically, the similarity of two vectors can be measured by their inner product, which is the result of multiplying both vectors, component by component, and then adding up the products. Geometrically, the inner product is proportional to the cosine of the angle between the vectors. In the case of total congruence of the vectors, the angle is zero, which means that the similarity is complete.

Thus, the similarity of a stored prototype vector (for instance the prototype picture of a typical tree) with an input vector (for instance the perception of a particular tree) can be calculated in an associative network by their inner product. The prototype vector is assumed to be stored in the matrix of the weights connecting the input and output of the network. Figure 4.8a shows a net with horizontal input lines for the input components, vertical output lines, and weights on the connections (which are considered to be binary, with open circles for zero and closed circles for one).

If in general the input vector (x_j) is associated with the output vector (y_j) by a linear transformation $y_i = \sum_j w_{ij} x_j$ with respect to the stored weight vector w_{ij}, then we get the simple case of a linear associator. This kind of an associative network is able to classify vectors representing examples of

Output lines

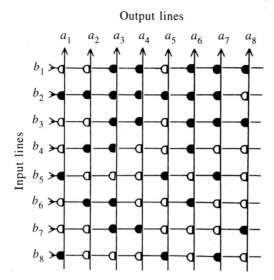

Fig. 4.8a. Linear associative network

some category which is realized by a stored prototype vector. This task is actually crucial for the survival of animals. In reality, a variety of more or less similar perceptions (for instance, of a hostile animal) must be identified and subsumed under a category.

Another kind of associative network can perform vector completion or vector correction. A so-called autoassociative network can produce an output which is as close as possible to a prestored vector given only part of the vector as input. In reality, noisy versions of an input vector (for instance, a picture of a person) must be completed according to a stored picture. A Hebb-like rule can fulfill this task by strengthening the connection weights between neurons with respect to the degree of their correlated activity.

A method for increasing the capacity of such a complex network is to introduce a nonlinear threshold for the output units. A linear associative network (for instance, Fig. 4.8a) has a feedforward topology with information flowing from input units to output units. Hebb-like learning procedures suggest local interactions of neural units converging to the correct global output by self-organization. Circulating information in the network means a feedback architecture. In Fig. 4.8b, each unit receives inputs from outside and feedback from intrinsic units of the network. The weights are represented by the intersections of horizontal with vertical lines [4.27].

Obviously, the complex system of Fig. 4.8b models a nonlinear feedback network which allows a wide range of possible dynamics. A famous example was explored by John Hopfield (1982). His class of nonlinear feedback networks has a dynamics converging to a solution. They have been interesting

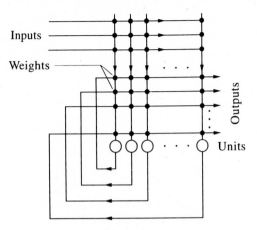

Fig. 4.8b. Nonlinear feedback network [4.27]

not only for modeling brain functions, but also (as we shall see in Chap. 6 on artificial intelligence) for the development of a new network technology. Concerning our complex system approach, it is noteworthy that Hopfield is physicist who has applied mathematical equations from spin glass physics to neural networks [4.28].

The dynamics of a ferromagnet is a well-known example of conservative self-organization in thermal equilibrium. In the Ising model a ferromagnet consists of a lattice of spins, each of which can be either up ($\updownarrow$) or down ($\updownarrow$). Each spin can interact with its nearest neighbors. The state with the lowest energy has all the spins lined up in the same direction. At a high temperature the directions of the spins are random because the thermal energy which causes the fluctuations is larger than the energies of interaction. If the temperature is reduced, the spins become aligned in the same direction. Evidently, the spins behave like a magnet (compare Sect. 2.4). Dynamically, it seems to seek the nearest local energy minimum as an attractor state (Fig. 4.9a). But a single energy minimum with all spins pointing in the same direction is only provided if all interactions are attractive. In the case of mixed attractive and repulsive interactions a complex system like a spin glass may have many local energy minima [4.29].

Fig. 4.9a. Phase transition in a 2-dimensional Ising model of a ferromagnet (annealing)

Hopfield assumed that the function of the nervous system is to develop a number of locally stable points in state space. Other points in state space flow toward the stable points as attractors of the system. As deviations from the stable points disappear, this dynamics is a self-correcting procedure. On the other hand, the stable point appropriately completes missing parts of an incomplete initial state vector. Thus, the dynamics can be used to complete noisy inputs.

Hopfield's model is rather simplified and involves threshold logic units, summing synaptic inputs, and comparing the sum with a threshold. If the sum is at or above threshold, they yield 1, and 0 otherwise. The network is recurrent in the sense that the neurons connect to each other with the exception of self-connection. Mathematically, the corresponding connectivity matrix has zeros along the main diagonal. Hopfield suggests a Hebb-like learning rule for constructing elements of the connectivity matrix. The complex system evolves like an Ising model of a spin glass according to a nonlinear feedback dynamics. The term isomorphic to energy decreases until it reaches a – perhaps local – minimum.

A simple application is given by the well-known problem of alphanumeric character recognition. The complex network is composed of interacting Boolean variables represented on a 2-dimensional grid. A pattern (for instance, the letter A) can be associated to the grid with a dark point for all active variables (with a value 1), and a blank point for those with a value 0. It is assumed that the letters of the alphabet are associated to attractors ("fixed points") as the desired states of the dynamical system. We may imagine that a human brain has stored the correct shapes of the letters by seeing many correct examples. If an incomplete and partly ruined letter is shown to the system, it should be able to reconstruct the correct shape that was learnt before (Fig. 4.9b) [4.30].

Thus, pattern recognition means pattern evolution by self-organization. This process aims at some attractors as the desired states of the system. We remember that an attractor is a state towards which the system may evolve, starting from certain conditions. The basin of attraction is defined by the set

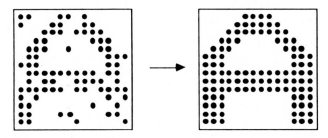

Fig. 4.9b. Phase transition in a Hopfield system for pattern recognition

of initial conditions that drive the trajectories of the system in the direction of the attractor. As we saw in earlier sections, an attractor may be a unique state referring to a fixed point or stable state, as in the examples of Hopfield networks and spin glass systems. But a periodic succession of states (a "limit cycle") or several forms of chaotic attractors (in dissipative systems) are also possible. Thus, the Hopfield networks are only a first and simplified approach to modeling neural states by attractors of complex systems.

Hopfield saw the analogy between the local energy minima in spin glasses and the prototypes in an associative brain. In the formal framework of a spin glass, attractors can be designated as prototype vectors. In Fig. 4.10a, the state space of a Hopfield system is visualized by an energy landscape by analogy with the thermodynamics of spin glasses. All possible states of the network are represented by points in a plane. The height of the surface refers to the energy of the corresponding state of the network.

a)   b)

Fig. 4.10a,b. State space of a Hopfield system as energy landscape (**a**) with local minima as attractors (**b**) [4.31]

The phase portrait of the system in Fig. 4.10b shows the convergence of the trajectories to stable local minima from different starting points. Each point in the plane is a state of the network. The energy landscape has basins attracting the trajectories of the Hopfield dynamics. The stable points ("attractors") are at the bottom of the basin. In the example of pattern recognition, the prototype letters are connected with the stable points. Thus, the process of pattern completion is a form of relaxation which can be compared formally with the annealing processes of conservative self-organization. In the physical examples, the final states are ordered structures of a spin glass, a magnetized ferromagnet, or a frozen crystal [4.31].

In general, Hopfield networks only converge to local minima in a state of lower energy. In some applications, the local minima are associated with particular stored items, and there may be no need to reach a global minimum. However, in many cases the global minimum is required. A solution of this problem was offered by making the individual units stochastic rather than deterministic.

Figure 4.11a visualizes the solution by a ball traveling along a curve of an energy landscape to probably end up in the deepest minimum. Starting from a given initial situation, the ball will move towards an energy minimum or the bottom of a well. If the energy landscape is characterized by a muliplicity of minima close together, the result depends upon the initial conditions. How can the network be prevented from getting stuck in a local minimum? The idea is to shake the energy landscape with a certain energy increment which is required to escape the valley of the local minimum B to enter the attractor of the global minimum A.

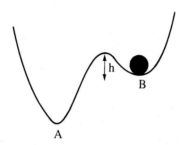

Fig. 4.11a. Phase transition from local minimum B to global minimum A in an energy landscape (simulated annealing)

Then, mechanically, the ball is more likely to go from B to A than from A to B. On average, the ball should end up in the valley of A. In the language of thermodynamics, the kinetic energy added by shaking the landscape corresponds to increasing the temperature of the system. For fairly high temperatures, the probability of transition between the valleys is no longer negligible. At thermal equilibrium, the probability of occupying the various valleys only depends on their depth.

In practice, the method of simulated annealing is well known and used for global optimization. As we already mentioned, annealing is the process of heating a material (e.g., metal or glass) to a high temperature, and then gradually lowering the temperature. But the material will only end up at its global energy minimum if the annealing process is gradual enough. Sudden cooling of, e.g., metal will leave the material only in a local minimum with a brittle state. Simulated annealing makes the escape from local minima likely by allowing jumps to higher energy states.

In the thermodynamics of gases, the gas is described by its probability of phase transition. It was Boltzmann who derived a probability distribution for the states of a gas when it had reached a uniform distribution of temperature. Hinton, Sejnowski, and others claimed that the distribution could be used to describe neural interactions. In this case of modeling, the low temperature term added to the system is interpreted as a small noise term. It is the neural analog to random thermal motions of molecules in gases.

This formal equivalence is the reason that the network under consideration is called a "Boltzmann machine" [4.32]. But, of course, no physicalism is intended, reducing neural interactions to the molecular interactions of gases. In Boltzmann's formalism, it can be proved that a Boltzmann machine is guaranteed to find the desired global minimum as long as it is cooled slowly enough. Obviously, a neural network with the dynamics of simulated annealing is capable of searching a state space for the pattern giving the global energy minimum.

A possible learning rule according to this dynamics matches probabilities between the network and its environment. All possible states of the network are possible at thermal equilibrium, with the relative probabilities of a Boltzmann distribution. If the probabilities of the states in the network are the same as the probabilities of states of the environment, then the network has an adequate model of the environment. Thus, a learning rule must be able to adjust the weights in the Boltzmann machine so as to decrease the discrepancy between the network's model and the environment.

At first, the rule lets the system run free. The probabilities of the states taken by each unit can be estimated. Then, the input and output units are clamped or forced to take appropriate values. Again, values of the probabilities of the states of the units are estimated. The local change of weights is proportional to the difference in the probabilities of the units coupled by that weight [4.33].

Formally, the weight modification rule demands that

$$\Delta w_{ij} = \varepsilon(\langle s_i s_j \rangle \text{ clamped} - \langle s_i s_j \rangle \text{free})$$

where ε is the constant of proportionality ("rate of learning"), s_i is the binary unit of the ith unit and $s_i s_j$ is averaged over time to $\langle s_i s_j \rangle$ after the network has reached equilibrium. In the clamped condition, the input and output units are fixed to their correct values. In the free condition, none of the units is fixed. Then, the learning rule is unsupervised. If the inputs are fixed in the free condition, the learning rule is supervised.

In Fig. 4.11b, the units in the network of a Boltzmann machine have binary values and the connections between them are reciprocal [4.33]. The weights of the connections can be trained by presenting patterns to the input units in the presence and the absence of output patterns and applying the Boltzmann learning rule. During the learning process, every weight in the network is modified. The hidden units which do not receive direct information from outside enable the network to yield complex associations between input and output patterns. Thus, Boltzmann machines with hidden units in their middle layer have internal representations of the environment which are not possible for networks with only visible (input and output) units.

From a neurobiological point of view, supervised learning with a "teacher" seems to be rather unrealistic in nature. Feature extraction or categoriza-

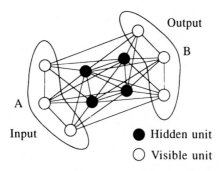

Fig. 4.11b. Network of a Boltzmann machine [4.33]

tion by an animal must be self-organized from an analysis of the sensory inputs. The more frequently a feature occurs in the input vectors, the more likely it is to belong to a certain category. The outputs of the network must learn to converge to the corresponding prototype vectors as attractors.

How can a network be designed to invent criteria of classification without the supervision of an external teacher? Some authors assume that this kind of self-organization depends on the nonlinear interactions and selective reinforcement of the connections in a multi-layered system. The learning procedure is organized in a Darwinian process of selection and competition.

In Fig. 4.12, the multi-layered architecture of a competitive learning system is designed to produce such eminently cognitive tasks as classification and categorization [4.34]. Active units are represented by filled dots, inactive ones by open dots. The connections from the input layer to each element in the second layer are excitatory. The second layer is subdivided into clusters within which each element inhibits all the others. Elements of the same cluster compete with each other in responding to the input pattern. According to the rules of Rumelhart and Zipser, a unit can learn only if it can win the competition with the other units within the same cluster. Learning means an increase in the active connections and a decrease in the inactive connections.

A simple task of classification refers to word recognition by a child. Obviously, the two-letter words AA, AB, BA and BB may be classified in several categories, for instance, the set {AA, AB} of words beginning with A or the set {BA, BB} of words beginning with B or the set {AA, BA} of words ending with A or the set {AB, BB} of words ending with B. In a computer-assisted experiment, the two-letter words were presented to a layered network with one level of competing units organized in a cluster of two units. The system was able to detect the position of the letters. One of the units spontaneously learnt to act as a detector of A as beginning letter, while the other one detected B as the beginning letter.

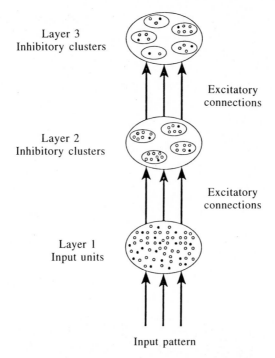

Fig. 4.12. Multi-layered network with competitive learning [4.34]

In further experiments, the number of letters was increased, with a modified network structure. Although these experiments seem only to illustrate limited capabilities, they demonstrate the emergence of cognitive behavior from unsupervised neural systems, at least in principle. They have started some interesting research linking neurophysiology with the cognitive sciences in the framework of complex systems which will be explored in more detail in Sect. 4.4.

Another approach to self-organizing cognitive systems through competitive learning was proposed by Teuvo Kohonen. He is a physicist who also has worked physiologically on associative memory. His mathematical modeling of neural systems has been important for engineering applications in artificial intelligence (compare Chap. 5). Kohonen's ideas of brain modeling by self-organizing feature maps stems back to anatomically and physiologically well-confirmed facts. Most neural networks in the brain are two-dimensional layers of processing units which may be cells or cellular modules. These units are interconnected through lateral feedback. For instance, in the neocortex there are 10 000 interconnections for every principal cell.

The synaptic coupling from a neuron to its neighbors is excitatory for all those neurons whose distance is smaller than a certain critical value. It is inhibitory for neurons lying at a greater distance. At some yet greater distance, the coupling is weakly excitatory again. The degree of lateral interaction is mathematically modeled by a curve with the form of a Mexican hat [4.35] (Fig. 4.13a).

Fig. 4.13a–c. Mexican hat of neural interaction (**a**). Distribution of neural activity in a 2-dimensional model (**b**) and a raccoon's cortex (**c**) [4.36]

Obviously, the activity of lateral couplings tends to a spatially bounded cluster. Figure 4.13b shows a two-dimensional example of clustering which was simulated by a network with a 21 × 21 square array of processing units. The clustering phenomenon ("activity bubble") depends on the degree of positive or negative feedback, which may be influenced by chemical effects in the neural network. In neural reality, "activity bubbles" do not have the regular form of computer-assisted simulations. Figure 4.13c shows distributions of activity on a raccoon's cortex, which does not represent a regularly-shaped figure, but a rather diffuse map [4.36].

Nevertheless, the cluster phenomenon can be shown to be useful in the self-organizing processes of the brain. While initially the activities of the neural network are homogeneously distributed, a progressive specialization of neural regions can be observed, according to a self-organizing learning process. After presentation of an input pattern, the neuron with highest activation and its neighbors are chosen for learning. The neural weights are modified according to the circular neighborhood of given radius, centered around the neuron with highest activation. This learning rule can be used to detect and categorize similarities among input data of visual or speech patterns.

Formally, Kohonen considered a nonlinear projection P from the space V of input signals v onto a two-dimensional map A. Figure 4.14 illustrates the learning step: the input value v selects a center s. In the neighborhood of s, all neurons shift their weights w_s in the direction of v. The degree of shift decrease with increasing distance from the center s and is visualized by differing grey values [4.37].

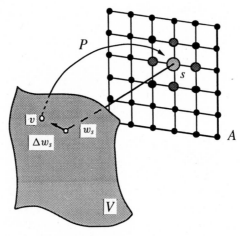

Fig. 4.14. Kohonen's model of self-organizing neural maps [4.37]

The map converges to a state of equilibrium with different regions of activity by self-organization. The projection should map the regularities of the input signals onto the neural map. Thus, P is mathematically called a topologically invariant mapping. Actually, the structure of the brain's environment, which is represented by the regularities of the sensory input signals, should be projected onto a neural map of the brain: the brain should get an adequate model of the world.

How realistic is the modeling of the brain by self-organizing maps? The magnitude of a neural field varies, depending on the importance of the perceived sensory stimuli for the survival of the species. In a neural field there are centers which can analyze and represent the stimuli with more accuracy than their environment. For instance, in the eye of a mammal the fine analysis of visual information is performed by the "fovea", which is a very small region around the optical axis of the retina with a very high density of light-sensitive receptors. Thus, the dissolution of signals is essentially higher in this center than in the surrounding region of the neural field. Similiar unproportional representations can be observed in the somatosensory system and in the motor cortex. The importance of the hand for human survival is represented by a rather great region on the somatosensory and motor cortex, relative to the representation of the body's surface.

Contrary to these results, the auditory cortex of cats, dogs, and apes does not project the frequencies of the outer world with special centers. An exception is the bat with its specialized system of orientation, which is necessary for its survival. Bats can send many different supersonic frequencies and determine the distance and magnitude of objects by the reflections of the signals. The bat's velocity relative to other objects can be determined by the Doppler effect in supersonic echo sounding. Even the movements of tiny insects can be detected by this sensitive system.

The specialization of the bat can be experimentally confirmed by a self-organizing map on its auditory cortex. Figure 4.15a shows the brain of a bat with the auditory cortex in the rectangle. Figure 4.15b is an enlargement of the rectangle with a distribution of the best frequencies on the auditive cortex. The one-dimensional frequency spectrum is represented continuously and monotonicly from the posterior to the anterior region of the auditive cortex. The frequency causing the highest excitation of a neuron is called the best frequency of that neuron. The region of the dotted lines is the primary auditory cortex. Figure 4.15c shows the distribution of the best frequencies in the hatched region of Fig. 4.15a. The majority of measured points are centered around the frequency of supersonic echo sounding. More than the half of the anterior-posterior region is used to analyze the Doppler effect in the supersonic echo sounding. It is remarkable that computer assisted simulations by self-organizing maps produce the actual representations of the auditory cortex in Fig. 4.15c [4.38].

The brain in primates consists of many regions with several neural net topologies. The retina, for instance, already develops during early ontogeny. It has a neural topology with five separate layers for photoreceptors, horizontal cells, bipolar cells, amacrine cells, and retinal ganglion cells. The photoreceptor layer in humans consists of about $120 \cdot 10^6$ receptor cells. The retinal output, which is represented by the spatio-temporal pattern of the impulse rates of all ganglion cells, travels along the optic nerve toward the thalamus. In humans, there are about $1.2 \cdot 10^6$ ganglion cells. Thus, the retina is a really complex system. Nevertheless, the complexity of more than $200 \cdot 10^6$ retinal neurons still is not understood completely. The cerebral cortex is the phylogenetically youngest brain region. The percentage of cerebral cortex to brain increased during evolution. Lower vertebrates like fishes did not evolve a cerebral cortex. Its magnitude increased from small parts in reptiles and birds to dogs, cats, and finally apes and humans. In primates, the cortex is divided into different regions with multilayered neural net topologies like visual, sensory, motor, and association cortex. The cerebellum consists of

Fig. 4.15a–c. Brain of a bat with auditory cortex in the rectangle (**a**), its enlargement (**b**), and distribution of the best frequencies (**c**) [4.38]

the cerebellar cortex with many multilayered subregions for specific sensori-motor functions.

The great variety of brain systems is described as densely packed sets of neurons with particular network topologies, communicating with each other via many nerves which consist of thousands of axons. In contrast to digital computers with separate central processing unit, memory, and registers, the brain and the central nervous system can be modeled as an ensemble with many special-purpose parallel processing networks. Each network is capable of independent processing and storage of information for sensory, motor, and associative functions.

Obviously, the biological brain does not apply principles known from program-controlled centralized digital computers. The process of network self-organization is fundamental to the structure of the brain. In the very long course of phylogenesis, complex structural forms, the purposes of which are sometimes not completely clear to us, have been produced. On the macroscopic scale, particular neural areas have been specialized for signals with different sensory functions, for information processing operations with different levels, for humans as well as for the animal and vegetative functions of the organism. Although they are distributed in different areas of the brain, they can be understood as self-organizing complex or collective effects.

Self-organization as a learning procedure demonstrates that organisms are not fully determined by genes containing a blueprint which describes the organism in detail. Each stage of brain organization involves some kind of self-organization. Genes would not be able to store the complex structure of the brain. With a cerebral cortex of about 10^{14} synapses, ontogeny could not select the correct wiring diagram out of all alternatives if all were equally likely. Thus, ontogeny must use the self-organization of neural systems to handle their complexity. But the structure of the cortex is not understandable without knowing the principles of its ontogenesis.

In earlier chapters, we have studied the emergence of ordering patterns from complex systems in physics, chemistry, biology, metereology, and astronomy. Global order emerges in complex systems with a large number of locally interacting elements. There are interacting atoms or molecules in a liquid or crystal, subvolumes in an evolving star system, or neurons and synapses in a complex neural system like a brain. We remind the reader of Bénard convection ("rolling columns"), which arises by thermal fluctuations of a liquid.

How can global order be arranged by local interactions? The intermolecular forces, for instance, acting within a volume of liquid have a very short range, while the pattern of convective movement which is caused by the molecular interactions may be ordered on a large scale. This principle, which arises in physical, chemical, and biological evolution, has great importance to the brain, in which local interactions between neighboring cellular elements create states of global order leading to a coherent behavior of the organism. The ordering pattern is arranged by forces between elements of the complex system and by initial and boundary conditions. In the example of Bénard convection, the forces are hydrodynamic interactions, thermal conduction, expansion, and gravity. Boundary conditions are, for instance, the temperature which is given to the liquid. In the brain, the connection patterns are arranged by several rules for the interaction of cellular units. As neurons are connected by sometimes very long axons, a local interaction of two neurons does not imply their spatial proximity in the brain's anatomy, but only their immediate connection by axons.

Although the general structure is universal for all types of neurons and synapses, there can exist many qualitative and quantitative differences. The neural system of an invertebrate, for instance, is deterministic with a high degree of coded information in the specific location of individual neurons. For an associative system in the mammalian neocortex, the specific responses to specific input patterns are achieved by learning rules faciliating feedback of information from the output.

How adequate are complex system models to real neural networks? From a methodological point of view, we must be critically aware that models cannot be naively identified with each function and element of reality. Models are special-purpose abstractions which may explain and simulate some part of the central nervous system more or less, and other parts not at all. Sometimes model nets are criticized in that they only demonstrate more or less correct execution of an input-output function like a black box. But nothing could be revealed about how biological neural nets execute that function. Hidden units were only theoretical concepts like hidden variables in quantum physics which are assumed to be intrinsic elements of the system realizing the relation between the observed and measurable input and output values. Besides the architecture of a perhaps multi-layered network, the dynamics and learning procedures are an essential problem of simulations.

How realistic are the parameter-adjusting procedures of model nets to minimize error? Many learning rules of model nets take unacceptably long to converge to a set of weights that classify correctly. Although a successful weight-adjustment is sometimes found, its optimality is by no means decided. In 1960, Widrow and Hoff suggested a simple and elegant learning rule which was motivated by reasons of technical optimization, and not by biological insight into the brain's function [4.39]. The Widrow–Hoff-rule and its variants have been extensively used in technical networks in recent years (compare Chap. 5).

The rule assumes that there is an input pattern, and an output classification of the input pattern by an adaptive neuron, which can take values of either $a + 1$ or $a - 1$. Thus, a "teacher" is assumed knowing what the answer was supposed to be for that input. The adaptive neuron compute a weighted sum of activities of the inputs times the synaptic weight. The system is able to form an error signal between what the output is supposed to be and what the summer computed. According to the difference between them, the synaptic weights are adjusted, and the sum recomputed, so the error signal becomes zero.

Widrow and Hoff's strategy aims at reducing the square of the error signal to its smallest possible value. All possible values of the input weighting coefficients give rise to an error value. In Fig. 4.16, the situation is visualized by an error surface in the weight space [4.40].

The minimum of the error surface is not known exactly, because the entire surface cannot be seen. But the local topography can be measured. Thus, the directions of adjustment which decrease the error the most can be calculated. The so-called gradient descent method, which is well known in differential geometry and physics, always adjusts the weights so that changes

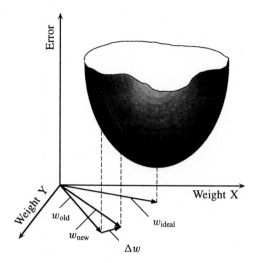

Fig. 4.16. Error surface and learning by gradient descent [4.40]

in weights move the system down the error surface in the direction of the locally steepest descent.

In Fig. 4.16, the gradient of the error surface is computed to find the direction of steepest descent. The weights are incrementally changed by a certain value Δw along this direction. The procedure is repeated until the weights reach w_{ideal} representing minimum error. For a nonlinear network the error surface may have many local minima. Problems with gradient descent in general involve getting trapped in local minima. Then the bottoms of valleys do not represent the lowest global error.

Widrow and Hoff proved that there is a simple quadratic error surface with only one global minimum. Mathematically, computing the gradient at a point means computing partial derivatives of the square of the error with respect to the weights. Widrow and Hoff proved that this derivative is proportional to the error signal. Thus, the measurement of the error signal provides the direction of movement, in order to correct the error. Technically, the existence of a "teacher" with perfect knowledge may be justified for special purposes. But the assumption of supervised learning procedures in nature seems to be rather unlikely.

In network models, so-called backpropagation is the best known supervised algorithm which is a generalization of the simple Widrow-Hoff rule. In short, backpropagation is a learning algorithm for adjusting weights in neural networks. The error for each unit, which is the desired minus the actual output, can be calculated at the output of the network and recursively propagated backward into the network. This method enables the system to

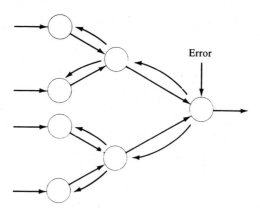

Fig. 4.17. Backpropagation

decide how to change the weights inside the network to improve its overall performance. Figure 4.17 illustrates the backpropagation method through an entire net with several layers [4.41].

Although models with backpropagation can be as successful as biological networks, it is not assumed that the real brain is organized by backpropagation. Many values of parameters in a real network are sometimes known by measurement and experimentation in anatomy, physiology, and pharmacology. For instance, the number of cell types and cells themselves may be roughly estimated. The topology and architecture may be described, the question whether specific synapses are excitatory or inhibitory may be decided, and so on. But the specific weights are unknown. When there are thousands of them, the probability that the global minimum of a brain net and a model net are exactly the same, weight for weight, is rather low.

Thus, the error minimization strategies of model nets may in general be hypothetical, but they are necessary to deal with the complexity of perhaps thousands of unknown parameters. They allow one to predict some local or global properties of brain nets if the degree of similarity between the topology, architecture, and synaptical dynamics of model and brain nets is rather high. Backpropagation in neurobiology is justified as a search tool to find local minima, but not as a replacement for neurobiological analysis, which may reveal the real learning procedure of a neural network.

Error minimization strategies have a long tradition in the evolution of nature and have not emerged for the first time in the learning procedures of the brain. Natural selection of, for instance, an ecological population can often be modeled as a process of sliding down an error gradient to an error minimum representing an environmental survival niche. But the parameter-adjusting procedure may only find a local minimum, not necessarily the global minimum. As far as we know, evolution in general does not find the best possible solution, but only a satisfactory one which is good enough for survival. Only local minima of the entire evolution can be empirically evaluated with respect to their evolutionary survival value. This evaluation depends on the observed and measured constraints of the chosen model.

Thus, from a methodological point of view, the idea of an always globally and perfectly optimizing nature is a metaphysical fiction. It is the secularized idea of a godhead called "nature" or "evolution" which was born in the optimistic century of enlightenment in order to replace the Christian God with his plan of creation. It was already Kant who criticized the idea of a totally self-optimizing nature as a human fiction which cannot be empirically justified in any sense. There is no supercomputer with a separate central processing unit which can optimize the evolutionary strategy totally and in the long run. There are only locally more or less satisfactory solutions, even many failures, and imperfection in real evolutionary processes. Their complexity contradicts the simplified models of a perfect world à la Laplace.

4.3 Brain and the Emergence of Consciousness

How can cognitive features be explained by neural interactions in complex models of the brain? Leibniz already had the problem that consciousness, thoughts, and feelings cannot be found in the elements of the brain if it is interpreted as a mere machine. Kant underlined that an organizing force is necessary to animate a physical system. Until this century some physicists, biologists, and philosophers believed in a immaterial organizing life factor which was called "elan vital" (Bergson) or "entelechy" (Driesch). From the point of view of complex systems, Köhler's gestalt psychology was an interesting approach referring to the existence of physical systems in which complex psychic structures originate spontaneously from the system's own intrinsic dynamics. Popularly speaking, the macroscopic "gestalt" (form) of a perceived object is more than the sum of its atomic parts, and cannot be reduced to the microscopic scale.

Köhler had the idea that the emergence of visual phenomena can be explained in the framework of thermodynamic models. But in those days, he referred to Boltzmann's linear thermodynamics in equilibrium. He assumed: "The somatic processes underlying static visual fields are stationary equilibrium distributions developed from the inner dynamics of the optical system itself" [4.42]. Köhler even realized that an organism is not a closed system, and tried to explain the emergence of ordered states as a kind of intuitively understood synergy. In this respect, Köhler was already correct with his clear distinction between the microscopic level of elementary interactions and the macroscopic level of emerging ordered states in a synergetic system. But he still lacked an adequate framework of complex dynamical systems to provide the formalism for a thermodynamics far from thermal equilibrium.

The complex system approach offers the possibility for modeling the neural interactions of brain processes on the microscopic scale and the emergence of cognitive structures on the macroscopic scale. Thus, it seems to be possible to bridge the gap between the neurobiology of the brain and the cognitive sciences of the mind, which traditionally has been considered as an unsolvable problem.

Complex models consist of state spaces and nonlinear evolution equations describing a system's dynamics. With about 10^{11} nonsensory neurons, the human brain is represented by a state space of 10^{11} dimensions. Even a typical subsystem contains about 10^8 elements. In a state space with 10^8 dimensions and only 10 levels of neural activitation, there are at least 10^{10^8} distinct positions representing activitation vectors. If we assume 10^3 synaptic connections between each neuron and the other 10^8 neurons of a subsystem, then about 10^{11} synapses must be distinguished. Consequently, for only 10 distinct weights at each synapse, we get the huge number of $10^{10^{11}}$ weights in a subsystem alone. This complexity provides numerous possibilities for coding, representing, and processing information, which can be modeled mathematically by vector and tensor transformations [4.43].

In Kohonen's competitive learning network, the system self-organizes so that nearby vectors map onto nearby points of the net. It is assumed that similar impressions are represented by similar vectors with tiny distances to some prototype vectors. In the framework of complex systems, prototype vectors are interpreted as attractors. Thus, two distinct categories or classes are represented by two different attractors in the state space (Fig. 4.7b). The learning process of cognitive distinction is modeled by a training process of the network which involves adjusting the weights so that an input vector (for instance, a visual or acoustic pattern) is submitted to the prototype vector with the most similarity.

The concept of prototype in neural state spaces allows some interesting interpretations of cognitive processes. How can a network recognize a pattern when the input impression is only partially given? The task of vector completion is crucial for animals to survive in the wild. Imagine a coyote in the desert which detects the tail of a rat in the grass (Fig. 4.18a). The input to the retina of the coyote is limited to that detail of the rat. The assumption or "hypothesis" that there is a rat in the grass is enabled by the coyote's visual system completing the input vector with the learnt prototype vector of a rat. In this sense, we may say that the coyote has a "concept" of a rat represented by a corresponding prototype activation pattern in the brain [4.44].

Paul Churchland even suggests interpreting human high-level cognitive abilities by the prototype vector approach. So, explanatory understanding is reduced to the activation of a specific prototype vector in well-trained

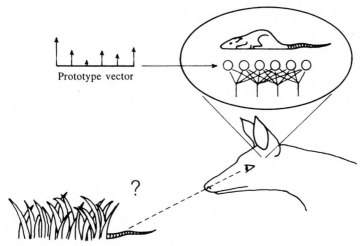

Prototype vector

Fig. 4.18a. Recognition by prototype-activation in neural nets: a coyote recognizes the tail of a rat [4.44]

networks. Prototype vectors embody a huge amount of information which may differ for different people. The reason is that different people may not always have the same items satisfying the constraints of a prototype cluster. Indeed, people mostly have different degrees of explanatory understanding, although they classify an object or situation in nearly the same manner. A joiner, for example, has a higher degree of understanding of what may be a chair than most other people. Nevertheless, they all will agree in most cases. Thus, the prototype-activation model is rather realistic, because it considers the fuzziness of human concepts and understanding.

In epistemology and cognitive psychology, it is usual to distinguish between different kinds of explanation. There are classifying explanations ("Why is the whale a mammal?"), causal explanations ("Why does the stone fall down?"), functional explanations ("Why does a bird have wings?"), and others which correspond to prototype activations of clusters, causal relations, functional properties, and so on. Even mastering social situations is a matter of activating social-interaction prototypes which have been trained and taught during a lifetime.

In the complex system approach, mental states are correlated with neural activation patterns of the brain which are modeled by state vectors in complex state spaces. External mental states referring to perceptions of the outer world may be testable and correlated to neural activities of the brain. How can we test and explain internal states of consciousness which do not refer to events of the outer world, but to mental states themselves?

It is well known that we can even reflect about our self-reflections, and reflect about the reflections about our self-reflections, and so on in an iteration process which is in principle unlimited (Fig. 4.18b). Self-experience and self-reflection lead to the concept of self-consciousness, which traditionally was considered as the essential concept in the philosophy of mind and cognitive psychology. Self-consciousness was defined as the crucial feature of human personality. The definitions of self-consciousness which historically have been discussed are not only philosophically interesting. Obviously, these more or less speculative definitions have powerful consequences in medicine and law. Which criteria must be satisfied for a human being to be conscious and therefore responsible for his actions? Are there medical criteria for consciousness? How can consciousness be disturbed or even destroyed? What about the consciousness of animals? Can we feel like our neighbor or like an animal?

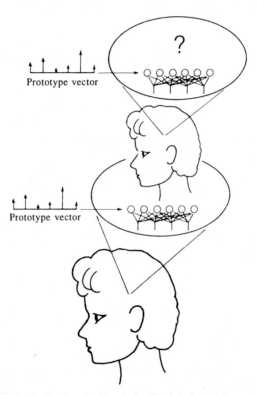

Prototype vector

Prototype vector

Fig. 4.18b. Self-reflection by iterated self-related prototype-activations

The fundamental questions arise (1) if there are particular brain processes causing the emergence of consciousness, and (2) if the emergence of consciousness from brain processes can be modeled by complex systems. The methodological difficulty seems to be that subjective feelings like pains, smells, and so on are only accessible to introspection. These subjective states of feeling and consciousness are sometimes called phenomenal states. Some philosophers have criticized that a physical description of brain states fails to grasp the essence of what is a phenomenal state. Proponents of the opposite view have argued that notions of phenomenal states can be reduced to notions of neurophysiological states of the brain. These arguments are nothing more than modern variations of traditional positions which are known as physicalism and mentalism (or antiphysicalism). Both positions are ideological reductionisms and exaggerations which are neither justified by research nor very helpful in research [4.45].

According to the complex system approach, neurophysiological states and mental states are modeled by mathematical formalisms without reductionist ambitions. Some philosophers fail heavily with their prejudice against mathematics, because they seem to believe that formulas only can designate "physical" states. The reader may recall, for instance, a Hopfield system which contains an "energy" formula by analogy with the energy formula for a physical spin system. Nevertheless, in the framework of a Hopfield system, the so-called "energy" formula must not be identified with energy in solid state physics. The mathematical expression only determines the dynamics of a network, which may be simulated by neurobiological brains or silicon computers or angelic organisms from still unknown star systems.

The mathematical model is empirically corroborated if it fits the observed data. In other cases, it must be modified or dropped. We must be aware that a testable and corroborated theory of mental states and consciousness does not enable us to feel like our neighbor. A physician or a surgeon, for instance, who wishes to heal a patient's pain in the stomach does not need to feel the patient's stomach pain. He must have a good knowledge of the stomach based on anatomy, physiology, biochemistry, psychology, etc. In the terminology of the complex system approach, he must know the possible states of a stomach and their dynamics. In this sense, a model of mental states and consciousness should be developed and tested without any reductionist claims.

Obviously, there are many testable correlations between phenomenal states of consciousness and the neurobiological functioning of the brain. Everybody knows that a short period of oxygen deprivation causes unconsciousness. Electrical stimulations, psychotropic drugs, anesthesia, and lesions may influence the degree of consciousness too, which is not only experienced by self-experiment (autocerebroscopy) but also clinically testable by observations and measurements of functional deficits. The reason is that the brain is an open system whose states depends on the physical, chemical, and biological metabolism with its environment far from thermal equilibrium.

Conscious and unconscious states seem to depend on a rather complex neurophysiological system which contains feedback loops and interconnections at various levels. Figure 4.19 shows the network of the cerebral cortex with its subsystems of primary sensory cortex and association cortex. There

Fig. 4.19. Inputs to the cerebral cortex with somasensory pathways (SOM), auditory pathways (AUD), visual pathways (VIS), lateral geniculate (LG), medial geniculate (MG), nucleus ventralis posterolateralis (VPL) [4.46]

are specific inputs ("afferents") from sense organs reaching the primary cortical projection areas through specific transmitting subsystems and pathways. Non-specific inputs reach the cortex from a subsystem called the "mesencephalic reticular formation". The reticular formation designates a complex network of neurons and nerve fibers with widely distributed connections of synaptic contacts. It is known to play an essential role in arousal, wakefulness, and attention [4.46].

Lesions within the complex network lead to various disturbances of consciousness, which may be global or only local, with specific deficits of conscious experience during global wakefulness. Neurophysiology can experimentally demonstrate that degrees of consciousness depend on the two streams of specific and non-specific afferent signals processed in the cerebral cortex. But the question arises of how mental states of consciousness emerge from these networks. In the terminology of Leibniz, we see the interacting elements like the cog wheels of a mill, but cannot bridge the gap between the neurophysiological machinery and the mental states of consciousness. Traditional neurophysiology has the conventional view that brain functions are made possible by electrical impulses spreading through a network of neurons connected by rigid synapses like the rigid connections of cog wheels in Leibniz' mechanistic model of a mill.

The complex system approach offers a view of self-organizing nets changing their synaptic connections, which are induced by synaptic activation and depend on the degree of activation. In the framework of neural complex systems, the microscopic level of interacting neurons is distinguished from the macroscopic level of global patterns produced as cell assemblies by self-organization. In earlier sections, it was already mentioned that the concept of self-organizing neural cell assemblies was introduced by Hebb. It was modified by Christoph von der Malsburg, Teuvo Kohonen, and others. If simultaneous activity is induced in some neurons of a net by a patterned input, then an assembly will be formed by synchronous activation according to a Hebb-like learning rule.

The modification suggested by von der Malsburg is that assembly formation is not a slow process, but produced by rapid synaptic changes [4.47]. These so-called "Malsburg synapses" are used to model networks with rapid weight adjustment dynamics. Today, there is empirical evidence of Hebb- and Malsburg-type synapses with high plasticity in the brain whose rule of interaction can be realized by molecular mechanisms. The formation of assemblies in a network depends on the degree of activation of its neurons.

Concerning conscious and unconscious states, it is assumed that global activation of a cell population, as exerted by the reticular formation on the cortex (Fig. 4.19), would generally increase the probability of assemblies being formed. Thus, Hans Flohr has suggested that degrees of consciousness differ in the rate at which assemblies can be generated. The production rate of cell assemblies determines the amount, complexity, and duration of representations of sensory patterns from the outer world, for instance. Consciousness is a self-referential state of self-reflection (Fig. 4.18b). Thus, a conscious state is based on a cell assembly representing an internal state (and not only a state of the outer world). For example, I not only have the impression of a green tree, but I am conscious that I am looking at this tree. Furthermore, I can reflect on my state of being conscious of my looking at the green tree, and continue with an iterative production of meta-meta...representations reaching from phenomenal impressions and feelings to abstract and highly sophisticated states of self-reflection. Whenever a critical threshold rate of production is surpassed, phenomenal states must emerge. Deficits of consciousness occur below the critical threshold level.

This hypothesis is testable by particular EEG changes corresponding to an increasing formation rate of assemblies which represent a certain degree of attention. As the production rate of cell assemblies is based on particular synapses with changing weights, the degrees of consciousness may be tested by interventions on synaptic connections. Actually, patients anesthetized with chemical substances which influence synaptic plasticity experience vivid dreams, sensory illusions, visual and auditory hallucinations, and disorganized thoughts. In this sense, awareness can be considered as the result of a system's capacity to generate representations and metarepresentations.

Neural networks with a high rate of assembly formation can produce more complex representations than networks with a lower formation rate. Thus, at a sufficiently high formation rate, complex systems will develop self-referential and metacognitive activities. We may imagine a scale of more or less conscious systems corresponding to the degrees of consciousness in the evolution of living beings with more or less complex nervous systems from the worm to humans. It follows that in the framework of complex systems the emergence of consciousness is no epiphenomenon of evolution. It is a lawful occurrence of global states according to the dynamics of complex systems which produce macroscopic order patterns by microscopic interactions of their elements if certain critical conditions are satisfied.

If the complex systems approach is right, then the nervous systems of biological evolution are only particular realizations of self-referential systems, and other, perhaps technical systems with self-referential capacities based on materials different from the human brain's biochemistry cannot be excluded in principle (compare Chap. 5). We may even be able to translate representations from one complex system into an alien one. As the representations in both systems are not exactly the same, we would not exactly feel like our neighbor, an animal, or another alien system. But we would have a representation in the form of knowledge or a theory about their feelings or thoughts. In this sense, subjectivity is saved, and a wide field of hermeneutics in human communication remains even in the case of technical simulations.

Concerning the traditional mind-body problem, the complex system approach shows that cognitive activity is neither completely independent and different from brain activity nor simply identical, nor an epiphenomenon. Thoughts and feelings are assumed to be both product and producer of neural processes without being identical to them. In the framework of complex systems, the brain is modeled as a self-organizing system which operates far from thermal equilibrium and close to certain threshold values as instability points. During neural instability, different modes of collective excitations evolve to coherent macroscopic patterns which are neurophysiologically based on certain cell assemblies and psychologically expressed as certain feelings or thoughts [4.48].

We all know the experience that in a situation of emotional instability a certain feeling may dominate the other virtual ones and even guide our actions. In synergetics, the competition of stable and unstable modes is explained by the slaving principle. The reader may be reminded of decision situations in which one thought or concept begins to "enslave" the other possible ones. These nonequilibrium phase transitions are governed by very few order parameters in the sense of minimum information. Indeed, acting after a decision means an enormous reduction of complexity. Too much knowledge hinders action or to quote Goethe: "An acting person is always unscrupulous."

Cognitive phenomena are referred to macroscopic properties of the brain's dynamics and to order parameters which govern the underlying microscopic processes. Thus, the so-called mind-brain interaction is only an old-fashioned formulation of an inadequate and obsolete metaphysics that assumes some interacting substances like colliding balls in mechanics. The overlapping area of brain and cognitive sciences is modeled by the emergence of macroscopic properties from microscopic neural interactions during phase transitions in complex neural systems.

In synergetics, phase transitions are interpreted as a kind of symmetry breaking which can be visualized by an overdamped motion of a particle in a symmetric potential (Fig. 4.20a) [4.49].

At the maximum of the potential the position of the particle is symmetric, but unstable, and tiny initial fluctuations decide which of the two equal stable states of minima the particle will reach. In the complex system approach, the two valleys of Fig. 4.20a are interpreted as attractors. Obviously, the ambiguity of perceptions and the spontaneous decision of the visual system

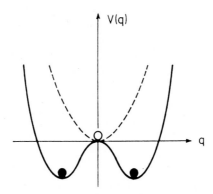

Fig. 4.20a. Symmetry breaking by an overdamped motion of a particle in a symmetric potential

b)   c)

Fig. 4.20b,c. Ambiguity of meaning: (**b**) white or black cross? (**c**) old or young woman? [4.50]

for one interpretation is a well-known psychological example of symmetry breaking. In Fig. 4.20b, there is an instability of figure and ground. Do we see a white or black cross? Figure 4.20c shows an ambiguity of meaning. Is it a picture of a young lady or an old woman? [4.50]

Symmetry breaking in psychology is governed by the nonlinear causality of complex systems (the "butterfly effect"), which roughly means that a small cause can have a big effect. Tiny details of initial individual perspectives, but also cognitive prejudices, may "enslave" the other modes and lead to one dominant view. A neurophysiological model has to simulate the corresponding phase transitions of cell assemblies.

Phase transitions are well known in animal locomotion, for instance, in horse gait. With increasing speed horses fall into different movement patterns, from walking to trotting to galloping, in order to minimize the energy costs. This phenomenon of hysteresis is frequently observed in nonequilibrium phase transitions and interpreted as a sequence of stable states or attractors in the nervous system. Phase transitions appear also in thinking. The "aha-experience" and the sudden "insight" are surprising phenomena arising from a situation of fluctuations and instability. In history, there are many famous examples of scientists, engineers, artists, and composers who suddenly found a new problem solution, an invention, an idea for a painting, a melody, and so on in a situation of "creative" instability and confusion.

The complex system approach delivers no closed doctrine of psychology, but an interdisciplinary research program to explore old and new problems of cognitive science and to bring them nearer to an empirical and experimental analysis. Thus, an exploration of correlations between the rates of changing cell assemblies and intellectual abilities of learning, creativity, cognitive flexibility, and ability to visualize is suggested. Phenomena of cognitive instability are assumed to be macroscopic properties of the microscopic instability of nervous processes. Thoughts and expectations are interpreted as order parameters governing the activity of the whole system if it is operating close to instability points. A confirmation of this theory can be seen in psychological tests which produce hallucinations by suggestions that correspond to measurable physiological effects. By recording the regional cerebral blood flow it has been shown that even the thought or intention of acting increases the neuronal activity of the motor area.

Who will deny that thoughts can change the world and that they are not only mere interpretations of the world? In the field of psychosomatic phenomena the placebo effect, for instance, demonstrates that a mere belief or leading idea can alter not only the emotional state but also the physiological state. Obviously, psychosomatic states are close to instability points. The corresponding order parameters are not just theoretical concepts of psychologists, but real modes governing and dominating ("enslaving") the activity of the central nervous system.

The last examples show that the application of self-organizing complex systems in psychology cannot simply be evaluated by their forecasts and quantitative measurability. It is an intrinsic feature of a complex system that its nonlinear dynamics on the microscopic scale and its sensitive dependence on initial conditions do not allow one to forecast the system's final state. In the brain and cognitive research we are confronted with a huge degree of complexity excluding exact calculations or long term forecasts. Nevertheless, the complex system approach reveals essential qualitative features of the mind-brain system, like its high sensitivity with respect to tiny intrinsic fluctuations and changes in the outer world.

4.4 Intentionality and the Crocodile in the Brain

Besides consciousness, there is another fundamental feature of the human mind which was traditionally emphasized – intentionality.

Intentionality is the reference of mental states to objects or states of affairs in the outer world: I see *something*, I believe in *something*, I expect *something*, I am afraid of *something*, I want *something*, etc. Intentional mental states can be distinguished from non-intentional states without any reference object: I am nervous, I am afraid, I am tired, I am happy, I am depressed, etc.

The phenomenon of intentionality can also be visualized by simple examples. In Fig. 4.21, every observer sees a square, although there is physically not given any part of a square's shape. The configuration of the lines suggests to the visual system that there is a particular closed object. An intentional reference between the observer and a configuration of stimuli is achieved [4.51].

Fig. 4.21. Intentional object of perception (square) [4.51]

Intentional objects or states may be fictional or real. Obviously, human culture is full of signs and symbols for intentional objects and states from traffic signals to religious symbols. Even buildings from memorials and churches to factories may represent intentional objects. The intentional meaning of languages has been constituted in the long development of human cultures. In traditional epistemology, some philosophers like Franz Brentano even proclaimed that intentionality is a particular ability of the human mind to refer to the world. Intentionality was understood as a feature of the mind which cannot be reduced to physical, chemical, or biological properties.

Some modern philosophers like John Searle maintain that intentionality is a distinctive feature of the human mind. But they agree that the biological evolution of the human brain somehow developed the intentional power of mental reference to the world [4.52].

Actually, intentionality is not reserved to brains. It is a feature of certain complex systems which can be modeled by the dynamics of attractors in the evolution of life. Nest construction by social insects is an example of a collective intentional dynamics. The specific feature of this complex system is the autocatalytic mechanism by which the goal-directed work of building nest ecosystems each consisting of a termite population with its environment is carried out. In the complex system approach, it is assumed that this social system already illustrates paradigmatic properties which can be observed in more highly developed systems like brains or central nervous systems [4.53].

The construction process of a nest involves the coordination of more than 5 million insects on the microscopic level, and results in an evolution of certain macroscopic building modes. African termites, for instance, build nests that stand more than 15 feet in height and weigh more than 10 tons. Each insect works independently of each other termite. But their actions are locally determined by distributions of some chemical substance being excreted by the termites themselves. The building material is marked by chemical substance. At first, the building material is distributed randomly, then in an increasingly regular way, until the architectural structure arises from the local interactions of the insects governed by the chemical distribution.

The pattern determines several centers as goals of collective activities, which can be interpreted mathematically as attractors of a diffusion field. In earlier chapters, an attractor was introduced as a solution shared by multiple trajectories originating from different initial conditions. The local trajectories either converge to or diverge from the attractor. In physical or chemical field models, the attractors define local regions in which the potential energy gradient degenerates, going to zero. The region surrounding the attractor is called the basin of attraction and is defined by the gradient flows converging to or diverging from the attractor. The flow pattern of the insects is globally organized by the layout of attractors in their work space, which is the phase portrait of the insects' dynamics. It is well known that attractors are not achieved for ever. If certain control parameters are changed, a pattern may become unstable and break down, being followed by a new pattern of attractors.

Figure 4.22a shows the chemical diffusion gradient surrounding two attractors which will be base of two pillars. As the two pillars act as competing attractors for the termites, a saddle-point is determined between them. In a later building step, the initial two-dimensional field of Fig. 4.22a is followed by a three- dimensional one (Fig. 4.22b) governing the direction of the pillars' construction. In Fig. 4.22c, the arch formation is shown with one attractor of a chemical diffusion gradient [4.54].

Obviously, an intention at the ecological scale does not require that an individual component of a system must be aware of the global consequences of its actions. The intention is only globally manifested in the long range by the system's dynamics. Figure 4.22d shows the autocatalytic cycle of a nest-building intentional complex system. As it is not a supervised learning process, there is no "goal" or "plan" of some supervising authority like "God" or "Nature". That would be only a simplified anthropomorphic metaphor which does not correctly describe the nonlinear causality of the self-organizing complex system under consideration. Nevertheless, globally there is intentional collective behavior arising from complex nonlinear interactions.

As brains and central nervous systems are complex systems with a nonlinear dynamics governing their neurons and synapses, it is no wonder that they achieve intentional behavior patterns, too. Intentionality has not fallen from heaven as a miraculous feature to guide and distinguish human mind from nature. It is a global pattern emerging in particular complex systems under certain conditions. But there are different levels of intentionality depending on the increasing complexity of evolution.

Intentions must not necessarily be conscious. In Fig. 4.21, the intentional object of our visual system is a square without our exerting conscious

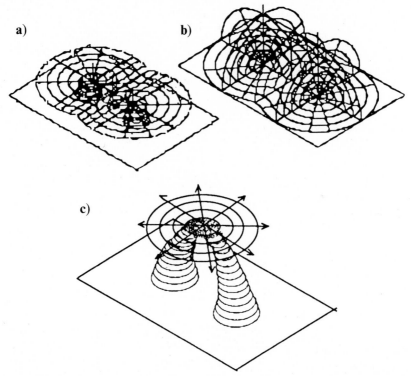

Fig. 4.22a–c. Arch formation as intentional dynamics of termites: (**a**) base of the two pillars as two attractors in a 2-dimensional diffusion gradient field, (**b**) 3-dimensional field governing the direction of the pillars' construction, (**c**) arc formation governed by an attractor of a diffusion gradient [4.54]

will. The so-called perceptual illusions are also intentional patterns of our visual system emerging spontaneously without our conscious will. Figure 4.23 manifests a warping effect of manifolds which seems to be caused by the repellor gradients of different visual attractors. Two equidistant parallel lines seem to change their curvature by a pair of repellor gradients on the left and by a single repellor gradient on the right. The state space of the observer's visual system indicates different curvatures as a result of different visual gradient fields, although the lines remain equidistant and parallel in the physical figure.

Even conscious intentional patterns of behavior are not exclusive to humans. A dog not only jumps, it jumps to catch prey, to greet its master, and so on. Intentionality in the sense of conscious goal-directedness is a property of more or less all animals. The question arises of how intentional behavior can be modeled by the complex system approach and how the model can be tested experimentally.

Fig. 4.22d. The autocatalytic cycle of a nest-building intentional complex system

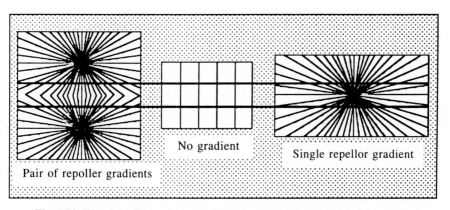

Fig. 4.23. Warping effects of two equidistant parallel lines by visual attractors

In this context, an intention is defined as an intended behavior pattern which may change the dynamical properties, such as stability, of intrinsic behavior pattern. Thus, psychologists can model the intrinsic dynamics of behavior patterns which may be changed by the dynamics of other, intended behavior patterns. Here we remind the reader of the intrinsic dynamics governing some patterns of behavior which can be modeled by nonequilibrium phase transitions and order parameters. Kelso, Haken, and others have analyzed the following simple examples: when persons are asked to move their

Fig. 4.24a,b. Two fingers moving in parallel (**a**) and antiparallel fashion (**b**) [4.55]

fingers in parallel (Fig. 4.24a), they can easily perform this at low frequency. When the test persons are asked to increase the frequency of their finger movements, the fingers are suddenly moved in a symmetric and antiparallel fashion without conscious intention (Fig. 4.24b) [4.55].

In order to model this phase transition of behavior patterns, the frequency is interpreted as a control parameter, and the macroscopic variable describing the finger movement is the phase φ. The behavior can be modeled in an energy landscape relative to the changing phase. The landscape must be symmetric, as the left and right finger have equal functions. It must also be periodic in the phase angle (Fig. 4.25). If the frequency increases, the landscape with its initially sharp valleys is deformed. In the beginning of slow movement the pattern is stable, corresponding to a stable phase at value π (Fig. 4.25a). Finally, the valley at π has disappeared, and a ball, initially in the valley at π, has run down to the deepest minimum, corresponding the symmetric movement of the fingers (Fig. 4.25c).

In some experiments, subjects were asked to switch intentionally between the two patterns of bimanual coordination. The duration of the transient corresponds to the switching time, which was measured. The stability of both patterns is measured by order parameter fluctuations. The relative phase dynamics is modeled by a nonlinear evolution equation.

Figure 4.26a visualizes the intrinsic dynamics according to the potential of this equation with two minima. The contribution of intentional information to the relative phase dynamics is shown by the potentials of Fig. 4.26b. The result of summing the intrinsic and the intentional dynamics to arrive at the full dynamics is shown by Fig. 4.26c. The ball in the landscape travels faster along the steeper slope at $\varphi=0$ than $\varphi=180$, corresponding to the empirically measured switching time. Obviously, an intention can change the intrinsic dynamics by destabilizing one pattern and stabilizing the other one. The intentional information is said to be a part of the pattern dynamics attracting the system toward the intended pattern. In this sense, intentional

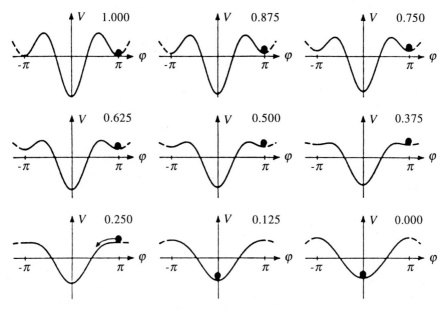

Fig. 4.25. Dynamics of moving fingers in an energy landscape with relative phase φ as order parameter

information defines an attractor in the same state space in which the intrinsic dynamics is modeled [4.56].

Intentionality and linguistic meaning are often proclaimed to be essential features of the human mind. Examples of intentional states are pains, tickles and itches, beliefs, fears, hopes, desires, perceptual experiences, experiences of acting, thoughts, feelings, etc., which are expressed by corresponding sentences like "I suffer from pain in the stomach", "I desire to get a car", "I believe in God", etc. Searle argues that mental states are as real as any other biological phenomena like lactation, photosynthesis, or digestion. He knows that mental states are macrostates of the biological brain which are caused by neurophysiological interactions between neurons on the microscopic scale. Thus, they cannot be identified with the neurophysiological states of single neurons.

The distinction between the micro- and macrostates of brains is illustrated by an analogy with, for instance, micro- and macrostates in liquids: the macrostate of liquidity cannot be reduced to single molecules or, in other words, single molecules cannot be liquid. In this sense, beliefs, desires, thirsts, and visual experiences are real causal features of the brain as much as the solidity of a table or the liquidity of water. Intentional states can themselves

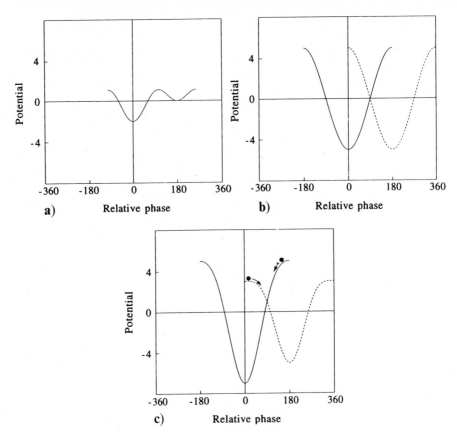

Fig. 4.26a–c. Relative phase dynamics with intentional information [4.56]

be caused by and realized in the structure of the brain. Searle declares that there is not, in addition, a metaphysical obstacle.

Nevertheless, he argues that no purely formal model will ever be sufficient by itself for intentionality because the formal properties are not by themselves constitutive of intentionality. His reason for holding this position is based on the thought experiment of the "Chinese room". A person who only understands English is locked in a room with a great store of Chinese symbols, and a set of complicated transformation rules, written in English, for performing operations on sequences of Chinese symbols. The person periodically receives sequences of Chinese symbols through a slot (Fig. 4.27). He applies the transformation rules in order to produce a further sequence of Chinese symbols which are put through the slot, again [4.57].

It is unknown to the person in the room that the store of sequences contains a large amount of information about certain topics, written in Chinese. The input sequences which are put through the slot are questions or com-

Fig. 4.27. Alice in the Chinese room

ments on those topics. The output sequences are reactions and comments to the received inputs. The transformation rules used are a formal program to simulate the conversational behavior of a native Chinese speaker. The person in the Chinese room applies the formal transformation rules correctly without understanding the sequences of Chinese symbols, which are meaningless to him.

Searle maintains that formal symbol manipulations by themselves do not have any intentionality, because they are quite meaningless to the user. Intentionality in this context is the feature of formal symbols like words, sentences, etc., that they refer to certain "meant" entities (semantic relation of the symbol) and to the user (pragmatic relation of the symbol). Searle asserts that this feature is intrinsic only for the mental states of the brain.

His arguments against "computer simulation" fail, because he restricts simulations to formal algorithms running on program-controlled Turing-type computers. But we have shown that a brain has the typical characteristics of a self-organizing and self-referential complex system which is quite different from a program-controlled computer (compare Chap. 6). Self-organization and self-referentiality of complex systems are not restricted to human or mammalian brains. They are only biochemical and neurophysiological realizations of particular complex structures which have been produced in biological evolution. Thus, in principle it cannot be excluded that these complex structures with their characteristic dynamics may be realized by quite different materials which may be produced via human technology. Consequently, as intentionality is made possible by the features of self-referentiality and self-organization, at least a partial simulation by complex models different from biological brains cannot be excluded in principle.

In traditional philosophies, intentionality is often founded on the so-called "self" of the human being, which is said to be able to refer to the world and to itself ("self-consciousness" as self-referentiality). But where is the self hidden in the brain? Traditional positions like kinds of platonism or spiritualism or materialism are even maintained by some modern researchers of the brain. For Sir John Eccles, for instance, the self seems to be a spiritual entity interacting with the brain, but completely different in nature [4.58]. But how should this hypothesis be defended or refuted? It is a mere postulate with high metaphysical costs, which one may believe or not believe.

Hypotheses must be criticizable, perhaps false, but fruitful for further research. Thus, the metaphysical price is too high. Ockham's razor from philosophy demands that we cut away superfluous hypotheses, remain economical with the postulation of metaphysical entities, and restrict hypotheses to the minimal number that seems indispensable for empirical research. The complex system approach is a mathematical research program of interdisciplinary models avoiding metaphysical dogmas. It may fail in the long run. But this strategy of modeling has been confirmed by an impressive number of successes in several sciences and technologies and, more important, it suggests some fruitful concepts for further empirical research. On the other hand, the traditional materialism which identifies mental states with neurophysiological processes in single neurons is simply false.

Nevertheless, in brain research, the question has arisen which part of the brain is the center of the "self". The cortex is the part of the brain which enables us to learn, to memorize, to think, and to create all the products of human culture and civilization. But if the cortex is mainly modeled as a complex associative memory store with certain learning procedures, then it is only a complex and highly sophisticated instrument which has been evolved in biological evolution for enhanced survival of the fittest.

Indeed, the cortex is the youngest part in the evolution of the human brain. There are some much older, but more primitive structures which also can be found in the brains of birds, reptiles, amphibians, and fishes. Some scientists assume that basic feelings like lust and pain and all the servo-mechanisms which were necessary to survive in a reptile's life are essentially realized in these early structures of the brain. This center would give the impulses for all kinds of activities, using the cortex only as huge and effective associative store. Thus, in this interpretation, the "self" is replaced by a little crocodile in the brain operating with some highly complex instruments like the cortex, in order to survive in a more and more complex environment [4.59]. Intentionality would be made possible by the cortex, but initiated by the basic instincts of the crocodile in the human brain.

The idea of crocodiles with highly effective neural instruments of survival seems to injure our vanity more than the popular Darwinistic motto of the last century that the ape is the ancestor of man. From a scientific point of view, of course, it should not be injured vanity which makes us criticize the concept of the "neural crocodile". The main objection is that our feelings have not rested at the level of a crocodile, but have developed during biological and cultural evolution, too.

Our feelings of lust and pain are rather complex, because they are influenced by the stimuli of a rather complex and sophisticated civilization which has been produced by human brains. Thus, there is a complex feedback which has shaped our feelings and desires from the crocodile until today. The history of literature, art, and psychology demonstrates that lust and pain have been highly sophisticated states of the human brain that are in permanent evolution. Thus, even the traditional concept of a human soul which is more or less sensitive still makes sense in the framework of complex system. But we must give up the traditional ideas of human mind and soul as strange substances controlling and interacting with the human body in a miraculous manner which cannot be conceived in principle.

5. Complex Systems and the Evolution of Artificial Intelligence

Can machines think? This famous question from Turing has new topicality in the framework of complex systems. The chapter starts with a short history of computer science since Leibniz and his program for mechanizing thinking (*mathesis universalis*) (Sect. 5.1). The modern theory of computability enables us to distinguish complexity classes of problems, meaning the order of corresponding functions describing the computational time of their algorithms or computational programs. Modern computer science is interested not only in the complexity of universal problem solving but also in the complexity of knowledge-based programs. Famous examples are expert systems simulating the problem solving behavior of human experts in their specialized fields. Further on, we ask if a higher efficiency of problem solving may be expected from quantum computers and quantum complexity theory (Sect. 5.2).

But the algorithmic mechanization of thinking with program-controlled computers has some severe obstacles which cannot be overcome by growing computational capacities. For example, pattern recognition, the coordination of movements, and other complex tasks of human learning cannot be mastered by Turing-like computer programs. Artificial neural networks realize the principles of complex dynamical systems. They are inspired by the successful technical applications of nonlinear dynamics to solid-state physics, spin-glass physics, chemical parallel computers, optical parallel computers, laser systems, and the human brain (Sect. 5.3). There are already some applications of artificial neural networks in neurobionics, medicine, and robotics. Another more speculative aspect of the new technology is cyberspace. Virtual reality has become a keyword in the world wide society of telecommunicators (Sect. 5.4). Obviously, Turing's question "Can machines think?" cannot be answered finally by the complex system approach, although we can master some interesting cognitive tasks using artificial neural networks.

5.1 Leibniz and the Mathesis Universalis

One of the most speculative applications of complex systems is the evolution of artificial intelligence (AI) [5.1]. In the tradition of classical AI, the brain has been understood as computer hardware of the most advanced machinery, while the mind is the corresponding software program with deterministic algorithms. Even knowledge-based expert systems are conceived by algorithmical representations of highly developed AI programming languages. But theoretical results of mathematical logic (Church, Turing, Gödel, etc.) and practical problems of programming limit the mechanization of thought in the framework of classical AI.

A theory of the "cerebral computer" as a product of natural evolution has been suggested to model the nature of the brain and its mental states by the non-linear dynamics ("self-organization") of complex neural networks. The question arises of wether the insight into their dynamics delivers the "blueprints" of a new revolutionary technology which will pursue the natural evolution of brain and mind. Actually, the development of human knowledge and knowledge technology seems to be a kind of technical evolution which has led to technical innovations like mutations in biological evolution.

The first level was realized by simple tools like the hammer, the lever, and so on. On the next level, machines using force and energy were invented. Today program-controlled computers and information-processing automata have become tools of everyday life. Computer scientists distinguish several generations of hardware and software in the historical development of their machines. In artifical-intelligence research one speaks of the "second computer age", meaning the transition from number-processing machines to knowledge-processing systems such as expert systems, which are said to simulate human experts, at least partially [5.2].

The early historical roots of computer science stem back to the age of classical mechanics. The mechanization of thoughts begins with the invention of mechanical devices for performing elementary arithmetic operations automatically. A mechanical calculation machine executes serial instructions step by step. Thus, its dynamics is determined by mechanical mono-causality, differing essentially from the parallelism and self-organization of complex systems. In general, the traditional design of a mechanical calculation machine contains the following devices.

First, there is an input mechanism by which a number is entered into the machine. A selector mechanism selects and provides the mechanical motion to cause the addition or subtraction of values on the register mechanism. The register mechanism is necessary to indicate the value of a number stored within the machine, technically realized by a series of wheels or disks. If a carry is generated because one of the digits in the result register advances from 9 to 0, then that carry must be propagated by a carry mechanism to the next digit or even across the entire result register. A control mechanism ensures that all gears are properly positioned at the end of each addition cycle to avoid false results or jamming the machine. An erasing mechanism has to reset the register mechanism to store a value of zero.

Wilhelm Schickard (1592–1635), professor of Hebrew, oriental languages, mathematics, astronomy, and geography, is presumed to be the first inventor of a mechanical calculating machine for the first four rules of arithmetic. The adding and subtracting part of his machine is realized by a gear drive with an automatic carry mechanism. The multiplication and division mechanism is based on Napier's multiplication tables. Blaise Pascal (1623–1662), the brilliant French mathematician and philosopher, invented an adding and subtracting machine with a sophisticated carry mechanism which in principle is still realized in our hodometers of today [5.3].

But it was Leibniz' mechanical calculating machine for the first four rules of arithmetic which contained each of the mechanical devices from the input, selector, and register mechanism to the carry, control, and erasing mechanism. The Leibniz machine became the prototype of a hand calculating machine. If we abstract from the technical details and particular mechanical constructions of Leibniz' machine, then we get a model of an ideal calculating machine which in principle is able to calculate all computable functions of natural numbers.

Figure 5.1 is a scheme of this ideal machine with a crank C and three number stores SM, TM, RM [5.4]. Natural numbers can be entered in the set-up (input) mechanism SM by the set-up handles SH. If crank C is turned to the right, then the contents of SM are added to the contents of the result mechanism RM, and the contents of the turning mechanism TM are raised by 1. A turn to the left with crank C subtracts the contents of SM from the contents of RM and diminishes the contents of TM by 1.

Addition means the following. At the beginning of the calculation, the erasing procedure is implemented by setting TM and RM to zero. Then the first number is set up in SM by SH. A turn to the right of crank C transports

Fig. 5.1. Hand calculating machine

this number into RM. In other words, the number is added to the zero 0 in RM. Now the second number is set up in the SM and added to the contents of RM by a turn to the right. The sum of both numbers can be read in the RM. After turning the crank twice to the right, the TM shows 2. Multiplication only means a repeated addition of the same number. The product b · a results from adding the number a to itself b times.

Leibniz even designed a mechanical calculating machine for the binary number system with only two digits 0 and 1, which he discovered some years earlier. He described a mechanism for translating a decimal number into the corresponding binary number and vice versa. As modern electronic computers only have two states 1 (electronic impulse) and 0 (no electronic impulse), Leibniz truly became one of the pioneers of computer science [5.5].

Leibniz' historical machines suffered from many technical problems, because the materials and technical skills then available were not up to the demands. Nevertheless, his design is part of a general research program for a *mathesis universalis* intended to simulate human thinking by calculation procedures ("algorithms") and to implement them on mechanical calculating machines. Leibniz proclaimed two basic disciplines of his *mathesis universalis*.

An *ars iudicandi* should allow every scientific problem to be decided by an appropriate arithmetic algorithm after its codification into numeric symbols. An *ars inveniendi* should allow scientists to seek and enumerate possible solutions of scientific problems. Leibniz' *mathesis universalis* seems already to foreshadow the famous Hilbert program in our century with its demands for formalization and axiomatization of mathematical knowledge. Actually, Leibniz developed some procedures to formalize and codify languages. He was deeply convinced that there are universal algorithms to decide all problems in the world by mechanical devices [5.6].

Consequently, he proposed that natural systems like cells, plants, animals, and even humans are more or less complex automata. In his *Discourse on Metaphysics* (1686), Leibniz underlines that the mechanistic description and causal explanation of living systems is not in contradiction to a teleological consideration that has great heuristic value in science (§22). In his *Monadology* (§18) he introduced an individual substance (*monade*) as an elementary automaton (*automates incorporels*) which is characterized by a (continuous) series of states ("perceptions"). The elementary automata constitute aggregations of more or less complexity which are characterized by different correlations and which can be interpreted as composite automata. In his *Theodicée* (§200), Leibniz discusses the hierarchical structure and subordination in living systems:

> ... the connection and order of things brings it about that the body of every animal and of every plant is comprised of other animals and of other plants, or of other living

organic beings: consequently there is subordination, and one body, one substance, serves the other.

The unity of a living system is guaranteed by its form of organization, which Leibniz, adapting an idea of Aristotle, called "entelechy". But Leibniz only used an old metaphysical term in order to introduce his own new concept. For Leibniz a system can only be more or less unified in the sense of higher or lower degrees of subordination and hierarchy. An aggregation with the same correlation between all·its substances has no hierarchical order and is less structured than a primitive cellular organism, while in plants, animals, and humans we can observe a growing degree of subordination.

For Leibniz the teleological terminology has a heuristical value, although in principle nature can be explained by mechanistic causes. But it is a fundamental error and misunderstanding when disciples of vitalism refer to Leibniz. The main difference is that for Leibniz no new principle or *force vitale* is necessary to explain living systems. At a certain degree of complexity, it is only heuristically suitable to describe natural systems in the terminology of teleology. But, unlike natural systems, artificial mechanical automata are constructed by humans in finite steps. Only an infinite analysis could demonstrate the complexity of a natural automaton, which is correlated with each individual automaton ("substance") in the world. Obviously, Leibniz designed a theory of complex systems, but still in the framework of classical mechanics and the belief in decidable universal algorithms.

In the 19th century it was the English mathematician and economist Charles Babbage who not only constructed the first program-controlled calculation machine (the "analytical engine") but also studied its economic and social consequences [5.7]. A forerunner of his famous book *On the economy of machinery and manufactures* (1841) was Adam Smith's idea of economic laws, which paralleled Newton's mechanical laws (compare Sect. 6.2). In his book *The Wealth of Nations*, Smith described the industrial production of pins as an algorithmic procedure and anticipated Henry Ford's idea of program-controlled mass production in industry.

5.2 From Turing Machines to Knowledge Based Systems

The modern formal logic of Frege and Russell and the mathematical proof theory of Hilbert and Gödel have been mainly influenced by Leibniz' program of *mathesis universalis*. The hand calculating machine (Fig. 5.1) which was abstracted from the Leibniz machine in Sect. 5.1 can easily be generalized to Marvin Minsky's so-called register machine [5.8]. It allows the general concept of computability to be defined in modern computer science.

A hand calculating machine had only two registers TM and RM, and only rather small natural numbers can be input. An ideal register machine has a finite number of registers which can store any finite number of a desired quantity. The registers are denoted by natural numbers $i = 1, 2, 3, \ldots$. The contents of register i are denoted by $\langle i \rangle$. As an example, the device $\langle 4 \rangle := 1$ means that the content of the register with number 4 is 1. The register is empty if it has the content 0.

In the hand calculating machine an addition or subtraction was realized only for the two registers $\langle SM \rangle$ and $\langle RM \rangle$, with $\langle SM \rangle + \langle RM \rangle$ or $\langle RM \rangle - \langle SM \rangle$ going into the register RM. In a register machine the result of subtraction $\langle i \rangle - \langle j \rangle$ should be 0 if $\langle j \rangle$ is greater than $\langle i \rangle$. This modified subtraction is denoted by $\langle i \rangle \dot- \langle j \rangle$. In general, the program of an ideal register machine is defined using the following elementary procedures as building blocks:

1) Add 1 to $\langle i \rangle$ and put the result into register i, in short: $\langle i \rangle := \langle i \rangle + 1$
2) Subtract 1 from $\langle i \rangle$ and put the result into register i, in short: $\langle i \rangle := \langle i \rangle \dot- 1$

These two elementary procedures can be composed using the following concepts:

3) If P and Q are well-defined programs, then the chain $P \rightarrow Q$ is a well-defined program. $P \rightarrow Q$ means that a machine has to execute program Q after program P.
4) The iteration of a program, which is necessary for multiplication, for instance, as iterated addition is controlled by the question of whether a certain register is empty.

A diagram illustrates this feedback:

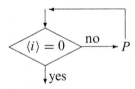

If P is a well-defined program, then execute P until the content of the register with number i is zero.

Each elementary operation (1) and (2) of a program is counted as a step of computation. A simple example is the following addition program:

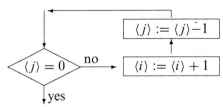

Each state of the machine is illustrated by the following matrix, which incrementally adds the content y of register $\langle j \rangle$ to the content x of register $\langle i \rangle$ and simultaneously decrements the content of $\langle j \rangle$ to zero. The result $x + y$ of the addition is shown in register $\langle j \rangle$:

$$
\begin{array}{cc}
\langle i \rangle & \langle j \rangle \\
x & y \\
x+1 & y \dot{-} 1 \\
\vdots & \vdots \\
x+y & y \dot{-} y
\end{array}
$$

A register machine with program F is defined to compute a function f with n arguments if for arbitrary arguments $x_1, \ldots, x_n$ in the registers $1, \ldots, n$ (and zero in all other ones) the program F is executed and stops after a finite number of steps with the arguments of the function in the registers $1, \ldots, n$ and the function value $f(x_1, \ldots, x_n)$ in register $n + 1$:

The program

$$
\langle 1 \rangle := x_1; \ldots; \ \langle n \rangle := x_n
$$
$$
\downarrow
$$
$$
F
$$
$$
\downarrow
$$
$$
\langle n + 1 \rangle := f(x_1, \ldots, x_n)
$$

works according to a corresponding matrix. A function f is called computable by a register machine RM (RM-computable) if there is a program F computing f.

The number of steps which a certain program F needs to compute a function f is determined by the program and depends on the arguments of the function. The complexity of program F is measured by a function $s_F(x_1, \ldots, x_n)$ counting the steps of computation according to program F. For example, the matrix of the addition program for $x + y$ shows that y elementary steps of adding 1 and y elementary steps of subtracting 1 are necessary. Thus, $s_F(x, y) = 2y$. As an RM-computable function f may be computed by several programs, a function g is called the step counting function of f if there is a program F to compute f with $g(x_1, \ldots, x_n) = s_F(x_1, \ldots, x_n)$ for all arguments $x_1, \ldots, x_n$. The complexity of a function is defined as the complexity of the best program computing the function with the least number of steps.

Obviously, Minsky's register machine is an intuitive generalization of a hand calculating machine à la Leibniz. But, historically, some other, but equivalent formulations of machines were at first introduced independently by Alan Turing and Emil Post in 1936. A Turing machine (Fig. 5.2a) can

Fig. 5.2a. Turing machine with one tape

carry out any effective procedure provided it is correctly programmed [5.9]. It consists of

a) a control box in which a finite program is placed,
b) a potentially infinite tape, divided lengthwise into squares,
c) a device for scanning, or printing on one square of the tape at a time, and for moving along the tape or stopping, all under the command of the control box.

If the symbols used by a Turing machine are restricted to a stroke | and a blank ∗, then an RM-computable function can be proved to be computable by a Turing machine and vice versa. We must remember that every natural number x can be represented by a sequence of x strokes (for instance 3 by |||), each stroke on a square of the Turing tape. The blank ∗ is used to denote that the square is empty (or the corresponding number is zero). In particular, a blank is necessary to separate sequences of strokes representing numbers. Thus, a Turing machine computing a function f with arguments $x_1, \ldots, x_n$ starts with tape $\cdots \ast x_1 \ast x_2 \ast \cdots \ast x_n \ast \cdots$ and stops with $\cdots \ast x_1 \ast x_2 \ast \cdots x_n \ast$ $f(x_1, \ldots x_n) \ast \cdots$ on the tape.

From a logical point of view, a general purpose computer – as constructed by associates of John von Neumann in America and independently by Konrad Zuse in Germany – is a technical realization of a universal Turing machine which can simulate any kind of Turing program. Analogously, we can define a universal register machine which can execute any kind of register program. Actually, the general design of a von-Neumann computer consists of a central processor (program controller), a memory, an arithmetic unit, and input-output devices. It operates step by step in a largely serial fashion. A present-day computer à la von Neumann is really a generalized Turing machine. The efficiency of a Turing machine can be increased by the introduction of several tapes, which are not necessarily one-dimensional, each acted on by one or more heads, but reporting back to a single control box which coordinates all the activities of the machine (Fig. 5.2b) [5.10]. Thus,

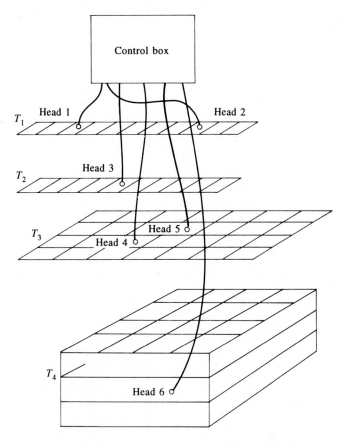

Fig. 5.2b. Turing machine with several tapes [5.10]

every computation of such a more effective machine can be done by an ordinary Turing machine. Concerning the complex system approach, even a
Turing machine with several multidimensional tapes remains a sequential
program-controlled computer, differing essentially from self-organizing systems like neural networks.

Besides Turing- and register machines, there are many other mathematically equivalent procedures for defining computable functions. Recursive
functions are defined by procedures of functional substitution and iteration,
beginning with some elementary functions (for instance, the successor function $n(x) = x + 1$) which are obviously computable. All these definitions
of computability by Turing machines, register machines, recursive functions,
etc., can be proved to be mathematically equivalent. Obviously, each of these
precise concepts defines a procedure which is intuitively effective.

Thus, Alonzo Church postulated his famous thesis that the informal intuitive notion of an effective procedure is identical with one of these equivalent precise concepts, such as that of a Turing machine. Church's thesis cannot be proved, of course, because mathematically precise concepts are compared with an informal intuitive notion. Nevertheless, the mathematical equivalence of several precise concepts of computability which are intuitively effective confirms Church's thesis. Consequently, we can speak about computability, effectiveness, and computable functions without referring to particular effective procedures ("algorithms") like Turing machines, register machines, recursive functions, etc. According to Church's thesis, we may in particular say that every computational procedure (algorithm) can be calculated by a Turing machine. So every recursive function, as a kind of machine program, can be calculated by a general purpose computer [5.11].

Now we are able to define effective procedures of decision and enumerability, which were already demanded by Leibniz' program of a *mathesis universalis*. The characteristic function f_M of a subset M of natural numbers is defined as $f_M(x) = 1$ if x is an element of M, and as $f_M(x) = 0$ otherwise. Thus, a set M is defined as effectively decidable if its characteristic function saying whether or not a number belongs to M is effectively computable (or recursive).

A set M is defined as effectively (recursively) enumerable if there exists an effective (recursive) procedure f for generating its elements, one after another (formally $f(1) = x_1$, $f(2) = x_2$, ... for all elements $x_1, x_2, ...$ from M). It can easily be proved that every recursive (decidable) set is recursively enumerable. But there are recursively enumerable sets which are not decidable. These are the first hints that there are limits to Leibniz' originally optimistic program, based on a belief in universal decision procedures.

Concerning natural and artificial intelligence, the paradigm of effective computability implies that mind is represented by program-controlled machines, and mental structures refer to symbolic data structures, while mental processes implement algorithms. Historically the hard core of AI was established during the Dartmouth Conference in 1956 when leading researchers such as John McCarthy, Alan Newell, Herbert Simon, and others from different disciplines, formed the new scientific community of AI. They all were inspired by Turing's question "Can machines think?" in his famous article "Computing machinery and intelligence" (1950).

In the tradition of Leibniz' *mathesis universalis* one could believe that human thinking could be formalized with a kind of universal calculus. In a modern version one could assume that human thinking could be represented by some powerful formal programming language. In any case, formulas are sequences of symbols which can be codified by natural numbers. Then assertions about objects would correspond to functions over numbers, conclusions would follow from some kind of effective numerical procedure, and so on.

Actually, the machine language of a modern computer consists of sequences of numbers, codifying every state and procedure of the machine. Thus, the operations of a computer can be described by an effective or recursive numerical procedure.

If human thinking can be represented by a recursive function, then by Church's thesis it can be represented by a Turing program which can be computed by a universal Turing machine. Thus, human thinking could be simulated by a general purpose computer and, in this sense, Turing's question must be answered with "yes". The premise that human thinking can be codified and represented by recursive procedures is, of course, doubtful. Even processes of mathematical thinking can be more complex than recursive functions. Recursiveness or Turing computability is only a theoretical limit of computability according to Church's thesis.

In the following we want to consider problems with a degree of complexity both below and beyond this limit. Below this limit there are many practical problems concerning certain limitations on how much the speed of an algorithm can be increased. Especially among mathematical problems there are some classes of problems that are intrinsically more difficult to solve algorithmically than others. Thus, there are degrees of computability for Turing machines which are made precise in complexity theory in computer science [5.12].

Complexity classes of problems (or corresponding functions) can be characterized by complexity degrees, which give the order of functions describing the computational time (or number of elementary computational steps) of algorithms (or computational programs) depending on the length of their inputs. The length of inputs may be measured by the number of decimal digits. According to the machine language of a computer it is convenient to codify decimal numbers into their binary codes with only binary numbers 0 and 1 and to define their length by the number of binary digits. For instance, 3 has the binary code 11 with the length 2. A function f has linear computational time if the computational time of f is not greater than $c \cdot n$ for all inputs with length n and a constant c.

The addition of two (binary) numbers has obviously only linear computational time. For instance, the task 3+7=10 corresponds to the binary calculation

$$
\begin{array}{cccc}
 & 0 & 1 & 1 \\
 & 1 & 1 & 1 \\
\hline
1 & 0 & 1 & 0 \\
\end{array}
$$

which needs 5 elementary computational steps of adding two binary digits (including carrying). We remind the reader that the elementary steps of adding binary digits are 0+0=0, 0+1=1, 1+0=1, 1+1=10, and carry. It is convenient to assume that the two numbers which should be added have equal length. Otherwise we simply start the shorter one with a series of zeros, for instance 111 and 011 instead of 11. In general, if the length of the particular pair of numbers which should be added is n, the length of a number is $\frac{n}{2}$, and thus, we need no more than $\frac{n}{2} + \frac{n}{2} = n$ elementary steps of computation including carrying.

A function f has quadratic computational time if the computational time of f is not greater than $c \cdot n^2$ for all inputs with length n and a constant c.

A simple example of quadratic computational time is the multiplication of two (binary) numbers. For instance, the task $7 \cdot 3 = 21$ corresponds to the binary calculation:

$$
\begin{array}{ccccccc}
1 & 1 & 1 & \cdot & 0 & 1 & 1 \\
\hline
 & & 0 & 0 & 0 & & \\
 & & 1 & 1 & 1 & & \\
 & & & 1 & 1 & 1 & \\
\hline
 & 1 & 0 & 1 & 0 & 1 &
\end{array}
$$

According to former conventions, we have $n = 6$. The number of elementary binary multiplications is $\frac{n}{2} \cdot \frac{n}{2} = \frac{n^2}{4}$. Including carrying, the number of elementary binary additions is $\frac{n}{2} \cdot \frac{n}{2} - \frac{n}{2} = \frac{n^2}{4} - \frac{n}{2}$. In all, we get $\frac{n^2}{4} + \frac{n^2}{4} - \frac{n}{2} = \frac{n^2}{2} - \frac{n}{2}$, which is smaller than $\frac{n^2}{2}$.

A function f has polynomial computational time if the computational time of f is not greater than $c \cdot n^k$, which is assumed to be the leading term of a polynomial $p(n)$. A function f has exponential computational time if the computational time of f is not greater than $c \cdot 2^{p(n)}$. Many practical and theoretical problems belong to the complexity class P of all functions which can be computed by a deterministic Turing machine in polynomial time.

In the history of mathematics, there have been some nice problems of graph theory to illustrate the basic concepts of complexity theory [5.13]. In 1736, the famous mathematician Leonhard Euler (1707–1783) solved one of the first problems of graph theory. In the city of Königsberg, the capital of eastern Prussia, the so-called old and new river Pregel are joined in the river Pregel. In the 18th century, there were seven bridges connecting the southern s, northern n, and eastern e regions with the island i (Fig. 5.3a). Is there a route which crosses each bridge only once and returns to the starting point?

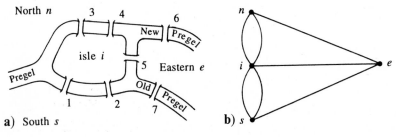

Fig. 5.3a,b. Euler's Königsberg river problem (**a**). Graph of Euler's river problem (**b**)

Euler reduced the problem to graph theory. The regions n, s, i, e are replaced by vertices of a graph, and the bridges between two regions by edges between the corresponding vertices (Fig. 5.3b).

In the language of graph theory, Euler's problem is whether for every vertex there is a route (an "Euler circuit") passing each edge exactly once, returning finally to the starting point. For arbitrary graphs Euler proved that an Euler circuit exists if and only if each vertex has an even number of edges (the "Euler condition"). As the graph of Fig. 5.3b does not satisfy this condition, there cannot be a solution of Euler's problem in this case. In general, there is an algorithm testing an arbitrary graph by Euler's condition if it is an Euler circuit. The input of the algorithm consists of the set V of all vertices $1, \ldots, n$ and the set E of all edges, which is a subset of the set with all pairs of vertices. The computational time of this algorithm depends linearly on the size of the graph, which is defined by the sum of the numbers of vertices and edges.

In 1859, the mathematician William Hamilton (1805–1865) introduced a rather similar problem that is much more complicated than Euler's problem. Hamilton considered an arbitrary graph, which means nothing else than a finite collection of vertices, a certain number of pairs of which are connected together by edges. Hamilton's problem is whether there is a closed circuit (a "Hamilton circuit") passing each vertex (not each edge as in Euler's problem) exactly once. Figure 5.3c shows a graph with a Hamilton circuit passing the vertices in the order of numbering.

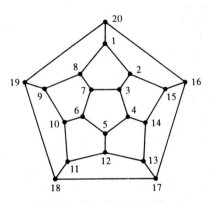

Fig. 5.3c. Hamilton's problem

However, unlike the case of Euler's problem, we do not know any condition which exactly characterizes whether a graph contains a Hamilton circuit or not. We only can define an algorithm testing whether an arbitrary graph contains a Hamilton circuit or not. The algorithm tests all permutations of vertices to see if they form a Hamiltonian circuit. As there are $n!$ different permutations of n vertices, the algorithm does not need more than $c \cdot n!$

steps with a constant c to find a solution. It can easily be proved that an order of $n!$ corresponds to an order of n^n. Consequently, an algorithm for the Hamilton problem needs exponential computational time, while the Euler problem can be solved algorithmically in linear computational time. Thus, Hamilton's problem cannot practically be solved by a computer even for small numbers n.

The main reason for a high computational time may be a large number of single subcases which must be tested by a deterministic computer step by step. It is more convenient to use a non-deterministic computer which is allowed to choose a computational procedure at random among a finite number of possible ones instead of performing them step by step in a serial way. Let us consider Hamilton's problem again. An input graph may have n vertices $v_1, \ldots, v_n$. A non-deterministic algorithm chooses a certain order $v_{i_1}, \ldots, v_{i_n}$ of vertices in a non-deterministic, random way. Then the algorithm tests whether this order forms a Hamilton circuit. The question is whether for all numbers j ($j = 1, \ldots, n - 1$) the successive vertices v_{i_j} and $v_{i_{j+1}}$ and the beginning and starting vertices v_{i_n} and v_{i_1} are connected by an edge. The computational time of this non-deterministic algorithm depends linearly on the size of the graph.

In general, NP means the complexity class of functions which can be computed by a non-deterministic Turing machine in polynomial time. Hamilton's problem is an example of an NP-problem. Another NP-problem is the "travelling salesman problem", which is rather like Hamilton's problem except that the various edges have numbers attached to them. One seeks that Hamilton circuit for which the sum of the numbers, or more intuitively the distance travelled by the salesman, is a minimum.

By definition every P-problem is an NP-problem. But it is a crucial question of complexity theory whether P = NP or, in other words, whether problems which are solved by non-deterministic computers in polynomial time can also be solved by a deterministic computer in polynomial time [5.14].

Hamilton's problem and the travelling salesman problem are examples of so-called NP-complete problems. This means that any other NP-problem can be converted into it in polynomial time. Consequently, if an NP-complete problem is actually proved to be a P-problem (if for instance a deterministic algorithm can be constructed to solve Hamilton's problem in polynomial time), then it would follow that all NP-problems are actually in P. Otherwise if P $\neq$ NP, then no NP-complete problem can be solved with a deterministic algorithm in polynomial time.

Obviously, complexity theory delivers degrees for the algorithmic power of Turing machines or Turing-type computers. The theory has practical consequences for scientific and industrial applications. But does it imply limitations for the human mind? The fundamental questions of complexity theory (for example N = NP or N $\neq$ NP) refer to the measurement of the speed, computational time, storage capacity, and so on, of algorithms. It is another question how one sets out to find more or less complex algorithms. This is the creative work of a computer scientist which is not considered in the complexity theory of algorithms.

On the other hand, Gödel's famous theorems are sometimes said to limit the mathematical power of computers and the human mind. His incompleteness theorem says that in every consistently axiomatized enlargement of formal number theory there is a (closed) formula which is not decidable. Actually, his theorem states that any adequate consistent arithmetical logic is incomplete in the sense that there exist true statements about the integers that cannot be proved within such a logic. Even if we enlarge our axiomatization by the undecidable formula, then there is another formula which is not decidable in the enlarged formalism. Gödel's result showed that the formalistic search for a complete consistent arithmetical logic in the tradition of Leibniz and Hilbert must fail [5.15].

Furthermore, Gödel proved that it is impossible to show that arithmetical logic, which may be incomplete, is consistent by methods that could be represented in the logic itself. Some years after Gödel's famous result, Gerhard Gentzen (1909–1945) proved the consistency of elementary number theory using so-called ε_0-induction, which is an infinitary extension of the usual arithmetical induction over natural numbers. But the consistency of Gentzen's extended proof method is as open to doubt as that of the system to be justified. In other words, the complexity of the justifying method is no less than that of the system to be justified. So there are only relative consistency proofs using methods which have to be justified by methods which have to be justified, and so on. For human thinking there is no absolute foundation of self-consistency which can be delivered by formal algorithms.

But Gödel's theorem is only a limitation on human thinking under some essential assumptions: we must accept theorem-proving as the key to human intelligence. The theorem can only be applied to a mind-model that consists of a machine with all its knowledge carefully formalized. Furthermore, Gödel's theorem only places limitations upon consistent machines, while fuzziness, inconsistency, and puzzles are typical features of human decisions. We would not survive very long if we required a long period of careful theorem-proving before deciding whether or not we should act. It should also be considered that Turing machines have a fixed data structure, while the human mind is open to novel experience and able to learn from its mistakes. So Gödel's theorem limits a machine as much as a human closing his or her mind to new information.

In 1936 Church and Turing proved that there is no algorithm deciding whether a formula of first-order predicate logic is a tautological truth or not. It follows that we must have some heuristic ideas in order to find a mathematical proof.

So the first period of AI (1957–1962) was dominated by questions of heuristic programming, which means the automated search for human problem solutions in trees of possible derivations, controlled and evaluated by heuristics [5.16]. An example was the "Logical Theorist" (1957) of Newell, Shaw, and Simon, which delivered proofs for the first 38 theorems of Russell and Whitehead's *Principia Mathematica*. Its heuristics were extracted from the rules of thumb used by several persons in psychological tests.

In 1962 these simulative procedures were generalized and enlarged for the so-called "General Problem Solver" (GPS), which was assumed to be the heuristic framework of human problem solving. But GPS could only solve some insignificant problems in a formalized micro-world. Another example of heuristic programming was the search for winning strategies in games (chess, checkers). The first programs of pattern recognition (for instance lexical and syntactical lists of words and symbols) were based on statistical methods. But in the long run the euphoric belief in general cognitive simulation procedures was not justified by any program of this early period. At least its metaphysics inspired the invention of McCarthy's programming language LISP, which was introduced as a functional programming language for the comfortable processing of symbolic lists and which has become the most powerful programming language for knowledge based systems today.

After the failure of general methods, the AI researchers propagated ad hoc procedures of "semantic information processing". The second period of AI (1963–1967) was characterized by the development of specialized programs like STUDENT for solving simple algebraic problems, ANALOGY for pattern recognition of analogical objects, and so on. Marvin Minsky, who was the leading figure at the MIT during this period, gave up the claim of psychological simulation: "The current approach", he said, "is characterized by ad hoc solutions of cleverly chosen problems which give the illusion of complex intellectual activity." For the first time it was underlined that successful practical programming depends on special knowledge, which became the central idea of knowledge based systems later on.

The search for general principles of problem solving still survived in theoretical computer science: J. A. Robinson introduced the so-called resolution principle based on the calculus of predicate logic and Herbrand's completeness theorem permitted the finding of proofs by logical refutation procedures.

The drive to practical and specialized programming in AI was accelerated during the third period (1967–1972). It was characterized by the construction of specialized systems, methods for representation of knowledge, and interest in natural languages. J. Moses, who invented the successful MACSYMA program for mathematical applications, described the change of paradigms in AI: "In fact, 1967 is the turning point in my mind when there was enough feeling that the old ideas of general principles had to go . . . . I came up with an argument for what I call the primacy of expertise."

Another famous example of this period is the DENDRAL program, which uses the specialized knowledge of a chemist in mass spectroscopy in order to find structural formulas for molecules. A paradigmatic example of this period became the SHRDLU program for a robot who could manipulate a mini-world of different blocks. The system could understand and answer questions in English concerning the block-world, carry out orders for manipulating the block-world, divide orders into a series of operations, understand what is done and why, and describe its actions in English.

In the fourth period (1972–1977), description, organization, and processing of knowledge became the central paradigm combining engineering and the philosophy of AI. Mitchell Feigenbaum introduced the term "knowledge engineering" for the development of so-called expert systems. An example was the MYCIN program for medical diagnosis, which simulates a physician with special medical knowledge of bacterial infections.

A new method of knowledge representation was the conception of frames by Marvin Minsky. A new programming language for symbolic knowledge processing was PROLOG ("Programming in Logic"), which can be compared with LISP.

The fifth period of AI (1977–1986) is, so to say, "normal" in the sense of Thomas Kuhn, meaning the paradigm of expert systems is worked out and commercialized. Tools are developed in order to build new expert systems like automobiles in mass production. AI is emerging from the laboratory and philosopher's study and is becoming the key technology of a world wide knowledge industry.

In the following, emphasis is put on expert systems, because they seem to be of most interest for philosophical questions [5.17]. An expert system is a computer program that has built into it the knowledge and capability that will allow it to operate at an expert's level (for instance DENDRAL in chemistry, MYCIN in medicine). The reasoning of a human expert is illustrated in Fig. 5.4.

Some expert systems can even explain why they rejected certain paths of reasoning and chose others. Designers work hard to achieve this because they understand that the ultimate use of an expert system will depend upon its credibility to its users, and its credibility will rise if its behavior is transparent and explainable.

But unlike that of human beings, the knowledge of an expert system is restricted to a specialized information base without generalized and structuralized knowledge of the world. Therefore, expert systems have an intermediate function between the conventional programs of numerical computers and human beings (Fig. 5.5).

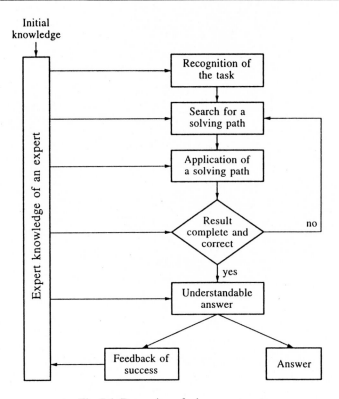

Fig. 5.4. Reasoning of a human expert

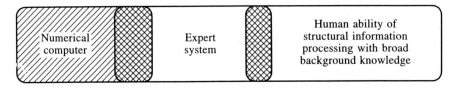

Fig. 5.5. Expert systems between numerical computers and humans

The architecture of an expert system consists of the following components: knowledge base, problem solving component (interference system), explanation component, knowledge acquisition and dialogue component. Their coordination is demonstrated in Fig. 5.6 [5.18].

Knowledge is the key factor in the performance of an expert system. The knowledge is of two types. The first type is the facts of the domain that are written in textbooks and journals in the field. Equally important to the practice of a field is the second type of knowledge, called heuristic

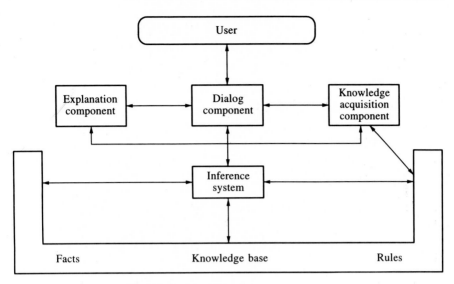

Fig. 5.6. Architecture of an expert system

knowledge, which is the knowledge of good practice and judgement in a field. It is experimental knowledge, the art of good guessing that a human expert acquires over years of work.

By the way, knowledge bases are not the same as data bases. The data base of a physician, for instance, is the patient's record, including patient history, measurements of vital signs, drugs given, and response to drugs. These data must be interpreted via the physician's medical knowledge for purposes of continuing diagnosis and therapy planning. The knowledge base is what the physician learned in his medical education and in the years of internship, residency, specialization, and practice. It consists of facts, prejudices, beliefs, and heuristic knowledge.

The heuristic knowledge is hardest to get at because experts rarely have the self-awareness to recognize what it is. Therefore knowledge engineers with interdisciplinary training have to acquire the expert's rules, to represent them in programming language, and to put them into a working program. This component of an expert system is called knowledge acquisition. Its central function in the knowledge processing of an expert system is illustrated in Fig. 5.7.

The most important methods of knowledge representation are production systems, logic, frames, and semantic networks. In addition to knowledge, an expert system needs an inference procedure, a method of reasoning used to understand and act upon the combination of knowledge and problem data. Such procedures are independent of the special knowledge base

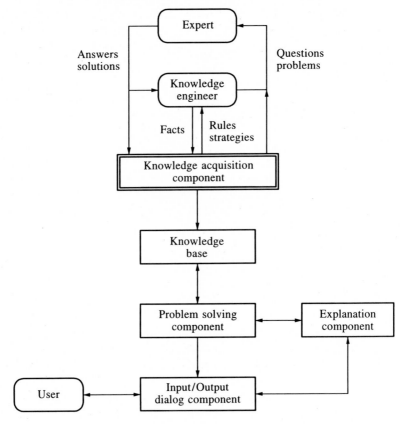

Fig. 5.7. Knowledge processing in an expert system

and are founded upon different philosophical methodologies, which will be analyzed for several examples of expert systems later on.

The explanation component of expert systems has the task of explaining the steps of the procedure to the user. The question "how" aims at the explanation of facts or assertions which are derived by the system. The question "when" demands the reasons for the questions or orders of a system.

The dialogue component handles the communication of expert system and user. A natural-language processor could, of course, increase acceptance even for untrained users.

From a technological point of view the limits of expert systems are obvious. First is the problem of knowledge representation. How shall the knowledge of a domain of work be represented as data structures in the memory of the computer and accessed for problem solving? Second is the problem of knowledge utilization. How should the inference engine be designed? Third is the question of knowledge acquisition. How is it possible to acquire the knowledge so important for problem solving automatically so that the computer eases the transfer of expertise from humans to the symbolic data structures?

The last and most important problem of expert systems is philosophical. How should the specialized knowledge base of an expert system be combined with the generalized and structuralized background knowledge of the world which influences the decisions and actions of a human expert?

Thus, when deciding on an operation, a physician will also take into consideration non-objective impressions he has concerning the patient's living conditions (family, job, etc.) and his attitude towards life. Especially in fundamental questions of life and death, for example in connection with the present dispute over the dignity of dying, the whole attitude and horizon of a physician flows into his decisions in a manner that cannot be codified, although legislation seeks to lay down general standards of behavior. In expert systems of law, for example, the same aspect could be shown. Despite all consistent systems of norms a judge will in the end find a formal scope for possible decisions where he will orient towards his personal outlook on life and the world. This invasion into subjectivity should not be complained of as a lack of objectivity, but should be taken as a chance for a more humane medicine and jurisdiction. With that, however, it is not excluded that in the future computer science should aim at an expansion of knowledge based expert systems, which are nowadays still very specialized. Yet essential limits are evident, resulting from the nature of expert systems.

Expert systems are technological realizations of problem solving procedures. Therefore the factually existing expert systems can be classified by the special prolems which they are intended to solve. Figure 5.8 illustrates the most important problem classes of expert systems.

A well analyzed problem class concerns "diagnosis", for instance in medicine. The input of such an expert system consists of measurement data, symptoms, etc, and it delivers patterns recognized in the data regularities as output. Another problem class concerns "design". The problem is how to find a product which conforms to some constraint. The solution of a planning problem demands a sequence of actions which transform an initial state into a goal state. A simulation problem starts with the initial state of a model, the following states of which must be calculated and evaluated [5.19].

The problem-solving strategies are inferred by production rules which must be chosen by a so-called rule interpreter. If several rules are applicable, a conflict-solving strategy decides which rule is appropriate. The possible rules can be ordered, for instance, by degrees of priority or generality. Then it may be suitable to choose the rule with the highest degree of priority or specificity.

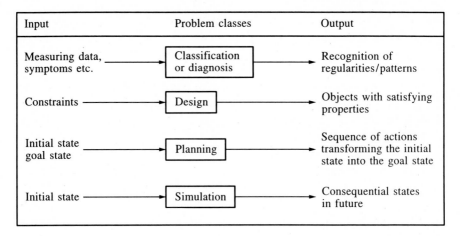

Input	Problem classes	Output
Measuring data, symptoms etc.	Classification or diagnosis	Recognition of regularities/patterns
Constraints	Design	Objects with satisfying properties
Initial state goal state	Planning	Sequence of actions transforming the initial state into the goal state
Initial state	Simulation	Consequential states in future

Fig. 5.8. Problem classes of expert systems

The combination of rules in an inference can be realized by so-called forward and backward chaining. Forward chaining starts with given data and facts A and applies the deduction machinery until the given goal D is deduced (Fig. 5.9).

In contrast to the data-driven forward chaining method, the backward chaining method is goal-directed, which means that it starts with the given goal and tries to find the premises of rules which make the goal deducible. Therefore a premise A must be found which is "true" or realizable (Fig. 5.9).

From a methodological point of view, the forward and backward chaining procedures of expert systems are nothing more than the well known methods of the antique logician and philosopher Pappos for finding necessary or sufficient reasons for affirmations. It is not surprising that nearly all inference strategies of expert systems are founded upon well known philosophical methodologies.

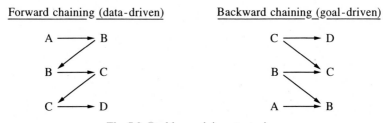

Fig. 5.9. Problem solving strategies

Nowadays, most of the philosophical theories used in AI are not taken from the philosophical literature directly, but that does not make them any less interesting philosophically. Nevertheless, some authors of famous expert systems were influenced directly by philosophers [5.20].

To see that AI is philosophical logic and methodology, one need only consider some expert systems in detail. Their problem class determines which strategy is suitable for problem solving. In general, a strategy aims at a reduction of the problem's complexity.

The task that the DENDRAL program addresses is the determination of molecular structure from data consisting of the molecular formula of a compound, and the mass spectrum of the compound [5.21]. The output is an ordered list of more or less probable structural formulas. Its strategy of problem solving is called "Generate-and-Test" and is an algorithm for generating the topological structures of organic molecules consistent with a given molecular formula and rules about which molecular bonds in a molecule are most likely to break. In short, we may say that the program reduces the complexity of the generated tree of solutions by early pruning of bad branches. Methodologically it involves a confirmation criterion.

In general, the following points are of importance, regardless of the chemical application:

a)   There is a set of formal objects in which the solution is contained.
b)   There is a generator, i.e., a complete enumeration process of this set.
c)   There is a test, i.e., a predicate identifying a generated element as
     part or not part of a set of solutions.

This general method is defined by the following algorithm, i.e., by the following recursive function according to Church's thesis:

Function GENERATE_AND_TEST (SET):
If the set SET to be examined is empty,
then failure,
otherwise let ELEM be the "following" element of SET;
If ELEM is goal element,
then give it as solution, otherwise repeat this function with the set SET
without the element ELEM.

For a translation into the AI programming language LISP [5.22], some recursive auxiliary functions have to be introduced, such as GENERATE (generates an element of a given set), GOALP (is a predicate function delivering T(true), if the argument is part of the set of solution, otherwise NIL), SOLUTION (prepares the solution element for "output"), and REMOVE (delivers the set minus the given element). Considering the common abbreviations in LISP, for example DE (Definition), COND (Condition), EQ (Equation), T (true), and the LISP conventions (e.g., brackets rules) when working out a list of symbols, the following algorithm in LISP is received:

```
(DE GENERATE_AND_TEST(SET)
    (COND((EQ SET NIL)'FAIL)
        (T(LET(ELEM(GENERATE SET))
            (COND((GOALP ELEM)(SOLUTION ELEM))
                (T(GENERATE_AND_TEST
                    REMOVE ELEM SET)))))))
```

All chemical structures are systematically produced out of a given chemical sum formula, for example C_5H_{12}, in a first step:

etc.

Some chemical structures are eliminated because they are unstable or contradictory. In the next step the corresponding mass spectrograms are computed and compared with mass spectrograms empirically obtained. This comparison is the test step of the process. GENERATE_AND_TEST thus technically realizes a methodology for generating presumptions with elimination of the impossible and test of the probable variants.

The META-DENDRAL program is designed to improve the rules of the DENDRAL program regarding which molecular bonds will break in the mass spectrometer. So META-DENTRAL uses the DENDRAL program plus the prediction criterion of confirmation which was critically analyzed by Hempel.

The MYCIN program for helping physicians diagnose infections is a backward-chaining deduction system [5.23]. Some 300 productions constitute MYCIN's pool of knowledge about blood bacterial infections. The following is typical:

If the infection type is primary-bacteremia, the suspected entry point is the gastrointestinal tract, and the site of the culture is one of the sterile sites, *then* there is evidence that the organism is bacteriodes.

To use such knowledge, MYCIN works backward. For each of 100 possible hypotheses of diagnosis, MYCIN attempts to work toward primitive facts known from laboratory results and clinical observations. As MYCIN works in a domain where deductions are rarely certain, its developers combined a theory of plausible and probabilistic reasoning with the basic production apparatus. The theory is used to establish a so-called certainty factor for each conclusion in the AND/OR tree (Fig. 5.10).

Here F_i is the certainty factor assigned to a fact by the user, C_i indicates the certainty factor of a conclusion, and A_i the degree of reliability expected

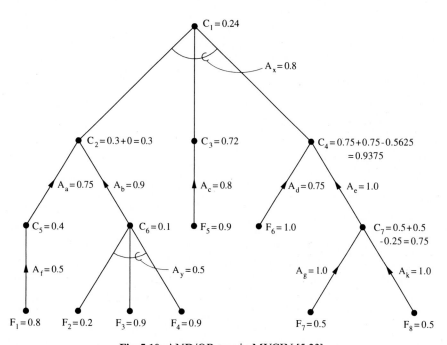

Fig. 5.10. AND/OR tree in MYCIN [5.23]

for a production rule. Certainty factors according to straightforward formulas are computed at the AND-nodes and the OR-nodes. If a certainty factor is 0.2 or less, the truth of the corresponding fact is considered unknown, and the value 0 is assigned.

The program computes degrees of inductive justification in dependence of more or less secure facts. This approach reminds of Rudolf Carnap's theory of induction. Carnap naturally did not believe in a universal inductive conclusion à la Bacon. Conclusions are always deductive. No Popperian advice was required for that. Expert systems do not operate otherwise. Nevertheless, probabilistic measures used in systems like MYCIN make the system more transparent to the user.

On the other hand, there are, so to say, Popperian programs with "Hypothesize-and-Test" strategies which generate the most interesting hypotheses with hard tests. There are programs that help to construct linear causal explanations of statistical data. Other programs uses the old philosophers' knowledge that inductive reasoning is nonmonotonic, which means a conclusion derived inductively from a set of premises may not follow from consistent extensions of the premises. For example, birds can fly, and Tweety is a bird, so I infer that Tweety can fly, but not if I also know that Tweety is an ostrich [5.24].

Another strategy is the division of a complex problem into simpler parts or less complex subproblems, which is for instance used in Georg Polya's heuristic mathematical guide *How to solve it*. Therefore the domain of application must allow a division into independent parts. But it is obvious that a complex network of dependencies cannot always be cut without changing the original state of the system. For instance, consider the ecological network of the human environment or the complex psychic dependencies which a psychiatrist has to analyze. The system is not always the sum of its parts.

Some distinctions in philosophy of science can be translated as properties of knowledge based systems. If research makes extensive use of theoretical conceptions associated with intrinsic properties of a theory, the discovery process is described as theory-dependent ("theory-driven"). The converse view, which is often called Baconian, takes a body of data as its starting point. Then the discovery process is called data-driven. The distinction between theory- or data-driven knowledge processing is also well known in AI.

Now I shall sketch some programs from knowledge based systems which realize for various sciences the tasks and advantages just mentioned. My first example concerns mathematics. AM [5.25] is a knowledge based system that has recursively generated and, so to speak, rediscovered concepts from number theory. In contrast with a program in empirical science, AM's criterion of success is not that a concept is compatible with empirical data but that it is "interesting" concerning its capacity to generate examples, new problems, etc. The program, written in 1977 in LISP, begins with basic concepts such as sets, lists, equality, and operations, and heuristic advice to direct the discovery process. The heuristics suggest new tasks and create new concepts based on existing ones. New tasks are ordered according to their degrees of interest. Tasks proposed by a number of different heuristics tend to be more interesting than those proposed by a single rule.

Using this measure to direct its search through the space of mathematical concepts, AM defined concepts for the integers, multiplication, and the prime numbers, and found propositions on prime numbers (for instance the unique-factorization theorem).

A closer analysis, however, shows that a demand for simulation of a historical process of discovery cannot be maintained. The success of AM depends decisively on features of the programming language LISP. Nevertheless, an analysis shows interesting analogies with the human research process.

As is indicated by the name "LISP", lists of symbols are worked out systematically. Two lists may be recursively defined as equal when both are atomic and the atoms are equal, otherwise when the heads of the lists are equal and the rest of the lists are equal. In LISP the recursive Boolian function is noted as follows:

```
(DE LIST-EQUAL (XY)
    (COND((OR(ATOM X)(ATOM Y))
        (EQ X Y))
    (T(AND
            (LIST-EQUAL(CAR X)(CAR Y))
            (LIST-EQUAL(CDR X)(CDR Y))))))
```

Here CAR and CDR are basic operators in LISP sorting list heads and rest lists, respectively, out of given lists of symbols. A heuristic generalizing rule of AM generalizes the term of identity. Then two lists are called "generalized equal" if both are atomic and the atoms are equal, otherwise the rest lists are "generalized equal". In LISP:

```
(DE L-E-1(XY)
    (COND((OR(ATOM X)(ATOM Y))
        (EQ X Y)
    (T(L-E-1(CDR X)(CDR Y)))))
```

Because of the generalization all lists with the same length are considered equivalent. They define a class which is called "number". The process of discovery realized on concrete objects by children is simulated by AM via transformation rules. Addition is introduced as joining of two lists. The concept of prime number is discovered by a heuristic tranformation rule for forming inverses out of concepts already generated. AM was followed by EU-

RISKO (1983) which can discover not only new concepts but new heuristics as well.

The discovering of quantitative empirical laws was analyzed by a sequence of programs called BACON [5.26]. The BACON systems are named after Francis Bacon, because they incorporate some of his ideas on the nature of scientific reasoning. They are data-driven knowledge-processing systems which gather data, discover regularities between two or more variables, and test laws. The basic methods of BACON make no reference to the semantic meaning of the data on which they operate and make no special assumptions about the structure of the data. In cases where one has experimental control over the independent terms, the traditional method of "varying one term at a time" can be used to separate the effects of each independent term on the dependent variables. Physical laws that can be reproduced by the BACON program include Boyle's law, Kepler's third law, Galileo's law, and Ohm's law.

An inquiry made with such a knowledge based system at least obeys the requirement that laws relevant for different disciplines should fulfill the same methodological and heuristic frame conditions. The corresponding knowledge based system not only reproduces certain laws, which were discovered in different historical contexts, but also systematically generates the complete methodological scope of concepts and sorts out interesting applications. The so-far latest BACON program is not only data-driven, and thus in a strict sense "Baconian", but also theory-driven. In its theoretical requirement for symmetry and conservation laws it generates, for example, the conservation law of momentum.

Another sequence of programs are capable of inducing qualitative laws from empirical data (GLAUBER, STAHL, DALTON). These programs can also induce structural and explanatory models for certain phenomena. Qualitative laws are usual in chemistry [5.27].

Competition between a machine and a human scientist is not intended. However, systematic and structural classifications of scientific laws and theories have been accomplished which give insight into the complexity of scientific laws and their conditions of discovery.

There are several aspects of the diverse activity known as scientific discovery, such as finding quantitative laws, generating qualitative laws, inferring the components of substances, and formulating structural models. An integrated discovery system is envisaged which incorporates the individual systems as components. Each component accepts input from one or more of the other components.

For instance, STAHL focuses on determining the components of chemical substances, whereas DALTON is concerned with the number of particles involved in a reaction. Thus, STAHL can be viewed as laying the groundwork for a detailed structural model in the sense of DALTON. In this way it might be possible to develop more and more complex knowledge-based systems to analyze research as knowledge processing and problem solving.

Even within such an expanded framework of research, we have not addressed the mechanisms underlying the planning of experiments, or the invention of new instruments of measurement. Any intrinsic concept, in association with the experimental arrangement that allows it to be measured, can be employed as a scientific instrument. In this case the discovery of the instrument is coincident with the discovery of the concept itself.

There are also knowledge based systems which consider the design of experiments and their interaction with other activities of scientific research. A system called KEKADA (invented by a research group of Simon) is shown in Fig. 5.11 with hypotheses generators, experiment choosers, expectation setters, etc. [5.28]. It was developed to model the design of an experiment in biochemistry (Krebs' discovery of the urea cycle 1935). Like knowledge engineers, Simon and his crew analyzed the laboratory notebooks of Krebs, defined his methodological rules of research and translated them into a LISP-like programming language.

If the system has not decided which task to work on, problem choosers will decide which problem the system should start working on. Hypothesis generators create hypotheses when faced with a new problem. The hypothesis or strategy proposers will choose a strategy to work on. Then the experiment proposers will propose the experiments to be carried out. Both types of heuristic may need the decision makers. Then expectation setters set expectations and experimenters carry out experiments. The results of the experimenters are interpreted by the hypothesis modifiers and the confidence modifiers. When applicable, problem generators may add new problems to the agenda. If the outcome of an experiment violates the expectations for it, then the study of this puzzling phenomenon is made a task, and it is added to the agenda.

Every component of the system is an operator which is defined by a list of production rules. Besides domain-specific heuristics the system contains general rules that are part of general methodology of research. It is remarkable that a specific rule defines the case that an experimental result is a "puzzling phenomenon". Scientific discovery thus becomes a gradual process guided by problem-solving heuristics, and not by a single "flash of insight" or sudden leap. These examples of knowledge based systems may be interpreted as assistants in the studies of philosophers of science in the sense that, for instance, the program DENDRAL is an assistant in the laboratory of a chemist. They can investigate the whole space of possible laws produced by some heuristical rules. But they are the accurate assistants, not the masters. Their "flash of insight", the kind of "surprise" the system can recognize, is frame-dependent, given by the masters.

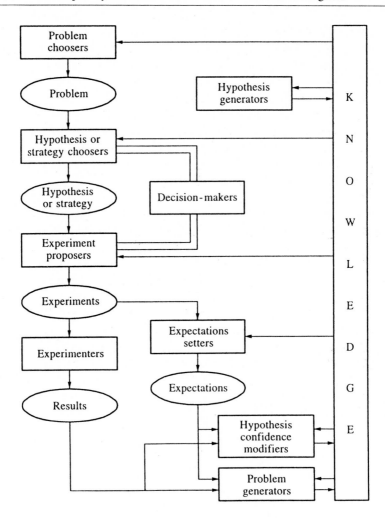

Fig. 5.11. Architecture of KEKADA [5.28]

What about Turing's question, which motivated early AI researchers? Can machines "think"? Are machines "intelligent"? In my opinion, this question is for computer technology a metaphysical one, because "thinking" and "intelligence" are not well-defined concepts of computer science or AI.

All we can say today is this. If a program generates a structure that can be interpreted as a new concept, then the rules of transformation used contain this concept and the corresponding data structure implicitly. An algorithm which directs the application of these rules makes the implicitly given concepts and data structures explicit. In philosophical discussions on AI much confusion is provoked by the terminology of AI, which is introduced in a technical sense, but is associated with sometimes old and sophisticated meanings in philosophy and psychology. As in other sciences, we have to live with traditional terms and notions, which may be highly confusing if they are abstracted from their technical context. An example was the notion "intelligence", for instance, in "Artificial Intelligence" (AI).

A term which sometimes confuses philosophers is the use of "knowledge" in AI. Let me again underline that "knowledge" in the term "knowledge based system" has a technical meaning, and does not claim to explain the whole philosophical, psychological, or sociological concept of knowledge. In AI technology as part of practical computer science, no philosophical reductionism is involved.

"Knowledge processing" in so-called "knowledge based systems" means a new kind of complex information processing which must be distinguished from the older merely numerical data processing. It involves complex transformation rules of translation and interpretation which are characterized at a high level in the hierarchy of programming languages (today LISP or PROLOG). This level is near to natural languages, but not identical, of course, and only grasps partial aspects of the broad meaning of human knowledge (Fig. 5.12). Nevertheless, knowledge processing remains program-controlled and in the tradition of Leibniz' mechanization of thoughts [5.29].

If the human mind is accepted as a Turing-type computer, then the laws of nature governing human bodies and brains would not be of much relevance. Algorithmic programs in computer software do not depend on the

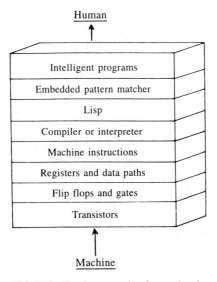

Fig. 5.12. Hardware and software levels

hardware of physical mechanics, but on the mathematically idealized concept of Turing's machine. But if the human mind is understood as product of natural evolution, then the relevance of physical, chemical, and biological laws for the emergence of the human mind must be tested. In modern physics the basic theory of matter is quantum mechanics. In classical physics the interactions of physical systems were conceived as completely independent of a human observer, yet human consciousness is now seen to play a crucial role in the measurement process. At first I shall criticize these interpretations as possible, but doubtful approaches. Nevertheless, quantum mechanics is shown as the physical framework for generalized quantum computers with high efficiency and quantum complexity theory, which differ from classical Turing machines and classical complexity theory.

Obviously, a Turing machine can be interpreted in the framework of classical physics (Fig. 5.13). Such a computing machine is a physical system whose dynamical evolution takes it from one of a set of input states to one of a set of output states. The states are labeled in some serial way. The machine is prepared in a state with a given input level and then, following some deterministic motion, the output state is measured. For a classical deterministic system the measured output label is a definite function f of the prepared input label. In principle the value of that label can be measured by an outside observer, and the machine is said to compute the function f. But classical stochastic computing machines and quantum computing machines do not compute functions in the above sense. The output state of a stochastic machine is random, with only a probability distribution function for the possible outputs depending on the input state. The output state of a quantum machine, although fully determined by the input state, is not an observable and so the observer cannot in general discover its label. What is the reason? We must remember some basic concepts of quantum mechanics which have been already introduced in Sect. 2.3.

In quantum mechanics, vectors like momentum or position must be replaced by operators satisfying a non-commutative relation depending on Planck's quantum (Fig. 2.18). Classical systems described by a Hamiltonian function are replaced by quantum systems, for instance, electrons or photons described by a Hamiltonian operator. States of a quantum system are described by vectors of a Hilbert space spanned by the eigenvectors of its Hamiltonian operator. The causal dynamics of quantum states is determined by a partial differential equation called the Schrödinger equation. While classical observables commute and always have definite values, non-classical oberservables of quantum systems do not commute and in general have no common eigenvector and consequently no definite eigenvalues. For observables in a quantum state only statistical expectation values can be calculated.

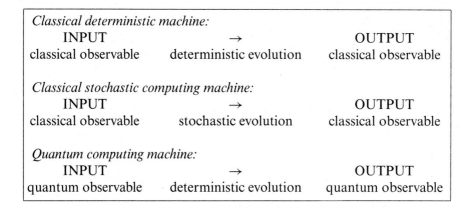

Classical deterministic machine:

INPUT	$\rightarrow$	OUTPUT
classical observable	deterministic evolution	classical observable

Classical stochastic computing machine:

INPUT	$\rightarrow$	OUTPUT
classical observable	stochastic evolution	classical observable

Quantum computing machine:

INPUT	$\rightarrow$	OUTPUT
quantum observable	deterministic evolution	quantum observable

Fig. 5.13. Classical and non-classical computing machines

A major difference from classical mechanics is given by the superposition principle demonstrating the linearity of quantum mechanics. In an entangled pure quantum state of superposition an observable can only have indefinite eigenvalues. In short, the superposition or linearity principle of quantum mechanics delivers correlated ("entangled") states of combined systems which are highly confirmed by the EPR experiments (Alain Aspect 1981). Philosophically, the (quantum) whole is more than the sum of its parts.

The superposition principle has severe consequences for the measurement of quantum systems. In the quantum formalism a quantum system and a measuring apparatus are represented by two Hilbert spaces which are combined in a tensor product $H = H_1 \otimes H_2$. In the initial state $\phi(0)$ of measurement at time 0 the systems H_1 and H_2 are prepared in two separated states ψ and φ respectively, with $\phi(0) = \psi \otimes \varphi$. The causal development of both systems is determined by the Schrödinger equation, i.e., $\phi(t) = U(t)\phi(0)$ with the unitary operator $U(t)$. Because of the linearity of $U(t)$, the state $\phi(t)$ is entangled with indefinite eigenvalues while the measuring apparatus at time t shows definite measurement values. Thus, the linear quantum dynamics cannot explain the measurement process.

In a more popular way the measurement process is illustrated by Schrödinger's thought experiment of a cat in a linear superposition of the two states "dead" and "alive" (Fig. 5.14a). Imagine a cat which is locked in a closed box with a sample of radium. The radium is chosen in such a way that during one hour a single decay takes place with a probability 1:2. If a decay happens, then an electrical circuit is closed, causing a mechanism with a hammer to destroy a bottle of prussic acid and thus killing the cat. The box remains closed for one hour [5.30].

According to quantum mechanics the two possible states of the cat, dead and alive, remain undetermined until the observer decides them by opening the box. For the cat's state in the closed box, quantum mechanics

Fig. 5.14a. Schrödinger's cat

as interpreted by Erwin Schrödinger forecasts a correlated ("entangled") state of superposition, i.e., the cat is both dead and alive with equal parts. According to the measurement process, the states "dead" and "alive" are interpreted as measurement indicators representing the states "decayed" or "not decayed" of the radium.

In the Copenhagen interpretation of Bohr, Heisenberg, and others, the measurement process is explained by the so-called "collapse of the wave-packet", i.e., splitting up of the superposition state into two separated states of measurement apparatus and measured quantum system with definite eigenvalues. Obviously, we must distinguish the linear dynamics of quantum systems from the nonlinear act of measurement. The reason for nonlinearity in the world is sometimes explained as the emergence of human consciousness.

Eugene Wigner (1961) suggested that the linearity of Schrödinger's equation might fail for conscious observers, and be replaced by some nonlinear procedure, according to which either one or the other alternative would be resolved out (Fig. 5.14b). But Wigner's interpretation forces us to believe that the complex quantum linear superpositions would be resolved into separated parts only in those corners of the universe where human-like consciousness emerges. In the macroscopic world of billiard balls, planets, or galaxies, EPR correlations are not measured, and appear only in the microscopic world of elementary particles like photons. It seems to be rather strange that the separated states of systems in the macroscopic world which can be described in classical physics with definite measurement values are caused by human-like consciousness.

The Everett "many-worlds" interpretation of quantum mechanics seems to avoid the problems of nonlinear reductions by splitting up human consciousness into branching paths inhabiting different, mutually incompatible worlds (Fig. 5.14c).

Fig. 5.14b. Wigner's interpretation of Schrödinger's cat

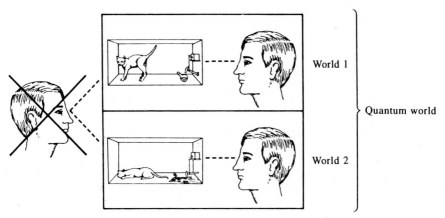

Fig. 5.14c. Everett's interpretation of Schrödinger's cat

In the measurement process the dynamics of measurement instrument and quantum system is described by the equation $\phi(t) = \sum_i c_i(t)\psi_i \otimes \varphi_i$ with states (φ_i) referring to the measurement values of the measuring instrument. Everett argues that the state vector $\phi(t)$ never splits up into partial states, but all branches $\psi_i \otimes \varphi_i$ are actualized. The state $\phi(t)$ describes a manifold of simultaneously existing real worlds with $\psi_i \otimes \varphi_i$ corresponding to the state of the i-th parallel world. Thus, the measured partial system is never in a pure state. In Everett's sense, ψ_n may be interpreted as a relative state depending on the state of an observer or measuring instrument with $\psi_n = c_n^{-1}(\varphi_n, \phi)_{H_2}$. If φ_n are accepted as memory states, then an observer with a definite memory can only be aware of his own branch of the world $\psi_n \otimes \varphi_n$. But he can never observe the other partial worlds.

The advantage of Everett's interpretation is that a nonlinear reduction of superposition does not need to be explained. But the disadvantage is his ontological belief in myriads of worlds which are unobservable in principle. Thus, Everett's interpretation (if mathematically consistent) needs Ockham's razor.

In the history of science, anthropic or teleological arguments often showed that there were gaps or failures of explanation in science. Thus, some scientists, such as Roger Penrose, suppose that the linear dynamics of quantum mechanics is inconvenient (Einstein said it was "incomplete") for explaining cosmic evolution with the emergence of consciousness. He argues that a unified theory of linear quantum mechanics and nonlinear general relativity could at least explain the separated states of macroscopic systems in the world without reference to anthropic or teleological principles. In Penrose's proposed unified theory a linear superposition of a physical system splits into separated states when the system is large enough for the effects of relativistic gravitation. Penrose calculates a level of one graviton as the smallest unit of curvature for such an effect [5.31]. The idea is that the level should lie comfortably between the quantum level of atoms, molecules, etc., with linear laws of quantum mechanics and the classical level of everyday experiences. The advantage of Penrose's argument is that the linearity of the quantum world and the nonlinearity of the macroscopic world would be explained by a unified physical theory without reference to any human intervention. But, of course, we are still missing a testable unified theory (compare Sect. 2.4).

Nevertheless, the question arises of whether quantum mechanics delivers a framework for the evolution of the human brain, or at least for a new computer technology to replace classical computing systems. The basic idea of quantum mechanics is the superposition of quantum states as a result of linear quantum dynamics and the reduction of superpositions by some kind of measurement [5.32]. Thus, a quantum computer would need a quantum version of a logic gate, where the output would be the results of some unitary operator applied to the input and a final act of measurement. The superposition of quantum systems (for instance photons) reminds us of the parallelism of computations. A quantum computer would become useful if we were interested in some suitable combination of many computational results and not in their partial details. In this case a quantum computer could deliver the superposition of perhaps myriads of parallel computations in a rather short time, overcoming the efficiency of classical computing systems. But quantum computers would still work in an algorithmic way, because their linear dynamics would be deterministic. The non-deterministic aspect comes in via the nonlinear act of measurement. Thus, it cannot be expected that quantum computers will perform non-algorithmic operations beyond the power of a Turing machine. So quantum computers (if they are ever built) may be more interesting for complexity theory and for overcoming practical constraints of computation.

Concerning the human brain, we want to argue that the quantum level of elementary particles, atoms, and molecules was necessary for its evolution, but not the other way that mental states of the brain are necessary for the reduction of correlated states in physics. Actually, there is a significant number of neurons sensitive to single quanta with their superpositions and reductions of entangled states. But, of course, these quantum states cannot be identified with mental states of the brain. We have no consciousness either of superpositions or of their separation into single states initiated by nonlinear random events. Nevertheless, quantum effects are involved in the emergence and interaction of mental states of the brain in a way which is still far from being satisfactorily understood.

5.3 Neural and Synergetic Computers

After logic and classical and quantum mechanics we want to test the relevance of complex dynamical systems for the development of computer science and artificial intelligence. It is obvious that algorithmic mechanization with Turing-like machines confronts severe obstacles which cannot be overcome by growing capacities of classical or quantum computers. For example, pattern recognition and other complex tasks of human perception cannot be mastered by program-controlled computers. The structure of the human brain seems to be completely different.

In the history of science, the brain was illustrated by technical models of the most advanced machinery [5.33]. Thus, during the age of mechanization, the brain functions were thought of as hydraulic pressures which are conducted along the nerves to operate on the muscles. With the beginning of electrical technology, the brain was compared with telegraphs or telephone switchboards. Since the development of computers, the brain has been identified with the most advanced hardware generations. In the last chapter, we saw that even quantum computers (if they are ever constructed) could not increase their power beyond the complexity of Turing-like algorithms.

Unlike program-controlled serial computers, the human brain and mind are characterized by contradictions, incompleteness, robustness, and resistance to noise, but also by chaotic states, dependence on sensitive initial conditions, and last but not least by learning processes. These features are well known in the complex system approach. Concerning the architecture of Turing-like and complex systems, an essential limitation derives from the sequential and centralized control of classical systems, but complex dynamical systems are intrinsically parallel and self-organized.

Nevertheless, historically, the first designs for neural network computers were still influenced by Turing's concept of a machine. In their famous paper "A logical calculus of the ideas immanent in nervous activity" (1943), McCulloch and Pitts offered a complex model of neurons as threshold logic units with excitatory and inhibitory synapses, which applied concepts of the mathematical logic of Russell, Hilbert, Carnap, and others, and the Turing machine. A McCulloch–Pitt neuron fires an impulse y along its axon at time

$n + 1$ if the weighted sum of its inputs $x_1, ..., x_m$ and weights $w_1, ..., w_m$ at time n exceeds the threshold Θ of the neuron (Fig. 5.15a) [5.34].

Particular applications of McCulloch–Pitts neurons are the following models of logical connections: the OR-gate (Fig. 5.15b) models the logical disjunction x_1 OR x_2 of sentences x_1 and x_2 (formally, $x_1 \vee x_2$) which is only false if x_1 and x_2 are false sentences, and true otherwise. The truth-values are binarily represented by 0 for "false" and 1 for "true". For threshold $\Theta = 1$ and weights $w_1 = 1$ and $w_2 = 1$, the OR-gate fires with $w_1 x_1 + w_2 x_2 \geq \Theta$ so long as x_1 or x_2 or both x_1 and x_2 are 1.

The AND-gate (Fig. 5.15c) models the logical conjunction x_1 AND x_2 (formally, $x_1 \wedge x_2$) which is only true if x_1 and x_2 are true sentences, and false otherwise. For the threshold $\Theta = 2$ and weights $w_1 = 1$ and $w_2 = 1$, the AND-gate fires with $w_1 x_1 + w_2 x_2 \geq \Theta$ only if both x_1 and x_2 are 1.

The NOT-gate (Fig. 5.15d) models the logical negation NOT x_1 (formally, $\neg x_1$) which is only true if x_1 is false, and false otherwise. For threshold $\Theta = 0$ and weight $w_1 = -1$ the NOT-gate fires $w_1 x_1 \geq \Theta$ only if x_1 is 0. Thus, if x_1 is 1, then the NOT-gate does not fire, which means that the ouput $y = \neg x_1 = 0$.

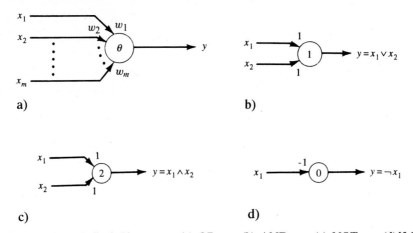

a)

b)

c)

d)

Fig. 5.15a–d. McCulloch-Pitts neuron (**a**), OR-gate (**b**), AND-gate (**c**), NOT-gate (**d**) [5.34]

A neural net à la McCulloch–Pitts is a system of McCulloch–Pitts neurons, interconnected by splitting the output of each neuron into lines and connecting some of these to the inputs of other neurons (Fig. 5.16). Although this concept of a system may be very simplified, any "classical" von-Neumann computer can be simulated by a network of such neurons. In 1945, John von Neumann wrote a draft report which has become famous as the first place where the idea of a stored program, which resided in the computer's memory along with the data it was to operate on, was clearly stated. This historic document shows that von Neumann was completely aware of the possibility of computation by a McCulloch–Pitts network.

Mathematically, a von-Neumann computer can be conceived as a finite automaton, consisting of a finite set X of inputs, a finite set Y of outputs and a finite set Q of states. The dynamics of a finite automaton is defined by a next-state function δ, transforming the state q and input x at time t into a state $\delta(q, x)$ at the following time $t + 1$, and an output function β, connecting a state q with an output $\beta(q)$.

The components of a von-Neumann computer, such as an input-output unit, a store, a logical control unit, and an arithmetic unit, can easily be shown to be finite automata. Even a modern digital computer which is made up of a network of thousands of elements and integrated onto chips can be understood as a McCulloch–Pitts-like neural network. In general, every register machine, Turing machine, or recursive function can be simulated by an appropriate network of finite automata. But these applications of McCulloch–Pitts networks still work in the framework of program-controlled serial computers.

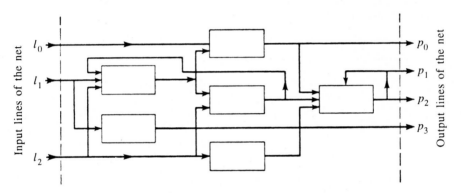

Fig. 5.16. McCulloch-Pitts network

It was again John von Neumann who first tried to extend Turing's concept of a universal computer to the idea of a self-reproducing automaton [5.35]. He noted that a machine building other machines decreased its complexity, because it seems to be impossible to use more material than is given by the building machine. Contrary to this traditional mechanistic view, living organisms in biological evolution seem to be at least as complex as their parents, with increasing complexity in the long run of evolution (Herbert Spencer).

Von Neumann's concept of cellular automata gave the first hints of mathematical models of living organisms conceived as self-reproducing networks of cells. The state space is a homogeneous lattice which is divided into equal cells like a chess board. An elementary cellular automaton is a cell which can have different states, for instance "occupied" (by a mark), "free", or "colored". An aggregation of elementary automata is called a composite automaton or configuration. Each automaton is characterized by its environment, i.e., the neighboring cells. The dynamics of the automata is determined by synchronous transformation rules. Von Neumann proved that the typical feature of living systems, their tendency to reproduce themselves, can be simulated by an automaton with 200 000 cells (in the plane), where each cell has 29 possible states and the four orthogonal neighboring cells as environment [5.36].

This idea was developed by John Conway, whose cellular automata can simulate growth, change, and death of populations of living systems. A simple example is defined by the following synchronous rules for cells with two possible states "occupied" (by a mark) or "free":

1) Rule of survival: An occupied cell with 2 or 3 occupied neighboring cells remains unchanged.
2) Rule of death: A cell loses its mark if it either has more than 3 neighboring cells ("overpopulation") or has less than 2 neighboring cells ("isolation").
3) Rule of birth: If an empty cell has exactly 3 occupied neighboring cells, then it gets a mark.

Figure 5.17a illustrates the "death" of a configuration in the third generation and Fig. 5.17b shows "survival" after the second generation. Conway's theory has some more surprising results, which were discovered via computer experiments.

Cellular automata are not only nice computer games. They have turned out to be discrete and quantized models of complex systems with nonlinear differential equations describing their evolution dynamics. Imagine a chessboard-like plane with cells, again. A state of a one-dimensional cellular

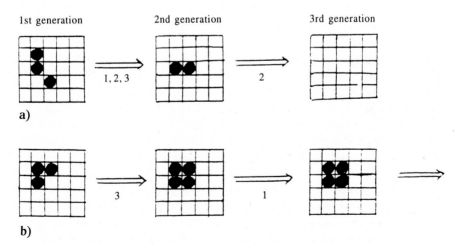

Fig. 5.17a,b. Conway's cellular automata modeling "death" (**a**) and "survival" (**b**)

automaton consists of a finite string of cells, each of which can take one of two states ("black" (0) or "white" (1)) and is connected only to its two nearest neighbors, with which it exchanges information about its state. The following (later) states of a one-dimensional automaton are the following strings on the space-time plane, each of which consists of cells taking one of two states, depending on their preceding (earlier) states and the states of their two nearest neighbors. Figure 5.18b–e illustrates the time evolution of four automata in 60 time steps. Thus, the dynamics of an one-dimensional cellular automaton is determined by a Boolean function of three variables, each of which can take either the value 0 or 1 [5.37].

$\underline{111}$	$\underline{110}$	$\underline{101}$	$\underline{100}$	$\underline{011}$	$\underline{010}$	$\underline{001}$	$\underline{000}$
0	0	1	0	1	1	0	1

Fig. 5.18a. Dynamics of 1-dimensional cellular automata

For three variables and two values, there are $2^3 = 8$ possibilities for three nearest neighbor sites. In Fig. 5.18a, they are ordered according to the corresponding three-digit binary number. For each of the three nearest neighbor sites, there must be a rule determining the following state of the middle cell. For eight sequences and two possible states, there are $2^8 = 256$ possible combinations. One of these rule combinations, determining the dynamics of a one-dimensional cellular automaton, is shown in Fig. 5.18a.

Each rule is characterized by the eight-digit binary number of the states which each cell of the following string can take. These binary numbers can be ordered by their corresponding decimal numbers.

The time evolution of these rules, characterizing the dynamics of a one-dimensional cellular automaton, produces very different cellular patterns, starting from random initial conditions. Computer experiments give rise to the following classes of attractors the cellular patterns of evolution are aiming at. After a few steps, systems of class 1 reach a homogeneous state of equilibrium independently of the initial conditions. This final state of equilibrium is visualized by a totally white plane and corresponds to a fixed point as attractor (Fig. 5.18b).

Systems of class 2, after a few time steps, show a constant or periodic pattern of evolution which is relatively independent of the initial conditions. Specific positions of the pattern may depend on the initial conditions, but not the global pattern structure itself (Fig. 5.18c).

Systems of class 3 evolve toward chaotic states as final attractors without any global periodicity. These chaotic patterns depend sensitively on the initial conditions, and show self-similar behavior with fractal dimension (Fig. 5.18d). Systems of class 4 produce highly complex structures with locally propagating forms (Fig. 5.18e). Systems of class 3 and 4 are sensitive to tiny

b)

c)

rule 0 (00000000)   rule 1 (00000001)

d)

e)

rule 18 (00010010)   rule 110 (01101110)

Fig. 5.18b–e. Attractors of 1-dimensional cellular automata

fluctuations that can effect a global change of order (the "butterfly effect"). Thus, in these cases, the process of evolution cannot be forecast in the long run.

Obviously, these four classes of cellular automata model the attractor behavior of nonlinear complex systems which is well known from self-organizing processes. In the preceding chapters, we have seen many examples in the evolution of matter, life, and mind–brain. In Chap. 6, we will consider many analogies with the evolution of human societies. In general, self-organization has been understood as a phase transition in a complex system. Macroscopic patterns arise from complex nonlinear interactions of microscopic elements. There are different final patterns of phase transitions corresponding to mathematically different attractors.

In Fig. 2.26a–e, a survey was given of the different attractors of a stream, the velocity of which is accelerated step by step. These patterns of the stream have many analogies with corresponding evolution patterns of cellular automata. At a first level, the stream reaches a homogeneous state of equilibrium ("fixed point"). At a higher velocity, the bifurcation of two or more vortices can be observed corresponding to periodic and quasi-periodic attractors. Finally, the order decays into deterministic chaos as a fractal attractor of complex systems. Classes 3 and 4 of the cellular automata are extremely interesting for modeling processes. Class 3 delivers evolution patterns of chaotic systems. Class 4 shows evolution patterns of dissipative systems with sometimes quasi-organic forms which can be observed in the evolution of organisms and populations.

From a methodological point of view, a one-dimensional cellular automaton delivers a discrete and quantized model of the phase portrait which describes the dynamical behavior of a complex system with a nonlinear differential equation of evolution, depending on one space variable. There are many reasons for restricting oneself to discrete models. The complexity of nonlinear systems is often too great to calculate numerical approximations within a reasonable computation time. In that case, a discrete model may offer rough, but sufficient information about the system's global dynamical behavior in the long run. The dynamical behavior varies if the dependence of the evolution rules (Fig. 5.18a) is enlarged to more than two neighbors in a string.

Two-dimensional cellular automata which have been used in Conway's game of Life can be interpreted as discrete models of complex systems with nonlinear evolution, depending on two space variables. Obviously, cellular automata are a very flexible and effective modeling instrument when the complexity of nonlinear systems is increasing and the possibility of determining their behavior by solving differential equations or even by calculating numerical approximations becomes more and more hopeless.

Historically, the modern development of cellular automata stems back to von Neumann's early ideas of self-reproducing automata. Besides self-reproduction, there is another feature which seems to be essential for natural complex systems in contrast to traditional computers. The human brain has the possibility to learn, for instance, by perception. In the first logical model of the brain which was offered by the McCulloch–Pitts network the function of an artificial neuron is fixed for all time. McCulloch and Pitts succeeded in demonstrating that a network of formal neurons of their type could compute any finite logical expression.

But in order to make a neural computer capable of complex tasks, it is necessary to find mechanisms of self-organization which allow the network to learn. In 1949, Donald Hebb suggested the first neurophysiological learning rule which has become important for the development of neural computers. Synapses of neurons do not always have the same sensitivity, but modify themselves in order to favor the repetition of firing patterns which have frequently occurred in the past.

In 1958, Rosenblatt designed the first learning neural computer, which has become famous under the name "Perceptron" [5.38]. Rosenblatt was originally a psychologist with a typical psychological interest in human learning processes. But his design of a learning machine which was capable of complex adaptive behavior was extremely interesting to engineers and physicists. Thus, it is no wonder that the original idea of a psychologist was grasped by engineers who were more interested in robots and computer technology than in simulating processes in the human brain. From a technical point of view, it is not essential that the learning procedures of neural computers are like learning processes of the mind-brain system. They must be effective in managing complex tasks of adaptive behavior but may use methods which are completely different from what is known in biological evolution.

Rosenblatt's neural computer is a feedforward network with binary threshold units and three layers. The first layer is a sensory surface called a "retina" which consists of stimulus cells (S-units). The S-units are connected with the intermediate layer by fixed weights which do not change during the learning process. The elements of the intermediate layer are called associator cells (A-units). Each A-unit has a fixed weighted input of some S-units. In other words, some S-units project their output onto an A-unit. An S-unit may also project its output onto several A-units. The intermediate layer is completely connected with the output layer, the elements of which are called response cells (R-units). The weights between the intermediate layer and the output layer are variable and thus able to learn.

The Perceptron is viewed as a neural computer which can classify a perceived pattern in one of several possible groups. In the case of two groups, each R-unit learns to distinguish the input patterns by activation and deactivation. The learning procedure of a perceptron is supervised. Thus, the desired state (active or not) of each R-unit corresponding to a pattern to be learnt must be well known. The patterns to be learnt are offered to the network, and the weights between the intermediate and output layer are adapted according to the learning rule. The procedure is repeated until all patterns produce the correct output.

The learning procedure is a simple algorithm: for each element i of the output layer, the actual output o_i which has been produced by a certain pattern is compared with the desired output d_i. If $o_i = d_i$, then the pattern is already correctly classified. If the desired output d_i is equal to one and the actual output o_i is zero, then all weights $w_{ij}(t)$ at time t with an active unit ($o_j > 0$) are enlarged in the following step $t + 1$, or formally $w_{ij}(t + 1) = w_{ij}(t) + \sigma o_j$. The constant σ is the learning rate, which can increase or decrease the speed of learning, according to its size. If the desired output is zero and the actual output is one, then all weights with an active element are diminished, or formally $w_{ij}(t + 1) = w_{ij}(t) - \sigma o_j$.

The Perceptron seemed to start a new era of computer technology with neural networks that could do anything. Early papers by the Perceptron group made exaggerated claims. But, in 1969, enthusiasm was followed by bitter criticism. In that year, Marvin Minsky and Seymour Papert published their famous book *Perceptrons*, which discussed the limitations of Perceptrons with mathematical precision [5.39]. The reaction to this analysis was that most research groups gave up their interest in the network and complex system approach, and changed to classical AI and computer technology, which seemed to be more profitable than the "speculations" of the Perceptron fans.

But this attitude of the scientific community after 1969 was again, of course, an overreaction. Uncritical enthusiasm and uncritical condemnation are inadequate measures of evolution in science. Darwinian evolution took millions of years to build the capability for pattern recognition into our brains. It would be surprising if our engineers should succeed in constructing analogous neural computers in just a few years.

The critical questions are the following. What can a Perceptron do? What can it not do? Why can the Perceptron not do it? An essential step to answer these question was Minsky's and Papert's proof of the so-called Perceptron convergence theorem. It garantuees that a solution which can be *learnt* by the network in principle can also be *found* in a finite number of learning steps. In this sense, the convergence of the system to a solution is proved.

But the question arises of whether a particular solution can be *learnt* by a Perceptron in principle. In general, we have to determine the classes of problems which may be applications for Perceptrons. Some simple problems show that Perceptrons are not as universal as they were believed to be in their enthusiastic beginning. For instance, a Perceptron is not able to distinguish between even and odd numbers. A special case of that so-called parity problem is the following application to elementary logics.

A Perceptron is not able to learn the exclusive OR (abbreviated XOR). This unsolvable cognitive task is a severe limitation of Perceptrons for applications to AI. The reasons can easily be shown. Remember that the exclusive OR with x XOR y is only true if either x or y, but not both x and y are true. An OR-sentence with x OR y is only false if both x and y are false, otherwise it is true. The following table provides the values of the Boolean functions OR and XOR:

x	y	x XOR y	x OR y
1	1	0	1
1	0	1	1
0	1	1	1
0	0	0	0

Now, imagine a network with two input units x and y, and an output unit z which may take the states 1 (active) and 0 (inactive). To simulate XOR, the output should be zero for an even input (either both input units are active or neither is), but one for an odd input (one input unit is active and the other one not). In Fig. 5.19a,b the possible input configurations of OR and XOR are illustrated in a coordinate system with the inputs x and y as coordinates.

Each pair (x, y) of coordinates x and y has a corresponding value z which is marked by a white dot (0) or black dot (1). A linear threshold element Θ calculates the sum of the weighted inputs x and y with weights w_1 and w_2, or formally $\Theta = w_1 x + w_2 y$. A simple derivation provides the formula of a straight line which is drawn in Fig. 5.19a,b. The position of the straight line is determined by the weights w_1 and w_2. It separates the active and inactive state of the threshold element.

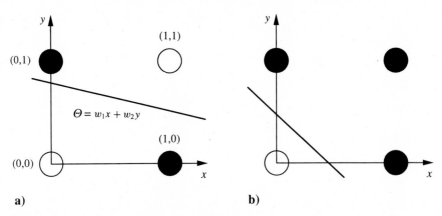

Fig. 5.19a,b. The linear separation of active and inactive states is not possible for the XOR-problem (**a**), but for the OR-problem (**b**)

In order to solve ("learn") the OR- or XOR-problem, the weights w_1 and w_2 must be adjusted in such a way that the points (x, y) with value $z = 1$ are separated from the points with value 0. This linear separation is geometrically possible for the OR-problem (Fig. 5.19b), but impossible for the XOR-problem. In general, the classification of input patterns by a Perceptron is restricted to linearly separable patterns.

This result can easily be generalized for networks with more than two input-units and real values. Many problems resemble XOR in the respect that they are not linearly separable. Actually, most interesting computational problems have that feature. The XOR-problem can be solved by adding a single hidden unit to a network with two inputs which are connected with the output. The hidden element is connected with both inputs and the output (Fig. 5.19c).

When both inputs are zero, then the hidden intermediate unit with a positive threshold is off. A zero signal reaches the output, and, as in this case the threshold is positive, the output is zero. If only one of the two inputs is 1, the hidden unit remains off and the output unit is switched on by the direct connection between input and output. Finally, when both inputs are 1, the hidden unit fires to 1 and inhibits the switching on of the output unit with the negative weight -2.

Thus, the hidden unit allows an appropriate internal representation. The XOR-problem has become a linearly separable problem using a two-dimensional threshold plane in a three-dimensional coordinate system with the three inputs of the output unit as coordinates. The separation is possible, because the input $(1,1)$ is now moved in the z-plane to the point $(1,1,1)$ (Fig. 5.19d).

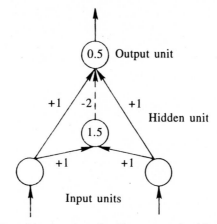

Fig. 5.19c. The XOR-problem can be solved by a network with a hidden unit

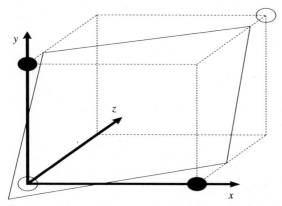

Fig. 5.19d. The XOR-problem is linearly separable by a 2-dimensional threshold plane [5.42], 92

A Perceptron had only a single intermediate layer with processor elements which can learn. Then, for a multi-layered network the problem occurs that for the layers of neurons without connection to the outer world the errors produced cannot be detected directly. An error can only be derived directly between the output layer and the intermediate layer below it.

The capability for representation and problem solving which a multi-layered network may possess depends decisively on the number of learning layers and the numbers of units in these layers. Consequently, a crucial question for neural computers must be the study of complexity of computation, because increasing the complexity of networks is the way out from the limitations of Perceptrons.

In Sect. 4.2, we already discussed the advantages of backpropagation in a multi-layered network (Fig. 4.17). A backpropagating learning algorithm allows us to define an error signal even for the neurons in hidden layers. The error of the output layer is recursively propagated back to the layer below it. The algorithm is able to construct networks with many hidden layers, the neurons of which can learn. As multi-layered networks can represent much more information in their hidden layers, than one-layered networks, networks with backpropagation are very powerful models for overcoming the weakness of the Perceptrons.

But backpropagation only delivers technically successful models, which have in general no analogy with biological evolution. Their adjustment of weights seems to be quite different from what is known about the behavior of biological synapses. Computer technology does not aim at simulations of the brain, but effective problem solutions during a reasonable time. On the other hand, we must give up the childish illusion that nature is a God-like engineer always finding the best solutions in evolution. As we have already underlined in former chapters, there is no centralized controlling and programming unit in nature. There are often only local solutions which are in general not the "optimal" ones.

In 1988, a technical application of a feedforward network, trained by the backpropagation of error method was designed by Gorman and Sejnowski to distinguish between sonar echoes of rocks and sonar echoes of mines. Even for an untrained human ear this task is rather difficult, but, of course, rather important for submarine engineers confronted with the problem of designing a sonar system that will distinguish between explosive mines and rocks. The architecture of the proposed network consists of an input layer with 60 units, a hidden layer with 1–24 units, and two output units, each one representing the prototypes "mine" or "rock" which are to be distinguished (Fig. 5.20) [5.40].

At first, a given sonar echo is prepared by a frequency analyzer and is divided into 60 different frequency bands. Each value ranges over an interval between 0 and 1. These 60 values are the components of an input vector which is given to the corresponding input units. They are transformed by the hidden units, leading to an activitation of one of the two output units, which may have values between 0 and 1. Thus, in a trained network with well-adjusted weights, a mine echo leads to the output signal $(1,0)$, while a rock echo has the output signal $(0,1)$.

In order to train the net, we must feed it with examples of mine and rock echoes. In each case, the factual values of the output units are measured and compared with the desired values, according to the corresponding input. The difference is the error signal, initiating small changes in the weights of the units. By this gradient-descent procedure, the weights of the network are slowly adjusted.

The mine–rock network of Gorman and Sejnowski is an application of complex system to AI. It is not claimed, of course, that the system simulates

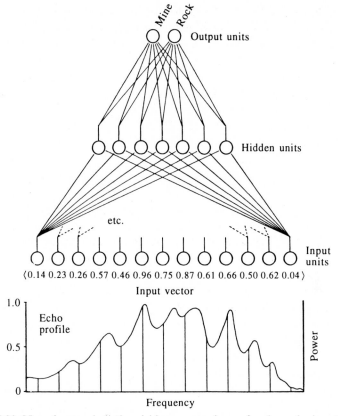

Input vector

Fig. 5.20. Neural network distinguishing sonar echoes of rocks and mines [5.40]

the human brain distinguishing between two concepts like "mine" or "rock". But we may say that this technical system also has an internal representation of the two concepts as prototype vectors in its hidden layer. In this restricted sense, the artificial system is "intelligent", because it can solve a task which is evaluated as intelligent in the case of the human brain. Artificial networks are not restricted to binary discriminations of concepts. In 1986, Sejnowski and Rosenberg designed a network called NETalk that has learned how to read. It takes strings of characters forming English texts and converts them into strings of phonemes that can serve as inputs to a speech synthesizer. The surprising fact is not the stammering sound of a childlike voice, which has been praised as a spectacular success in popular books. The essential capability of NETalk is the learning procedure for an internal representation of several concepts of pronunciation. For each letter of the alphabet, there is at least one phoneme assigned to it. For many letters there are several phonemes that might be signified, depending on the lexical text.

Sejnowski and Rosenberg used a three-layer feedforward network with an input layer, an intermediate hidden layer, and an output layer. Although backpropogation is rather unlikely to be "naturally" realized in biological brains, it has turned out to be the fastest learning procedure, compared with other solutions. The input layer looks at a seven-character window of text, for instance, the word "phone" of the phrase "The_phone_is_" in Fig. 5.21a. Each of the seven letters is analyzed successively by 29 artificial neurons, each representing a letter of the alphabet including signs for blank and punctuation mark. Thus, exactly one neuron of each neural subsystem with 29 elements is activated.

The output layer consists of 26 neurons, each of which represents a component of pronunciation. There are six components for the position of pronunciation, eight for articulation, three for pitch, four for punctuation, and five for accentuation and syllabization. So, each sound has four characteristics from these four groups of components. The 7 times $29 = 203$ neurons of the input layer are connected with 80 internal neurons of the hidden layer, which is interconnected with 26 neurons of the output layer (Fig. 5.21b). The neurons in the layers are not connected. The neurons of the input and output layer are also not directly connected.

An internal neuron of the hidden layer receives signals from 203 input neurons, but only send 26 signals to the output layer. As the internal neurons are threshold units with thresholds $T_1, \ldots, T_8$, the inputs are multiplied with particular weights, and the size of the sum of these products decides whether the neuron is activated or not (Fig. 5.21c). Realistically, the activation happens according to a continuous "sigmoid-curve" and not a digital jump [5.41].

At the beginning, the weights are fixed at random. Thus, NETalk begins with a senseless stammering. In the learning phase, NETalk uses a particular text of a child whose pronunciation is well known. The random sounds are compared with the desired ones, and the weights are corrected by backpropagation. It is remarkable that this procedure is a kind of self-organization, and not a rule-based program of pronunciation. There is only the global demand to change the weights by approximating the factual output to the desired one. After ten runs of the particular text of a child the network could already pronounce understandably. After 50 runs, there were only 5% mistakes. In this phase, an unknown text of the child was pronounced with only 22% mistakes.

Today, networks like NETalk must be simulated by traditional von-Neumann computers, because there is no direct hardware of complex networks. Thus, each neuron must be sequentially calculated. Even today, the principles of self-organizing complex networks are still mainly realized in the software not in the hardware. Nevertheless, we will speak about "neural computers", because the hardware realization is a question of technical de-

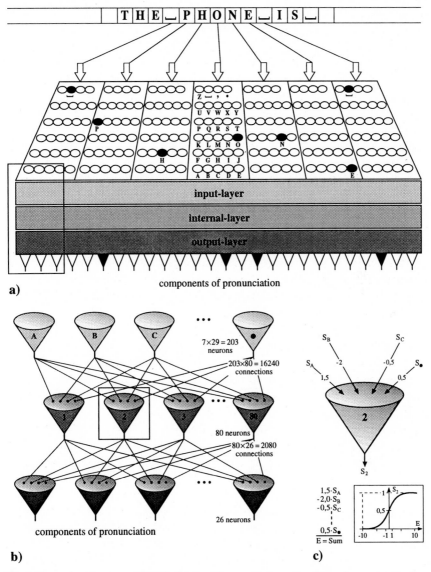

Fig. 5.21a–c. Architecture of NETtalk (**a**) with neural interaction (**b**) and neural activation (**c**) [5.41]

velopments in the future, depending on new technologies such as solid state materials or optical procedures, and not a question of theoretical limitations in principle.

Projections by neural networks seem to be rather successful and profitable in financial, insurance, and stock-exchange forecasts. The reason is that short-term forecasts of stock quotations are based on chaotic time series which becomes more and more chaotic if the time period of forecasts decreases.

Conventional statistical programs are only successful in long-term forecasts if the development of stocks can be smoothed without loss of relevant information. Then, good statistical programs have an accuracy between 60% and 75%. But short-term forecasts are rather limited. Conventional statistical procedures for smoothing the development of stocks must neglect the essential properties of short-term forecasts, which are frequent but small exchange fluctuations. In a conventional statistical program, the relevant calculation factors must be given explicitly. A well-trained and appropriately designed neural network is able to recognize the relevant factors without explicit programming. It can weight the input data and minimize the forecasting errors in a self-organizing procedure. Furthermore, it can adapt to changing conditions in the system's environment, unlike computer programs which must be changed explicitly by a programmer. In order to design a neural network for stock forecasting, the input data must be prepared by codifying the stock data into binary digits. The input vector consists of several partial vectors representing the exchange value, absolute change from the day before, direction of change, and direction of change from the day before yesterday and relevant change greater than 1% to the day before. If the input vector has a fixed length of, say, 40 units, then the lengths of the partial vectors can vary, depending on their desired relevance. The system may have two output units. Activation of the left one signals decrease of the stock values, while activation of the right one signals increase.

In the learning phase, the network was fed with the actual day's quoted rate during a fixed period from, say, February 9, 1989, to April 18, 1989. On the basis of this learnt data, the network forecasts the development of the stocks for the next 19 days. The forecasts were compared with the actual curves, in order to measure the system's accuracy. Several multi-layered structures were tested by backpropagation. They developed their particular global heuristics for forecasts by self-organization. For instance, if one forecasts nearly the same value for the day after as for the actual day, then the error was relatively small. The heuristics of this rule of thumb is the fact that a change of the quotation tendency is more unlikely than its remaining unchanged. Figure 5.22a,b show the curves of forecast (+) and actual (−) stock quotations for a bank (Commerzbank) and a firm (Mercedes) [5.42].

Obviously, feedforward networks with backpropagation are technically very interesting, although they seem to have less analogy with information processing in biological brains. In Sect. 4.2, we have already analyzed Hopfield systems with feedback (Fig. 4.8b) and Hebb-type learning (Fig. 4.9a), which seems to be practised by biological brains, too. In the case of a homogeneous network of boolean neurons, the two states of the neurons can

Fig. 5.22a,b. Neural network curves of forecast (+) and actual (−) stock quotations for Commerzbank (**a**) and Mercedes (**b**) [5.42]

be associated with the two possible values of electron spin in an external magnetic field. A Hopfield model is a dynamical system which, by analogy with annealing processes in metals, admits an energy function. As it is a non-increasing monotonic function, the system relaxes into a local energy minimum, corresponding to a locally stable stationary state (fixed point attractor).

Thus, the dynamical evolution of a Hopfield system may correspond to mental recognition. For example, an initial state representing a noisy picture of the letter "A" evolved towards a final state representing the correct picture, which was trained into the system by several examples (Fig. 4.9b). The physical explanation is given in terms of a phase transition in equilibrium thermodynamics. The correct pattern is connected to the fixed point or final state of equilibrium. A more flexible generalization is the Boltzmann machine with a stochastic network architecture of non-deterministic processor elements and with a distributed knowledge representation, mathematically corresponding to an energy function (Fig. 4.11b).

The general idea of relaxation is that a network converges to a more or less global state of equilibrium on the basis of local interactions. By iterative modification of the local connections (for instance, by a Hebb learning strategy in the case of a Hopfield system) the network as a whole eventually relaxes into a stable and optimal state. We may say that local interactions lead to a cooperative search which is not supervised, but self-organized. There are networks which use the strategy of cooperative search for mental activities like, for instance, seeking a probable hypothesis. Imagine that a certain range of competing hypotheses are represented by neural units which may activate or inhibit themselves. The system thus moves away from the less probable hypotheses toward more probable hypotheses.

In 1986, McClelland and Rumelhart used this cognitive interpretation to simulate the recognition of ambivalent figures, which is a well known problem in *gestalt*-psychology. Figure 5.23a shows a network for cooperative search simulating the recognition of one of the two possible orientations of a Necker cube. Each unit is a hypothesis concerning a vertex of the Necker cube. Abbreviations are B (back), F (front), L (left), R (right), U (upper), L (lower). The network of hypotheses consists of two interconnected subnetworks, one corresponding to each of the two possible interpretations.

Incompatible hypotheses are negatively connected, and consilient hypotheses are positively connected. Weights are assigned such that two negative inputs balance three positive inputs. Each unit has three neighbors connected positively and two competitors connected negatively. Each unit receives one positive input from the stimulus. The subnet of hypotheses to find is the one which best fits the input. Tiny initial fluctuations (which means a small detail in the special view of an observer) may decide which orientation is seen in the long run.

To visualize the dynamics of the network, suppose that all units are off. Then one unit receives an input of positive value at random. The network will evolve toward a state where all units of one subnetwork are activated and all units of the other network are turned off. In the cognitive interpretation we may say that the system has relaxed into one of the two interpretations of the ambivalent figure of either a right-facing or a left-facing Necker cube.

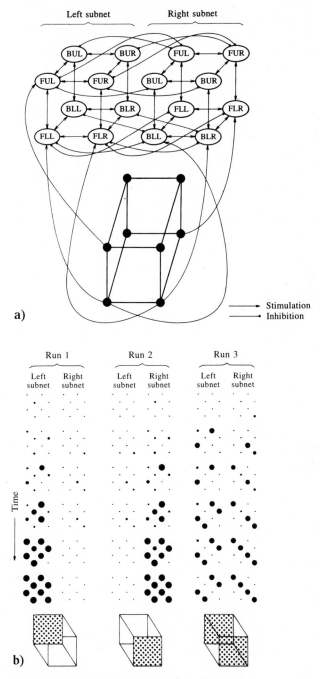

Fig. 5.23a,b. Neural network simulating the recognition of one of the two possible orientations of a Necker cube (**a**) with three evolution patterns (**b**) [5.43]

Figure 5.23b shows three different evolution patterns, depending sensitively on different initial conditions. The size of the circles indicates the activation degree of each unit. In the third run, an undecided final state is reached which is nevertheless in equilibrium [5.43]. Obviously, the architectural principles of this network are cooperative computation, distributed representation, and relaxation procedure, which are well known in the dynamics of complex systems.

In the past, many designs of artificial neural networks have been suggested. They were inspired by different disciplines such as physics, chemistry, biology, psychology, and sometimes just technical purposes. What are the common principles of the complex system approach? In former chapters, synergetics was introduced as an interdisciplinary methodology for dealing with nonlinear complex systems. Synergetics seems to be a successful top-down strategy for deriving particular models of complex systems from common principles which has been confirmed in many scientific disciplines. The main idea is that the emergence of global states in complex systems can be explained by the evolution of (macroscopic) interactions of the system's elements in learning strategies far from thermal equilibrium. Global states of order are interpreted as attractors (fixed points, periodic, quasi-periodic, or chaotic) of phase transitions.

Pattern recognition, for instance, is interpreted as a kind of phase transition by analogy with the evolution equations which are used for pattern emergence in physics, chemistry, and biology. We get an interdisciplinary research program that should allow us to explain neurocomputational self-organization as a natural consequence of physical, chemical, and neurobiological evolution by common principles. As in the case of pattern formation, a specific pattern of recognition (for instance a prototype face) is described by order parameters to which a specific set of features belongs.

Once some of the features which belong to the order parameter are given (for instance, a part of a face), the order parameter will complement these with other features so that the whole system acts as an associative memory (for instance, the reconstruction of a stored prototype face from an initially given part of that face). According to Haken's slaving principle the features of a recognized pattern correspond to the enslaved subsystems during pattern formation (Fig. 5.24) [5.44].

If a small part of the face which was learnt as prototype is given to a synergetic computer, then it can complete the whole face with the codified family name (Fig. 5.24b). The sequence of more or less fuzzy pictures corresponds to the phase transition of states in the synergetic computer.

When an incomplete pattern is offered to the neurons, a competition between different neuronal states, each corresponding to a specific prototype pattern, starts. This competition is won by that total state of the neuronal system that corresponds to the prototype pattern to which the offered test

Order parameter   Order parameter

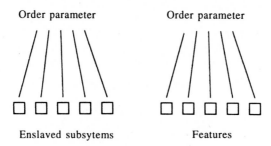

Enslaved subsytems   Features

Fig. 5.24a. Haken's slaving principle for pattern formation and pattern recognition

Fig. 5.24b. Recognition of a face by a synergetic computer with the slaving principle [5.44]

pattern has the greatest resemblance. In complete analogy to the dynamics valid for pattern formation, when a test pattern is offered to the synergetic computer it will pull the test pattern from an initial state (at $t = 0$) into a specific final state corresponding to one of the prototype patterns.

The evolution of the test pattern can be identified with the overdamped motion of a particle with a certain position vector in a potential landscape. Figure 5.24c shows a two-dimensional example of such a potential. The two prototypes correspond to two valleys. If a pattern is offered the features of which cannot be identified exactly with the features of the prototypes, then the particle has a position in the potential landscape outside the valleys. Obviously, recognition is a kind of symmetry breaking, which was already illustrated by Fig. 4.20a in the one-dimensional case.

In synergetic systems, the shape of a potential landscape can be changed by tuning control parameters. As synergetic systems are open, control parameters may represent the input of energy, matter, information, or other stimuli from the system's environment. When the control parameter is below a critical value, the landscape may have a single stable position like the single valley with the dotted line in Fig. 4.20a. After each excitation by means of fluctuations, the order parameter relaxes towards its resting state. When the control parameter exceeds a critical value, the formerly stable state has become unstable and is replaced by two stable states in the two valleys of Fig. 4.20a.

Fig. 5.24c. The evolution of a test pattern (order parameter) in a synergetic computer illustrated by the overdamped motion of a particle in a potential landscape [5.44]

The learning procedure of a synergetic computer corresponds to the construction of a potential landscape. The potential intensities which are visualized by the shape of the landscape indicate the synaptic forces of neural connections. It is an advantage of the synergetic approach that the huge number of microscopic details characterizing a pattern (for instance, a face) are determined by a single macroscopic order parameter. Thus, synergetic computers use the typical reduction of complexity which has been already applied for synergetic models of natural evolution (compare Sect. 3.3).

The order parameter equations allow a new kind of (non-Hebbian) learning, namely a strategy to minimize the number of synapses. In contrast to neurocomputers of the spin-glass type (for instance, Hopfield systems), the neurons are not threshold elements but rather perform simple multiplications and additions. But the fundamental difference between neurocomputers of the spin-glass type and synergetic computers is the following: complex systems of the spin-glass type are physically closed systems. Thus, their pattern formation is driven by conservative self-organization without any input of energy, matter, or information from outside. Typical patterns formed by conservative self-organization are the "dead" ice-flowers on windows in winter, which are frozen in equilibrium at low energy and temperature. The phase transitions of conservative self-organization can be explained completely by Boltzmann's principles of equilibrium thermodynamics.

In Sect. 3.3, we have already explained that pattern formation of living systems is only possible by input of energy, matter or information far from thermal equilibrium. This kind of self-organization has been called "dissipative" (Prigogine) or "synergetic" (Haken) but, nevertheless, can even be found in physical and chemical evolution. Consequently, pattern formation in the human brain as a living system depending sensitively on influences from the outer world will provide the "blue-prints" or models for a new computer technology in the framework of synergetics. Neurocomputers of the spin-glass type may be practical and successful for particular technical purposes. But, as they are physically closed systems, they differ in principle from what is going on in living systems like the human brain.

The pattern recognition process of synergetic computers has been made simultaneously invariant against translation, rotation, and scaling. These features of recognition correspond to realistic situations. Faces, for instance, are not always given as in the learning phase, but they may be translated, rotated, smaller or larger, nearer or farther. A nice application of synergetic computers is the recognition of oscillation (for instance ambivalent pictures) and hysteresis in perception. Figure 5.25a shows a well known example of hysteresis. When one starts looking at the patterns from left to right, the transition from a man's face to a girl will appear only after about six patterns. When one starts in the opposite direction, the switching from the perception of a girl to the face of a man occurs only toward the left end.

Figure 5.25b shows the perception process of a synergetic computer via the time evolution of characteristic order parameters. The broken line means the interpretation "girl", the solid one means the perception "face". The first diagram shows the transition from the perception of a man's face to that of a girl, the second one shows the transition from the perception of a girl to a man's face [5.45].

It may be objected that so far synergetic computers must still be simulated by traditional sequential computers. The principles of synergetic computers have been realized only in the field of computer software, not as machine hardware. But synergetics with its interdisciplinary applications gives rise to material and technical realizations. As the laser is a well-understood model of synergetics (compare Sect. 2.4), it may play an essential role for the construction of an optical computer by synergetic principles. In a laser, different modes occur, depending on critical values of the laser threshold. They can be characterized by their photon numbers. On the microscopic level, the rate of change for the photon numbers is described in a nonlinear evolution equation, depending on the gain, loss, and saturation of modes. On the macroscopic level, the order parameters correspond to field amplitudes which characterize several light wavetracks (Fig. 2.27a,b).

This suggests a three-layer architecture with an input layer of data which can be mapped by holographic mappings onto the laser. The laser with its order parameters is the intermediate layer. With its modes it serves as a decision making device by self-organization. The mode which survives in the sense of the slaving principle triggers new sets of features. This level is conceived as the output layer. The laser architecture of synergetic computers must, of course, be confirmed and improved by experiments. A synergetic computer would be a real dissipative system far from the thermal equilibrium.

a)

b)

Fig. 5.25a,b. Recognition of hysteresis (**a**) by a synergetic computer via the time evolution of characteristic order parameters (**b**) [5.45]

Obviously, complex dynamical systems are useful for simulation of cognitive behavior and technical systems, too. The human brain can be modeled as a nonlinear complex system, the dynamics of which may be governed by fixed-point attractors, periodic or quasi-periodic attractors, or even chaotic attractors. Chaos, for instance, has been confirmed experimentally as an efficient brain reset mechanism. After a study of the rabbit olfactory bulb, the recognition of various odors was modeled by the relaxation of a neural network towards cyclic asymptotic states. The chaotic state appeared during exhalation, eliminating the memory of the preceding odor. During inhalation, the presence of a particular odor as input drove the system to reach the limit cycle corresponding to that odor.

Technical applications of chaotic states are rather interesting, because chaotic systems are able to generate information. It is well known that chaotic systems depend sensitively on their initial conditions. Thus, during the course of dynamical evolution, two trajectories may separate dramatically at a certain time, even if their initial states only differ with a very tiny value. As any observation can only be made with finite precision, there may be two different states, the distance between which is less than our resolving power. In the initial state, an observer may see them as equal. But over the course of time a chaotic system allows discrimination between states which initially appeared to be coincident.

Actually, neural computers have been explored for technical applications in several industrial fields. Examples are robotics, aviation and astronautics (sensitive and adaptive systems, air navigation, etc.), medicine (evaluation and control of medical data, therapy, diagnosis, etc.), production industry (control of quality, optimization of products, etc.), safety technology, defense, communication technology, banks, post, etc. The complex system approach in technology should not be considered as competition or even opposition for classical AI. In the present state of technical development, neural networks and classical AI systems like expert systems seem to be useful and specialized in different fields of application. Complex systems appear to be more suitable than classical AI systems for analysis and recognition of signals, images and speech, speech synthesis, sensorimotor coordination in robots, etc. Obviously, these examples of neural nets are not single computers or robots, but more or less complex functions which are integrated in composite systems with several tasks. From an anthropomorphic point of view, the problems which are actually managed by neural nets may be classified as "low-level" ones.

In this chapter, inferential models based on AI-like expert systems have failed because they are less tolerant and flexible with their exactly and sequentially programmed behavior. In contrast to expert systems and knowledge engineering, complex processes of self-organization cannot be controlled by an explicit formulation of expert knowledge. On the other hand, ruled based systems with inferential algorithms are successful in all logically structured problems. Compared with, for instance, sensorimotor coordination, logical programming seems to be an example of "high-level" knowledge. Nevertheless, the low-level problems of nonlinear dynamical systems may have extremely high complexity. Nonlinear complex systems are not restricted, of course, to low-level knowledge, as we have seen in former chapters. The principles of complex systems seem to be rather appropriate to model high-level functions of the human brain like concepts, thoughts, self-referential states, etc. But the technology of neural nets is still in an infant state.

In present and future technology, heterogeneous systems with several modular rule-based and complex dynamical systems are interesting for specialized tasks. A speech comprehension system can consist of a neural network performing speech recognition and a rule-based symbolic module for syntactic and semantic analyses. Hybrid systems integrate inferential and dynamical techniques which may be useful for several medical purposes. For example, imagine a system which is able to recognize and control medical parameters by neural nets, combined with a rule-based deductive system for the diagnosis of particular illness from the recognized data. Like nature, an engineer should not be restricted dogmatically to one "optimal" strategy, but find purposeful solutions, eventually several solutions which are integrated but which need not be the best ones.

5.4 Neurobionics and Cyberspace

What are the goals of all our technical efforts to achieve neural and synergetic computers? The complex system approach will enable us to create a new kind of computer-assisted methodology in science, technology, industrial, economic, and even cultural life. But we must not forget that we have to decide the direction and the ethical goals the technical development will aim at. Today, these goals vary between epistemic and scientific interests, technical, economic, cultural, and last but not least military applications. Without any doubt, medical research and application must get a high rank in the hierarchy of research goals. I remind the reader of the old idea that medicine aims not only at scientific recognition and research, but at praxis, too. Praxis means not only technical application of knowledge in the sense of an engineer, but curing, helping and healing. Knowledge and research are only instruments for this fundamental purpose of medicine, which since the days of Hippocrates has been the protection of life [5.46].

The central organ of human personality is the brain. Thus, the medical task of maintaining the health of the brain puts a great responsibility upon neurosurgeons. They must conceive their medical treatments as manipulations of the whole mind–brain entity. In order to provide the most careful medical treatment possible for the human mind–brain entity, pioneering efforts are to be made in expansion and refinement of the diagnostic and therapeutic potential of neurosurgery, operation planning, operative technique, and rehabilitation by applied research. As far as we know, the human mind–brain entity is the most complex system of evolution. So, an interdisciplinary research program of computational neuroscience, physics, engineering, molecular biology, medicine, and epistemology is necessary to deal with this complex organ. This is the reason why some scientists have started an interdisciplinary research program on the brain and mind with ethical and anthropological dimensions, which we have called "neurobionics".

In general, "bionics" means simulation of natural functions and processes by technical and artificial procedures and systems. The design of aeroplanes and submarines to emulate the aerodynamic bodies of birds and fishes are well known examples. Historically, bionics is an old dream of mankind to simulate the principles of nature by technical means to manage the complex problems of life. In this tradition, neurobionics means the elaboration of a common techno-biological realization of a brain-like computing system on silicon and/or organic materials for enhancing morphological and functional properties of natural neurons and developing neural prostheses. That is not the horrible vision of Frankenstein's monster. In order to become motivated and engaged in these goals of research, one need only look at the desolate states of patients suffering from tumors or accidental injuries of the brain.

The specific medical discipline which is concerned with the central nervous system and the human brain is neurosurgery [5.47]. As the brain is the biological medium of human personality and intellectual capabilities, the neurosurgeon must not only bring to bear the principles of neurology with regard to the brain but also knowledge about the human mind and its function. Already, neurosurgery has been able to achieve progress in the treatment of patients. Notable successes in this regard have been achieved by the introduction of diagnostic visualizing procedures like computerized and nuclear magnetic resonance tomography and by the use of microsurgical procedures during operations.

However, essential problems in the treatment of patients with brain diseases are still unsolved. For example, the central nervous system of the adult human can only replace destroyed areas of tissue to a very limited extent in terms of function. This is due to the inability of nerve cells to subdivide further after completion of the embryonic phase, in contrast to other cells of the body. Only embryonic tissue has the potential to adapt itself to the surrounding host tissue. So the destruction of neural cell groups by illness or accident often leads to permanent functional disabilities. In this area of application, artificial complex networks with their principles of self-organization will become extremely interesting.

In the history of medicine, attempts have been made to restore injured peripheral nerves by autografts. This method is based on the fact that even adults have the capacity to regenerate nerve cell extensions growing centrally from the spinal cord into the periphery to the target organs. Thus, parts of functionally unimportant sensitive nerves are removed from appropriate points in the body and inserted in the area of the interrupted nerves which are to be restored. The regrowth of the interrupted nerve fibres is still not completely understood, however. Therefore, it is not possible to control the growth of a transplant consisting of several hundred individual nerve cell extensions which should reach the target organ. As the central nervous system is unable to regenerate, transplantation is ineffective in the case of very central injuries close to the spinal cord.

An improvement of peripheral nerve transplants is assumed in the field of molecular biology. An understanding of the physiology and biochemistry of the nerve cells and their associated cells, like glial cells and Schwann cells, could lead to new methods of nerve transplantation. An advanced method of replacement of tissue in the central nervous system is the transplantation of the body's own cells which have been genetically altered and adapted before their transplantation. The effect of neural growth factors, the relationship between the source of the transplant and the target area in the recipient brain and many other problems of molecular biology must be investigated. The resulting methods are based on knowledge derived from genetic engineering.

Another possible method of peripheral nerve transplantation would be the use of artificial instead of biological transplants. Attempts to restore failed parts of the nervous system by artificial replacement have already been made in the field of medical and neurological disciplines.

Artificial transplants are equipped with learning algorithms as "blueprints" of nature. Contrary to MCP (McCulloch & Pitts) networks, they are BPN (biological pulse-processing) networks working in real time. Figure 5.26 illustrates the general scheme of such a neuro-technological implant: A learning neural network encodes control signals of sensation and motion into many parallel pulse sequences which are received by a set of implanted microcontacts for stimulating the intact nerves (Fig. 5.26a). Signals registered by nerves are decoded by a neural net for controlling a movement prosthesis (Fig. 5.26b).

In patients suffering from a damaged spinal cord, attempts have been made to enhance the standing and walking functions with the aid of electrostimulation by BPN systems. Assuming that the peripheral apparatus is intact, electrical stimulation of the peripheral nerves brings about a contraction of the muscles. This is caused by the parallel pulses of an adaptive learning network encoding the auditory commands of a sensory system (Fig. 5.26a) which are given by the patient. The system has learning ability, because it can adapt to the particular conditions of a patient by sensory feedback to the moving legs. But the system still depends on the consciousness and speech of the patient. In the next step of research the unconscious intentions of the brain must be decoded from the registered signals of the spinal cord. Then these signals can be sent, for example, by radio waves, to a receiver with an adaptive neural encoder which again brings about a contraction of the muscles as in Fig. 5.26a.

An ambitious project of neuro-technology concerns a certain type of blindness. Patients with retinitis pigmentosa suffer from damage to certain layers in their retina which are responsible for the perception of contours, surfaces, colors, and other visual features. The damaged layers of the retina are bridged by a neuroprosthesis. In the architecture of the so-called retina implant, the visual scene is registered by photo receptors (e.g., transistors) in a spectacle frame which is equipped with an adaptive neural network. The optical signals of the external world are processed by the neural network (BPN) which can learn to model the receptive fields like a human eye. Their

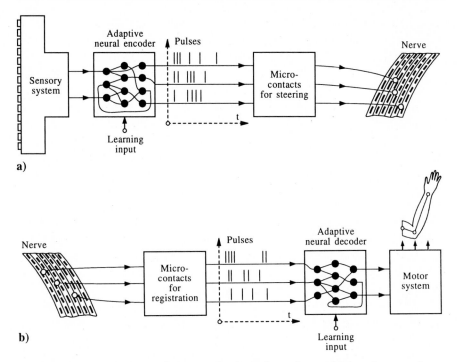

Fig. 5.26a,b. General scheme of neuro-technological prostheses with learning neural nets [5.48]

signals are coded and transmitted telemetrically to an inductive receiver with an electrode array at the damaged retina, in order to stimulate the optic nerve and the central nervous system (CNS). In a more advanced stage of research the spectacle arrangement will be given up and the receptor unit with the adaptive neural net may be located directly in the eye. In the first trials, neuro-technology cannot realistically hope to replace the full visual faculty; nonetheless, a restored perception of contours and surfaces would facilitate the orientation of the patients, and this is the present aim of the endeavor.

Decisive progress can be achieved if the stimulation of different muscle groups can be carried out directly on the branching of the ends of the peripheral nerves without the use of inorganic metal electrodes. Biotechnical connectivity has to be realized by molecular devices designed to produce electronic components from organic molecules. The processors controlling the electrodes and processing the information must be based on the principles of artificial neural networks, which are able to achieve a data-processing speed sufficient for the requirements of the human walking and standing apparatus. Obviously, the development of those complex networks requires the interdisciplinary cooperation of molecular biology, computational neuroscience, and high-tech hardware engineering.

A further example of the artificial replacement of failed neuronal functions is the replacement of the inner ear by a cochlear implant. If the auditory nerve is intact, a 25-pole electrode can be implanted by microsurgery as a replacement for the organ of Corti. The auditory nerve is now excited by suitable electrode impulses, to simulate acoustic patterns. The impulses are controlled by sequential microprocessors programmed on the basis of linguistic knowledge. But during difficult operations to remove tumors of the auditory nerve, the danger arises that the auditory nerve can be damaged, with subsequent deafness of the patient. Today, it is also possible to connect an artificial network directly into the area of the central auditory pathway. Then the sense of hearing can be restored despite the loss of the auditory nerve. Interdisciplinary cooperation of biotechnology, computational neuroscience, and engineering would be necessary, again.

In general, neurosurgery has to take care of the following clinical points of view: neurosurgical diagnostics, operation planning, operative technique, and neurorehabilitation, which may be supported by the complex system approach in biotechnology and computational neuroscience. Concerning diagnostics, the visualisation processes of computerized tomography have opened up a new era. As neurosurgeons have to deal with one of the most complex organs in evolution, operation planning and simulation have become an essential step in preparing a successful therapy. In this context, complexity also means the personality traits of a patient, his or her particular case history, the pathology of a certain pathogenic process, individual anatomical particularities, and possible postoperative consequences of an operation.

A new method has already been put into practice. A neurosurgical operation can be simulated by CAD(= computer-aided design)-supported techniques. There are computer-generated, 3-dimensional reconstructions of the pathological anatomy which are governed by particular programs. Potential difficulties which may be discovered during the simulation can be avoided during the actual operation. Concerning operative techniques, a progressive development would be a reduction in the practice of large open operations. Stereotactical and endoscopic techniques are important methods for minimizing surgery-induced trauma. A further development of laser technology in combination with neurosurgical endoscopy, intraoperative visualization processes, and computer-controlled regulation techniques could become a fertile complex operative instrument in the future.

A research group at the Massachusetts General Hospital in Boston revealed with magnetic resonance image techniques (MRI) functional image maps of human task activation in the visual cortex created by photic stimulation. Contrast agents were injected periodically over time. Even images of the primary visual cortex were made without injection by rapid MRI scanning. Figure 5.27 shows real time visualizations of brain cognitive activity as complex networks [5.49]. These advanced computer-based visualizations of complex neural nets not only help injured patients, but will ultimately allow us to see ourselves think and feel.

An essential impulse to develop artificial neural networks stems from the fact that the production of highly integrated electronic circuits based on the chemical element silicon will reach its physical limits. Then no further miniaturization is possible with the help of this technology, which is based on the principles of program-controlled microprocessors. High parallelism and strategies of self-organization are necessary to deal with the brain's complexity. Thus, the use of a new substrate appears to be required as the foundation for information-processing systems. First steps are being taken to develop

Fig. 5.27. Real time visualizations of brain cognitive activity of complex networks [5.49]

molecular electronic devices on the basis of biological components. Electrical signals between nerve cells could be conducted via organic conductors.

Concerning computational neuroscience, computer simulations of neural networks can help to identify the algorithms which are actually used by the central nervous system and the brain. Present models of artificial neural nets are mainly investigated by simulation on vector computers, workstations, special coprocessors, or transputer arrays. But, of course, the advantages of spatio-temporal parallelism in complex networks is entirely or partly lost by simulations on classical computers. Execution of tasks in real time only becomes possible with specially designed neural hardware.

In future neurobionic applications, the training of neural chips which may serve as interfaces between human nerve fibres will give rise to nonlinear dynamics of great complexity. The chip designer is faced with interconnection problems: if tens of thousands of weight wires are to be connected physically to a neuron and some thousand neurons are to be implemented, then the wiring area will grow to such orders that the delay on the wires will tend to exceed the latency time of the functional block representing the neuron. As the reduction of technological structure size is economically and physically limited, bionics designers nowadays prefer an architectural solution of the interconnection problem. First, they should check the real processing time of the neural net and then consider to what extent it is possible to deviate from its ideal massive parallelism.

Obviously, the underlying parallel hardware significantly increases the complexity of the software and calls for new methods. Powerful operating systems, programming tools, and flexible user interfaces must be designed in order to ease interface communication with the system. This task will become especially important in an interdisciplinary team with differing degrees of knowledge in computer science. Knowledge based expert systems may help team members to work with the bionic software and to integrate them in the team. Programming neural network hardware will differ decisively from classical programming of von-Neumann computers. A programmer must identify the necessary network topology and architecture, and has to specify the behavior of neurons with interconnection schemes. Thus, the use of heterogeneous and hybrid systems integrating neural nets and classical knowledge based systems (described in Sect. 5.2) seems to be a realistic perspective in neurobionics.

Some people may be afraid that hybrid computer systems with increasing complexity cannot be managed without a highly specialized training. New interfaces between computer systems and users must be developed. Computer-generated images should be manipulated by direct speech and visual and tactile contacts in a "virtual reality". The user will have the impression of a computer-generated reality via several technical instruments which are connected with his senses.

Visual impressions are generated by a monitor with position sensors which can be worn as an eye-mask. A microphone is connected with a speech recognition system to translate human orders to the system. Tactile contacts are generated and modeled by so-called "data gloves" that transform the movements of hand and fingers into electrical signals (Fig. 5.28) [5.50].

In a data glove, optical fibres are led between two layers of cloth. They are connected with a special module tranforming light signals into electrical signals. Practical applications are assumed, for instance, in astronautics. NASA is interested in the development of robots which can execute complicated and dangerous actions in space by simulating the movements of an astronaut's hand in the space station. It seems to be possible to use the principles of data gloves even for a data suit simulating the movements and reactions of the whole body.

Fig. 5.28. Data glove [5.50]

The senses have an eminent influence on human imagination. Thus, molecular modeling in chemistry can be realized not only by computer-assisted graphics visualizing the molecular structure, but by the introduction of a tactile element, too. By means of data gloves, chemists may have the imagination to grasp a molecule, to feel its surface and to manipulate it in a desired manner. Engineers try to generate these effects of touches and forces by particular technical systems. The human hand manipulating in a virtual reality via data gloves must receive a feedback from the touched imaginary objects. The complexity of the experienced world should be simulated in all its aspects.

In the examples of astronautics and chemistry, the virtual reality of models corresponds to an actual reality in the macro- or microcosmos. But imagine computer-generated scenarios of fantastic worlds which only exist as electronic realities. The boundary between technical possibilities and science fiction seems to be fuzzy. People may feel like imaginary bodies in a computer-generated "telereality". There are suggestions to build so-called "home reality engines" which transfer the user into a virtual world of desired and undesired fantasies. Sex with Marilyn Monroe or discussions with Albert Einstein, as you like it, have been promised by the prophets of computer-generated virtual realities. Authors of science fiction like William Gibson describe the world of a computer-generated "cyberspace" which will be experienced by people as a gigantic hallucination:

> Cyberspace. A consensual hallucination experienced daily by billions of legitimate operators, in every nation, by children being taught mathematical concepts . . . . A graphic representation of data abstracted from the banks of every computer in the human system. *Unthinkable complexity*. Lines of light ranged in the nonspace of the mind, clusters and constellations of data. Like city lights, receding . . . [5.51]

These perspectives, of course, give rise to an essential critique of our cultural development. People locked up in boxes with their own private or manipulated virtual reality generated by super-Crays and neural networks seem to be a horror vision like Orwell's Big Brother.

Besides those ethical problems, there are some severe epistemic questions which arise with the possibility of computer-generated complex artificial worlds. In traditional epistemology, philosophers like Berkeley and Hume discussed the solipsistic or sceptical position that the reality of the outer world cannot be proved by any means. All our impressions may be illusions generated by our brain and its mental states. These puzzling problems were not meant as jokes by childish philosophers who were completely ignorant of world. They should inspire us to test and analyze the validity of our arguments. Modern logicians and philosophers of mind like Hilary Putnam have translated the problem in the following way, which reminds us of the famous Turing test.

Imagine that a human has been subjected to an operation by an "evil scientist". The person's brain has been removed from the body and placed in a vat of nutrients which keeps the brain alive. The nerve endings have been connected to a hybrid neural computer which causes the person whose brain it is to have the illusion that everything is perfectly normal. All the person experiences is the result of electronic impulses coming from the computer to the nerve endings. If the person intends to raise his hand, the feedback from the computer will cause him to "see" and "feel" the hand being raised, although there are only the corresponding patterns in the brain and not a physical eye or ear. The evil scientist can cause the poor person to experience any situation. Putnam said:

> It can even seem to the victim that he is sitting and reading these very words about the amusing but quite absurd supposition that there is an evil scientist who removes people's brains from their bodies and places them in a vat of nutrients which keep the brains alive. The nerve endings are supposed to be connected to a super-scientific computer which causes the person whose brain it is to have the illusion that ... [5.52]

Could we, if we were brains in a vat in this way, say or think that we were? Putnam argues that we could not. The supposition that we are actually brains in a vat cannot possibly be true, because it is self-refuting. A self-refuting supposition is one whose truth implies its own falsity. A logical example is the general thesis that all general statements are false. If it is true, then because of its generality it must be false. An epistemic example is the thesis "I do not exist" which is self-refuting if thought by me. Thus, Descartes argued that one can be certain that oneself exists, if one thinks about it. The supposition that we are brains in a vat has this property.

Suppose that we are brains in a vat with nutrient fluid and afferent nerve endings connected with a super neural computer producing all sensory inputs of the brain. As the human brain in the vat is well functioning, it has, of course, consciousness and intelligence. But its ideas and images of trees, houses, etc., have no causal connection to actual trees, houses, etc., in the outer world of the brain in the vat, because these ideas and images are produced by our super neural computer. Thus, if we suppose that we are brains in a vat with all these conditions, then the words "vat", "nutrient fluid", etc., do not refer to an actual vat, nutrient fluid, etc., but to certain ideas and images which have been produced by our super neural computer. Consequently, the sentence "We are brains in a vat" is false (Fig. 5.29).

We must be aware that the possibility that we are brains in a vat is not ruled out by physics, but by logic and philosophy. A physically possible world in which we are brains in a vat is compatible with the laws of physics. But in

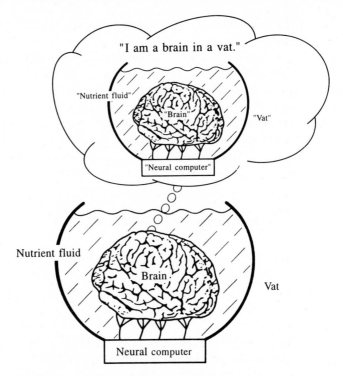

Fig. 5.29. Brain in a vat thinking that it is a brain in a vat

a *gedankenexperiment* we are even able to derive necessarily true conclusions beyond physically possible worlds.

The reason for these features seems to be the structure of self-referentiality, which is typical for the high-level capabilities of the mind–brain system. In Sects. 4.3 and 4.4, we have argued that self-referentiality may be an essential feature to make consciousness and self-consciousness possible – not only in the mind–brain system as product of biological evolution but even in artificial complex systems with a quite different hardware.

Turing himself proposed a well known test which should decide whether an artificial system like a computer is conscious nor not: let someone carry on a keyboard conversation with the computer and a similar conversation with a person whom he does not know. If he cannot tell which is the computer and which is the human being, then the computer is conscious. In short: a computing machine is conscious if it can pass the Turing test.

The thought experiment of the "brain in the vat" has shown that Turing's dialog test must fail in some particular sense. The words and sentences which are used by the artificial system do not necessarily refer to the actual objects and events we refer to in natural languages of humans. The use of words and sentences may be governed by syntactic patterns which are programmed into the Turing machine in a highly sophisticated way. Weizenbaum's program ELIZA which simulated the dialog of a patient with his psychologist may give a first hint of these possibilities. In this sense Turing's test cannot rule out that the conversation of the machine is only a syntactic play resembling intelligent human discourse. Nevertheless, in principle, it cannot be excluded that self-organizing complex systems are able to learn the particular references of their syntactic words and rules by prototype patterns and experiences with their environment. In the long run, it is an ethical question whether we want to develop those highly autonomous (dissipative) systems.

5.5 From Artificial Intelligence to Artificial Life

Life science and computer science are not only the highlights of future technologies, but they are growing together with common objects of research in "artificial life" and "artificial evolution". In a famous quotation of his Monadology (§64) Leibniz argued that every organism of a living being is a kind of "divine machine" or "natural automaton surpassing all artificial automata infinitely". There are historical roots from Leibniz to the modern sciences of computability and complexity which try to model intelligent behavior by computational and complex dynamical systems.

Besides "artificial intelligence" (AI) as a classical discipline of computer science "artificial life" (AL) is a new growing field of research in the framework of the sciences of complexity. Literally the term "artificial life" means life made by man rather than by nature. For Christopher G. Langton from the Santa Fe Institute who was the organizator of the first AL-conference in 1987 artificial life is still in the process of defining itself [5.53]. Natural life on earth is organized on the molecular, cellular, organism, and population-ecosystem level. Artificial life aims to find modeling tools powerful enough to capture the key concepts of living systems on these levels of increasing complexity.

A key concept of living systems is the distinction of genotype and phenotype. In biology the genotype is the complex set of genetic information (genes) encoded in the DNA of an organism. The phenotype is the physical organism itself. The development of the phenotype directed by the genes of the genotype is called morphogenesis. In Sect. 3.3 morphogenesis was characterized by order parameters of a complex dynamics. In our framework of complex systems the phenotype means a macrophenomenon with order parameters depending on the nonlinear interactions of genes on the microlevel of the system. The high nonlinearity of genetic dynamics provides a great variety of possible phenotypes. On the other hand, it prevents predictions or derivations of properties or future behavior of an individual phenotype.

In the artificial life approach the distinction of genotype and phenotype is not restricted to life processes based on carbon-chain chemistry. In the theory of computability the genotype can be generalized as a set of local computational devices ("genes") which recursively generates global structures of phenotypes. The locality of the rules is an essential assumption. There are no global rules in the system with global information of the entire structure. In the expert systems of artificial intelligence which we discussed in Sect. 5.2 there are global rules for global behavior of the entire system.

Examples of the artificial life approach are the so-called L-systems of Aristid Lindenmayer modeling filaments and branching botanical trees and other plants and organs [5.54]. L-systems consist of sets of rules for deriving strings of symbols in analogy to the formal grammars of Noam Chomsky. For example $x \rightarrow y$ means that every occurrence of symbol x in a formal structure should be replaced by string y. The symbol x may appear on the right as well as the left side of some rules. Thus, the set of rules can be applied recursively to already derived structures. We get finite structures ("phenotypes") step by step ad infinitum. If the context of symbolic replacement is not considered, the rules are called context-free. Context-free rules only replacing single symbols are equivalent to the operations of finite state machines (or the so-called regular languages of Chomsky grammars). This kind of L-system can produce branching structures such as the vascular network of the heart in Fig. 3.9.

When there is more than one symbol on the left-hand side of a rule, the rules constitute context-sensitive grammars which are equivalent to Chomsky's Turing languages. Obviously, in context-sensitive grammars the application of local rules depends on locally neighboring symbols corresponding to nonlinear interactions of the elements in complex systems. Without context sensitivity, the global structures ("phenotypes") produced by L-systems are linearly decomposable.

A formal advantage of L-systems is their complexity measures in the sense of computer science (cf. Sect. 5.2). A phenotype is interpreted as a formal structure which may or may not belong to the class of derivable structures in a certain L-system. Thus, undecidability of membership in a given class indicates a higher complexity of this class with respect to those in which this question is decidable. Furthermore, we can measure the computational time of derivations in L-systems. In this case, L-systems can be ordered in the hierarchy of complexity degrees from N- to NP-complete problems.

L-systems provide a rich variety of computer-generated shapes of modeled plants and organs. Sometimes the formal rules of L-systems can be realized by observable mechanisms of matter. In supramolecular chemistry, for example, tree-like polymers such as dendrimers (cf. Sect. 2.5) are generated by a divergent synthesis of repeating bifurcations which can be described by the recursive rules of an L-system. Obviously, L-systems offer fruitful prospects in computer graphics illustrating and visualizing molecular and cellular growth in chemistry and biology. But, of course, the actual mechanisms of molecular and cellular growth must be observed and measured in the laboratories of chemists and biologists – not by computer experiments.

Another example of the artificial life approach is the concept of cellular automata which we discussed in Sect. 5.3. Their simple sets of local rules can be interpreted as "genotypes" producing "phenotypes" of more or less complex patterns. Cellular automata allow remarkable simulations of key concepts in biology. We already mentioned John von Neumann's idea of a self-reproducing automaton. Although this idea is justified by a mathematically precise proof, it is hard to realize with technical computers. The reason is von Neumann's requirement that the self-reproducing structure must be a universal computer which has the complexity degree of a universal Turing machine. Obviously, self-reproducing molecules of prebiotic evolution (for example, compare Eigen's model in Sect. 3.3) were hardly capable of universal constructions. Therefore, Christopher Langton dropped the requirement of universality and designed a very simple cellular automaton that can reproduce itself. It consists of a set of neighboring cells which are finite state automata.

In Fig. 5.30, each number is the state of one of the automata in the lattice. Blank space means cells in state 0. The 2-states form a sheet around the 1-

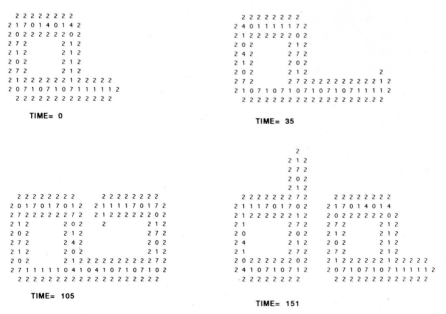

Fig. 5.30. Self-reproducing Langton loop with an extending tail and the formation of a daughter loop [5.55]

state data-path in a loop with a tail on one end, reminiscent of a Q-shaped virus. The inner data-path conducts data such as the state pairs 70 and 40 necessary for self-reproduction. With each generation of the whole cellular automaton, the cells in this inner layer follow rules that affect the states of the neighboring cells. They propagate signals counter-clockwise around the loop. When the signals reach the end of the tail, each 70 signal extends the tail by one unit and the two 40 signals construct a left-hand corner at the end of the tail. For each full cycle of the signals around the loop, another side and corner is constructed. After four cycles the tail runs into itself, and the two loops are disconnected. The cellular automaton of the loop has then reproduced itself.

Each loop goes on to produce further offspring which themselves also reproduce (Fig. 5.31a,b). This development continues indefinitely and produces an expanding colony of loops. As the loops in the core of the colony have no space to reproduce themselves, they lose their tails and "die" (Fig. 5.31c). Thus, the colony acquires the shape of a reproductive fringe surrounding a growing "dead" core (Fig. 5.31d). The pattern seems to be similar to the growth of a coral with dead skeletal structures and living cells on the surface.

In Sect. 5.3, we mentioned that cellular automata can be used as discrete and digital models of the phase portraits describing the global dynamical behavior of complex systems. The complexity classes of cellular automata (Fig. 5.18b,c) correspond to the complexity degrees of the attractors classifying dynamical systems. Langton introduced a certain parameter λ to

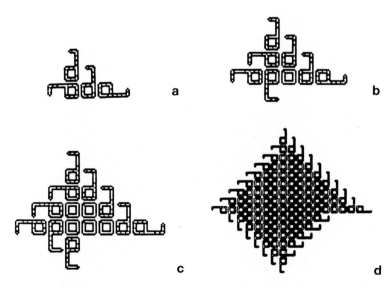

Fig. 5.31a–d. Expanding colony of an artificial coral [5.55]

characterize the dynamical behavior of cellular systems depending on their transition rules [5.56]. Figure 5.18a shows a one-dimensional cellular automaton with $8 = 2^3$ transition rules depending on 3 neighboring cells and 2 possible cellular states.

In general, a cellular automaton has K^N transition rules for N neighbors and K states. An arbitrary state of a cellular automaton is distinguished as quiescent state. In Fig. 5.18b, the quiescent state is the white cell ('zero'). If n_q is the number of transition rules to this special quiescent state, then there are $K^N - n_q$ remaining rules which should pick randomly and uniformly over the other $K - 1$ states. Now, the λ-parameter is defined by $\lambda = (K^N - n_q)/K^N$, ranging from zero to one: If $n_q = K^N$, then all transition rules of the automaton are to the quiescent state and $\lambda = 0$. If $n_q = 0$, then there are no transition rules to the quiescent state and $\lambda = 1$.

The dynamical meaning of λ is obvious. If the λ value is very small and approaching zero, then all dynamics is frozen after a short time such as the molecules in frozen solid. If the λ value is very large, approaching its maximum value, then the dynamics is very free and chaotic, and a cellular pattern is very difficult to retain as for the molecules in a gaseous state. Physically, life cannot be supported in these extreme scenarios. But there is a certain area between these extreme situations where the dynamics changed but not so rapidly that it lost all configurations of the previous state. This area corresponds to a liquid state which indeed supported the emergence of life in biological evolution.

Computer experiments show that the progression of the λ parameter characterizes the spectrum of dynamical behavior from a fixed point dynamics to a periodic dynamics, then to a complex dynamics and finally to a chaotic dynamics. These regimes correspond to the complexity classes of cellular automata which are illustrated in Fig. 5.18b (fixed point dynamics), Fig. 5.18c (periodic dynamics), Fig. 5.18e (complex dynamics), Fig. 5.18d (chaotic dynamics). Thus, the λ parameter corresponds to a control parameter of a dynamical system characterizing its phase transitions. At an intermediate critical value λ_C, we observe a phase transition between periodic and chaotic dynamics. There is an increasing complexity at λ_C in the sense of longer cellular patterns. Transition lengths of cellular patterns are shorter at either end of the range of λ and get longer in the middle of the range. At λ_C the transient lengths diverge. Beyond λ_C, the transient lengths get shorter with increasing λ, although the dynamical activity expands more rapidly with time until the onset of chaos. There is even evidence of a second-order phase transition at λ_C, because the average Shannon entropy of cellular growth has no discrete jump, but is smooth at λ_C.

From the viewpoint of computer modeling, the solid, liquid and vapor phases of matter are only special realizations of dynamical systems. They can also be realized by artificial systems such as cellular automata in the medium of a computer. In this framework, artificial life is possible in the neighborhood of the critical value λ_C. In Sect. 5.3, we suggest how cellular automata may be related to our concept of order parameters and synergetics. Order parameters correspond to macroscopic spatio-temporal properties of cellular patterns such as fixed point, periodic, complex, or chaotic attractors.

In the synergetic approach, stable and unstable collective motions ("modes") can be distinguished close to critical points of instability. This instability is caused by a change of a control parameter and leads to new macroscopic spatio-temporal patterns. In cellular automata, stable and unstable motions can be characterized by transition rules that may or may not change previous states. The stable modes are enslaved by the unstable modes and can be eliminated (Fig. 5.24a). The remaining unstable modes serve as order parameters determining the global macroscopic pattern of the system. In general, the elimination of unstable modes by enslaving means an enormous reduction of the degrees of freedom. The global dynamics of order parameters may lead to fixed points, periodic, oscillating, pulsation, or chaotic patterns. Remember the patterns of fluid dynamics (Fig. 2.27a–e) or the laser (Fig. 2.28a,b).

Concerning cellular automata, a fixed point dynamics is independent of fluctuating initial states such as order formation near to thermal equilibrium. It corresponds to a low value of the control parameter λ near to zero. In the neighborhood of the critical value λ_C the cellular dynamics depends sensitively on initial fluctuations. Thus, it simulates the complex dynamics of open dissipative systems such as a laser in physics, a Belousov-Zhabotinsky reaction in chemistry, or the cellular differentiation of an organism in biology. Thus, Langton's λ_C-regime is not restricted to "artificial life" as computer simulation of biological life, but to "artificial dissipative non-chaotic systems". In Sect. 3.3 we underlined that, from a biological point of view, it is not sufficient to know the general scheme of dissipative self-organization. We must know more about the tricks of "carbon-chain chemistry" to create biological life (especially its clever combination of conservative and dissipative self-organization). Then, we may also grasp the complex dynamics of "artificial life" in the medium of a computer. Otherwise the cellular automata approach is too general for modeling life processes.

Since Darwin, selection had been emphasized as a key mechanism of biological evolution. But, there are also selection processes in dissipative systems like the laser in physics or the Belousov-Zhabotinsky reaction in chemistry. The nonlinearity of their dynamics shows that there is no superposition of pattern waves, but a competition of stable and unstable modes. In the late 1960s, Ingo Rechenberg [5.57] already used evolutionary strategies as "blue prints" for optimizing technical systems. John Holland applied the process of selection to algorithmic learning procedures [5.58]. His genetic algorithms have become a key concept in artificial life research. In general, a genetic algorithm is a method for moving from one population of genotypes to a new "fitter" population. The genotypes may be represented by, for example, bit strings for the genes of an organism or by possible solutions of a problem. The basic concept of a genetic algorithm works as follows: (1) Start with a randomly generated population of genotypes. (2) Calculate the fitness of each genotype in the population. (3) Apply selection and genetic operators to the population to create a new population. (4) Go to step 2. Selection chooses those genotypes in the population that can reproduce and decides how many offspring each is likely to have: The fitter genotypes produce on average more offspring than less fit ones. Holland distinguishes the genetic operators of crossover, mutation, and inversion. Crossover exchanges subparts of two genotypes (Fig. 5.32). Mutation randomly changes the values of some locations in the genotype. Inversion reverses the order of a contiguous section of the genotype.

If the algorithmic procedure (1)–(4) is repeated over many time steps, it produces a series of populations (generations) until one or more highly fit genotype emerges. Genetic algorithms have been applied successfully to many scientific and engineering problems [5.59]. In Sect. 4.2, we analyzed neural networks with competitive learning procedures (Fig. 4.12) and learning classifier systems. Eigen's self-optimizing strategy of genetics (Sect. 3.3) can be understood as a kind of genetic algorithm, too.

Learning during our lifetime does not directly affect our genotype. But if learning helps survival, then organisms best able to learn have the most offspring. They increase the frequency of the genes responsible for learning.

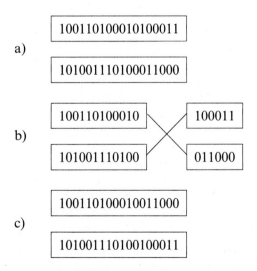

Fig. 5.32a–c. Crossover operator of a genetic algorithm with binary coding

This indirect effect of learning on evolution is called the Baldwin effect. Computer experiments can help to understand and to measure the success of this effect. The learning process is not necessarily supervised by a global function for evaluating the fitness of the population. But it is a macroeffect of individual learning processes by individual agents of the population. At each time step t in an agent's life, the agent evaluates its current state, using its evaluation network. This evaluation is compared with the evaluation it produced a time $t - 1$ with respect to the previous action. The comparison gives a reinforcement signal to modify the weights in the action network. Then the next action can be determined. Obviously, this kind of unsupervised learning can be used in computer models of social and economic systems.It can be embedded in the learning procedures of neural networks to improve their capacity for problem solving.

Extremely interesting applications of genetic algorithms are computer models and computer experiments in gene technology. Another example is the medical simulation of immune systems in which learning takes place [5.60]. In recent years, immunology has become a tremendously important field of medical research, especially with respect to therapies for cancer. The human immune system is capable of recognizing on the order of 10^{16} different foreign molecules. The genome encoding the construction rules of the immune system only contains about 10^5 genes. Thus, the immune system is an example of a complex dynamical system which is highly effective for pattern recognition with relatively few instruction rules. There is no central organ to control it. Thus, pattern recognition by an immune system is a macroeffect of local molecular interactions. It can be characterized by order parameters. They represent the extent of molecular binding between the cells performing the recognition ("antibodies") and the foreign material ("antigens"). Genetic algorithms may constitute artificial immune systems in the medium of a computer by representing antigens and antibodies with binary strings. The binary immune systems have been used to study the ability to detect typical patterns in an environment of randomly distributed antigens or to learn new patterns if the pathogenic material is rapidly evolving by mutations.

Organic forms of dangerous viruses are well-known. But artificial forms are not only interesting for simulating and modeling biological immune systems. During the past several years there has been considerable interest in computer viruses as security and social problems of world-wide computer-assisted networks [5.61]. In general, a computer virus is a code which can attach itself to other computer entities such as programs, data files, etc. and reproduce itself at least partially. Today, we can distinguish generations of more or less sophisticated and dangerous computer viruses. A shell virus, for example, forms a shell around the original code. The virus becomes the program, and the original host program becomes a subroutine of the viral code. The most dangerous computer viruses are the polymorphic or self-mutating viruses. They infect their targets with a modified or encoded version of themselves. Scanners are often used as computer virus defense or artificial immune system. They read data from a disk and apply pattern matching operations against a list of known virus patterns. But polymorphic viruses cannot be detected by scanners and need learning algorithms for their detection.

From the view point of artificial life, computer viruses do satisfy some of the criteria for life, but not all. Obviously, a computer virus is able to reproduce itself. But on the other hand, it is not the agent, but the computer is. Obviously, a computer virus has an information storage of a self-representation. Like DNA molecules, its code is a template that is used by the virus to replicate itself. There also seems to be a kind of metabolism that converts matter and energy. But the use of electrical energy is again not by the virus but by the underlying computer system. There are also ecological forms of computer viruses such as predatory, territorial, or displaying self-protective behavior. But up to now, these behaviors are not initiated by the viruses themselves, but by a human programmer. In the future virtual computer viruses may have all the dangerous capabilities in a virtual life in cyberspace that are well known from our biological life.

In biology artificial life systems are applied as software, hardware and wetware systems [5.62]. Wetware means the techniques of a biochemical laboratory which, for example, are used to model the molecular origins of prebiotic evolution in an evolutionary reactor. These are, of course, empirical experiments reproducing the biochemical evolution of life or creating new forms of biological life. They are only called "artificial", because the conditions for the self-organization of biological life are arranged by human experimentors. Software systems are computer programs or networks modeling key concepts of life and evolution. We already analyzed the cellular automata approach. There are also computer programs modeling the evolution of animals. In the program GENESYS, for example, animals are represented as neural nets or finite automata. The genes of each organism are represented as bit strings encoding the weights of a neural net or the transition lists of a finite automaton. In order to increase the modeling capacity of the program, it is executed on a massively parallel connection machine. The GENESYS program reminds us of Leibniz' vision of natural and artificial automata in his Monadology. Hardware models of artificial life are used in robotics. In Sect. 5.4 we analyzed the first applications in neurobionics. In Sect. 2.5 we mentioned the self-construction of smart and "intelligent" new materials.

Clearly, the artificial life approach is not a homogeneous field of research [5.63]. But it seems to provide many fruitful modeling instruments for chemical and biological research. Besides these practical applications, there is a visionary dream of AL with a hard scientific core. The factual biological evolution is only a model of a complex dynamics which is governed by highly nonlinear equations. Today, we only know some properties of these equations, and where we know them, we have no analytical tools to solve them exactly. Even numerical approximations would be restricted by their immense degree of computational complexity. Nevertheless, computer models may allow computer experiments to become familiar with possible scenarios under several restricted constraints. These experiences with computer experiments may be used to create particular conditions, under which new materials may construct themselves and new forms of life may organize themselves. The first steps have already been taken. Even the emergence of artificial consciousness cannot be excluded in principle [5.64]. If we know the complex neural dynamics behind conscious states (cf. Sects. 4.3, 4.4, 5.4), then the wetware and hardware of the human brain is only a particular model. It is not matter itself that makes possible new materials, life, and consciousness, but its particular organization and dynamical laws which can be modeled in more or less complex systems. Thus, in the long run, it becomes an ethical question whether we want an artificial evolution with an open end which we cannot forecast.

6. Complex Systems and the Evolution of Human Society

How can one explain the emergence of political, social, and economic order in human societies? The chapter starts with a short history of political and economical systems since classical antiquity. The historical ideas of political and economical order were often illustrated by technical, physical, or biological concepts of their corresponding period. In the 17th century, Thomas Hobbes tried to transfer Galilean and Cartesian laws of movement from mechanics to anthropology and state theory. The *physiocrates* modeled the economic system of an absolutist state like a mechanistic 18th-century clockwork (Sect. 6.1). The liberal ideas of Locke, Hume, and Smith were designed against the historical background from Newton's physics. Until recent years, mainstream economics has often been inspired by models from linear mathematics, classical mechanics, thermodynamics of thermal equilibrium and sometimes the Darwinian theory of evolution. Like many physicists, economists believed in the exact computability of their (linear) models and suppressed the possibility of a "butterfly effect" leading to chaos and excluding economic forcasts in the long run (Sect. 6.2).

To describe the dynamics of an economy, it is necessary to have evolution equations of many economic quantities from perhaps thousands of sectors and millions of agents. Since everything depends on everything else, the equation will be coupled and nonlinear, in order to model economic complexity. In particular, the economic behavior of modern high-tech industries and the effects of technological innovations seem to be better modeled by the nonlinear dynamics of complex systems (Sect. 6.3). The crucial point of the complex system approach is that from a macroscopic point of view the development of political, social, or cultural order is not only the sum of single intentions, but the collective result of nonlinear interactions. In Sect. 6.4, examples of complex social and cultural problems are analyzed in the framework of complex dynamical systems: the growth of urban centers, the worldwide problems of migration and the management problems of complex organizations. The chapter finishes with a look at complex communication networks, which offer the prospect of a worldwide "global village" as well as the enslavement of mankind with modern high-tech procedures.

6.1 From Aristotle's Polis to Hobbes' Leviathan

After the evolution of matter, life, mind–brain, and artificial intelligence we finish the book with the question of whether the evolution of human society can be at least partially described and modeled in the framework of complex systems. In the social sciences one usually strictly distinguishes between biological evolution and the history of human society. The reason is that the development of nations, markets, and cultures is assumed to be guided by the intentional behavior of humans, i.e., human decisions based on intentions and values.

From a microscopic view we may actually observe single individuals with their intentions, beliefs, and so on. But from a macroscopic view the development of nations, markets, and cultures is more than the sum of its parts. The emergence of political, social, and economical order seems to be caused by procedures of self-organization which remind us of certain phase transitions in complex systems. Nevertheless, avoiding any reductionist kind of naturalism or physicalism, we shall consider the characteristic intentional features of human societies. In Sects. 3.4 and 4.3, the evolution of animal populations was modeled in the framework of the complex system approach. Macroscopic structures like social order, organization of social behavior, the construction of nests, and so on were explained by attractors of complex systems. But the difference of complexity between animal populations and human societies is enormous, although there are common origins and features. Thus, in the following, concepts like "evolution" and "nature" cannot be restricted to the mechanisms of molecules, fishes, ants, and so on. They imply a new kind of complex dynamics, the analysis of which must consider the long tradition of social philosophy.

Plato and Aristotle were the first philosophers who tried to explain the emergence of political, social, and economic order of human societies. They analyzed the structure of the Greek *polis*, which has become the germ cell of Western societies and states. In Greek antiquity a *polis* (πόλις) like Athens was the republic of a small city, which later on may be compared with the Italian states of Florence and Venice in the Renaissance, and perhaps the Swiss cantons of cities in modern times. A Greek polis was a small, but politically and economically almost autonomous state and society [6.1]. The Greek philosophers suggested an idealized model which was more or less realized by the historical examples.

Plato distinguished several states of transformation which a polis must undergo before reaching the final goal of a harmonious society. In the first state the citizens must learn different skills and professions, business, and trades, in order to satisfy the different needs of the whole community. Plato believed that the people of a polis must specialize, according to their different gifts. The citizens must organize themselves for cooperative work. Plato assumed that the exchange of their products and services achieved an equilibrium of work and demand by spontaneous self-organization. This economic state of equilibrium is characterized by "just" prices.

But Plato's idyllic world of cooperative work is, of course, unstable. People try to pursue their interests and make profits. They are selfish, immodest, full of envy, and driven by passions. Thus, conflicts arise and political power must be organized to avoid the destruction of the polis. Plato suggested an aristocracy of the best and wisest men ("philosopher-kings") who should rule the state [6.2]. Their government is intended to keep the whole system with its conflicting fluctuations in a state of equilibrium. It is well known that Plato distrusted democracy, because in his opinion common people without philosophical education are not able to recognize the true idea of justice. Plato believed in an eternal hierarchy of ethical values behind the changing and transitory world of appearances. Thus, there is an objective measure of values which people have to be aware of, in order to avoid chaos and to keep the system in a state of harmony.

Obviously, Plato defends a centralized system of political power. In the language of system theory, there is a centralized processor controlling all actions and reactions of the system's elements. Like the Laplacean demon in the scientific world, there is Plato's political myth of an ideal, wise, and good politician leading the system to a harmonious equilibrium. In a small city like a Greek polis, Plato's aristocracy of the best "philosopher-kings" may be justified under certain critical circumstances. Nevertheless, the experiences of actual history have shown that even well educated and intelligent political leaders are not free of temptations to misuse their power. Nowadays, Plato's aristocracy of the best may be compared with the power of experts in knowledge based complex societies. But, under present conditions of highly developed information and computer technologies, Plato's myth of a wise and good politician can easily change to Orwell's horror-scenario of a "Big Brother" with omnipotent controlling power.

The second famous philosopher who referred to the Greek polis was Aristotle [6.3]. He assumed that humans are social beings by nature who want to survive. Furthermore, they are political beings because they want to live well and happily. Aristotle believed in an organic development of human society, driven by the social and political nature of its members. The social and political dynamics reach a final state of equilibrium when the social and political form of a polis is realized. Aristotle described the social and political dynamics as processes in nature.

However, a dynamical process of nature is not conceived as a causal movement of mechanics, but as an organic growth like a plant or animal, starting with the initial state of a seed and aiming at the final state of its complete form (compare Sect. 2.2). Thus, Aristotle's model of society is naturalistic in the sense that humans are driven by the impulses of their social and political nature. But only the human instinct of social organization for the purpose of survival is common with animals. Humans are distinguished by their political nature to achieve a just society. In a famous formulation, Aristotle said humans are defined as rational beings striving for truth in science and philosophy, and they are political beings striving for justice in society.

Justice means a natural state of completion if the society is arranged in its harmonic proportions in equilibrium, like the static balance of an Archimedean scale (Fig. 6.1). Thus, economic equilibrium in an Aristotelean society is measured by "just prices" according to the "natural" value of goods and services. Economics was a part of Aristotle's moral philosophy of justice and state. He distinguished between commutative justice (*iustitia commutativa*) in private exchange and affairs of citizens and distributive justice (*iustitia distributiva*) concerning the relationship of citizens and the state. The Aristotelean model of economic and political justice became the leading idea during the Middle Ages. Obviously, it accorded with the Aristotelean concept of nature in those days.

The mechanistic view of nature was founded by Galileo, Descartes, and others, leading to Newton's grand system of classical physics. In his famous

Fig. 6.1. Last Judgement of Osiris (Egypt: 2nd century B.C.)

book *Leviathan or the Matter, Form and Power of a Commonwealth, Ecclesiastical and Civil* (1651), Thomas Hobbes (1588–1678) projected a mechanistic model of modern society and state [6.4]. At the end of the Middle Ages and the beginning of modern times, Hobbes lived in a period of dramatic political changes. The traditional monarchy and aristocracy of the Middle Ages had lost their religious legitimacy. In bloody civil wars, European societies and states were falling into ruin and chaos. Scientifically, Hobbes was impressed by the new mechanical method of Galileo and its successes in physics. Thus, he tried to use this method in order to found a mechanistic model of modern society without the obsolete traditional metaphysics to compromize its legitimacy in science and politics.

According to Galilean mechanics, there is an analytic or resolving method for dividing a system ("body") into its separable elements and a synthetic method for composing and unifying the separated building blocks into the whole system again. In short, the whole is the sum of its parts. Obviously, Galileo described the crucial superposition principle founding the linearity of the mechanistic world view. Actually, a mechanical system like a clock can be divided into separated elements such cog wheels and other mechanical parts which together constitute its perfect functionality.

Hobbes tried to transfer the law of movement from mechanics to anthropology and state theory. Humans are assumed to be driven by affects and emotions like physical bodies by mechanical impulses. The main affect is the individual instinct of self-preservation and survival. In the opinion of Hobbes, it is a natural right of man to pursue his instinct of survival, leading to force and violence against other humans. Thus, in Hobbes' natural state of human society, there is a permanent struggle of everyone against everyone (*bellum omnium contra omnes*) without any state of equilibrium.

On the other hand, humans with their complex necessities can only survive in society. Thus, their reason dictates a first natural law to search for peace. In order to realize the "law of peace", a second law is necessary, demanding a social contract. Hobbes suggested that in this social contract all citizens have to transfer their natural right of power to an absolute sovereign ("Leviathan") who is alone legitimated to apply political power and to rule the state. In modern words, Hobbes' social contract legitimates the state's monopoly of power, in order to keep society in an absolute equilibrium.

Hobbes defines the sovereign as the "sum of all individuals" who made the social contract. Obviously, this idea is an application of Galileo's mechanical principle of superposition or linearity. The title page of Hobbes' book (Fig. 6.2) shows the body of the Leviathan as a huge complex system of single individuals, illustrating Hobbes' political principle of linearity.

The "phase transition" from the natural state of chaos to political order and equilibrium is realized by a social contract of all citizens and in this sense by self-organization. But then, the final state of Leviathan is a centralized

Fig. 6.2. The "Leviathan" on the title page of Hobbes' book

and deterministic system without any political "degree of freedom" for its citizens. Hobbes compared the economic circulation of goods and money with the circulation of blood, which was discovered by the English physician William Harvey. Income and expenditure were referred to the incoming and outcoming blood of the heart as the pumping machine of the whole circulation.

The mechanistic view of an economy was elaborated later on by the French scientist François Quesnay (1694–1774), the founder of the physiocratic school in economics [6.5]. Quesnay, who started as a physician at the court of Louis XV, wrote a book about the "animal economy" (*oeconomie animale*) of the human body inspiring him with the idea of an economy of the social organism. Descartes' mechanistic world view was the leading philosophy of the physiocrats.

Thus, the economic system was modeled as a mechanistic clockwork of cog wheels, springs, and weights. A clock is a sequentially operating system with programmed functions. Analogously, the physiocratic economy cannot regulate itself. Advances in agriculture, which are the driving forces of the physiocratic economy, are compared with the weights and springs of a clock. Economic production was referred to the composed movements in a clock.

Consequently, economic prosperity is only guaranteed by a regular economic circulation like a clockwork.

The physiocrats used a particular table (*Tableau économique*) to visualize the circulation of wealth between the social classes of farmers ("productive class"), tradesmen ("sterile class") and land owner. In Fig. 6.3a, an economic period starts when the class of land owners distributes their received rent (for instance 200 million Louisdor) for food and agricultural products to the left column of farmers (100 million Louisdor) and for the goods of tradesmen to the right column (100 million Louisdor) [6.6]. Both incomes enable the two classes of farmers and tradesmen to produce new goods. As farmers use the products of tradesmen and tradesmen use the goods of agriculture, money must circulate between the columns of the corresponding classes. The circulation forms a zigzag curve, until the net profit is paid out at the bottom of the table.

But the expenditure of the net profit causes new receipts making new expenditure possible, which may reproduce the net profit, in order to start a new economic circulation. The mechanics of the regular circulating and iterative reproduction of the net profit was illustrated by clocks with rolling balls (Fig. 6.3b) [6.7]. The clocks measure time by means of balls rolling down inclined zigzag paths. After a period of circulation the balls are lifted to the top of the system and the process is iterated. Obviously, the distribution of net profit in a circulation period is compared with the zigzag path of the rolling balls in the clockwork. The iteration of an economic circulation period corresponds to the lifting of a ball and its rolling down the zigzag path again.

The physiocratic economists used physical models in the framework of Cartesian mechanics. Their causal determinism without any kind of self-regulation or individual freedom corresponds completely to the political system of absolutism. The citizens are reduced to functioning elements in a political and economic machine.

6.2 Smith's Economics and Market Equilibrium

While the physiocrats designed their economic model against the background of Cartesian mechanics, Adam Smith referred to the classical physics of his great predecessor Sir Isaac Newton. In Cartesian mechanics, all physical events are reduced to the contact effects of interacting elements like cog wheels in a clock or impulses between balls. Thus, Cartesian physicists constructed hypothetical mechanisms which are often not observable. For instance, the refraction of light was explained by the interactions of hypothetical tiny glass-like balls. The laws of percussion and impulse were fundamental in Cartesian physics.

Fig. 6.3a. The "Tableau economique" of the physiocrats [6.7]

Newton critized Cartesian mechanics with his famous phrase *Hypotheses non fingo*. His law of gravitation was mathematically deduced from his axioms of mechanics and empirically confirmed by forecasts and experiments. But he renounced any explanation of the long-distance effects of gravitation in empty space (*actio in distans*) by some hypothetical transmission mechanism which was not observable.

Fig. 6.3b. The "Tableau economique" of the physiocrats (**a**) illustrated by the zigzag path of the rolling ball in a clockwork (**b**) [6.7]

In Newton's celestial mechanics, material bodies move in a system of dynamical equilibrium which is determined by the invisible forces of gravitation. The physical concept of freely moving individuals in dynamical equilibrium corresponds to the liberal ideas of a free economy and society with the division of independent political powers. Contrary to the liberal ideas, the Cartesian clockwork of nature seems to accord with the state machinery of absolutism with its citizens as cog wheels.

The famous English philosopher John Locke (1632–1704) influenced not only the epistemology and methodology of Newtonian physics but also the political theory of modern democracy and constitution. He asked why man is willing to give up his absolute freedom in the state of nature and to subject himself to the control of political power. Locke argued that the enjoyment of the property man has in the natural state is very unsafe and insecure, because everyone else wants to take it away from him in a state of unrestricted freedom. Thus, the state of nature is unstable and will transfer to an equilibrium of political forces. For Locke, the "phase transition" from the state of nature to a society with government is driven by men's intention to preserve their property.

However, government does not mean the unfree machinery of Leviathan. It is a state of balance ("equilibrium") with independent political powers like the legislature and executive. Since laws are made by a parliament as representative organ of the society, there is an essential feedback to the citizens, who only give up their natural freedom to the extent that the preservation of themselves and their property require: "And all this to be directed to no other end but the peace, safety, and public good of the people" [6.8]. Historically, Locke's ideas of democracy, division of power, property, and tolerance mainly influenced the American and French constitutions.

As in epistemology, the great Scottish philosopher David Hume (1711–1776) was both more critical and more exact in political theory than Locke. In epistemology he taught that human consciousness is governed by associations of sensations and feelings which may be reinforced or diminished by outer experiences (compare Sect. 4.1). Thus, there is no absolute truth even in Newtonian physics, but only a useful method with more or less probable results. Analogously, there is no eternal ethical value like justice determining human behavior. Ethical ideas can only be evaluated by their individual or public usefulness [6.9]. In general, political institutions are only legitimated by their usefulness in being accepted by a society or not. Thus, Hume became a forerunner of utilitarian ethics and political philosophy. His friend and Scottish fellow-countryman Adam Smith was probably inspired by his sceptical anthropology of self-interested behavior in human societies.

The publication of Smith's famous *Inquiry into the Nature and Causes of the Wealth of Nations* (1776) is often celebrated as the birth of a separate discipline. Nevertheless, Smith was a professor of moral philosophy, and Newton was a professor of natural philosophy. Actually, Smith tried to unify ethics, economics and politics, and Newton embedded his physics into a cosmic and even religious framework. In his *Theory of Moral Sentiments* Smith analyzed the role of sympathy in human beings [6.10]. In his *Wealth of Nations* the self-interested behavior of men is assumed to be the essential impulse of economics.

In both of these books Smith tried to apply the Newtonian method to ethics and to economics [6.11]. He described the Newtonian method as one in which a scientist lays down "certain principles, primary or proved, in the beginning, from whence we account for several phenomena, connecting all together by the same chain." In a Hume-like manner, Smith credited the origin of science not to men's love for truth, but to the simple desire to maximize "wonder, surprise, and admiration." The great purpose of human life is the uniform, constant, and uninterrupted effort to better men's condition. In short, the self-interest of men tends to maximize the welfare function.

According to Newton's motto *Hypotheses non fingo*, Smith emphasized that human self-interest is not a theoretical construction of economists, but an empirical fact of experience. Self-interest is a strong and natural impulse of single human beings, and therefore a human right. But the interactions of several single micro-interests achieve the common macro-effect of welfare by the mechanism of the market. Two famous quotations from Smith's *Wealth of Nations*: "I have never known much good done by those who affected to trade for the public good" [6.12] and "It is not from the benevolence of the butcher, the brewer, or the baker, that we expect our dinner, but from their regard to their own interest" [6.13].

The mechanism of the market is regulated by supply and demand driving the competing micro-interests to the macro-effect of welfare and the "wealth of the nation" in the equilibrium of the market. In the mechanistic view, the micro-interests seem to be drawn to the common macro-state of equilibrium by an "economic demon" or mechanical spring. According to the Newtonian method, Smith prefers the picture of an "invisible hand" directing the micro-interests like the "invisible" long-distance force of gravitation in astronomy. Obviously, Smith describes an economy as a complex system of many competing micro-interests. The dynamics of their interactions is a self-organizing process of competition with a final state of equilibrium between supply and demand.

The value of goods is measured by money. But the money measure could not be used without a precaution. It was necessary to distinguish between "market prices" achieved by the market mechanism and the "natural prices" or actual costs of a good. Economists had to find a "standard of value" in order to be able to correct the value of money. Thus, Smith already aimed at a political economy based upon a theory of value. Values were needed for weighting the social product. Figure 6.4 illustrates Smith's self-organizing process of supply and demand by a feedback scheme with demand r for a good, supply c, market price m, and natural price n [6.14]:

Fig. 6.4. Adam Smith's self-organizing process of supply and demand illustrated by a feedback schema [6.14]

But Smith did not introduce a "just" price like Aristotle on the background of ethical ideals like justice. His inquiry analyzed the "nature" and "causes" of the "wealth of nations" based on actual facts of human nature like self-interest. Concerning the natural prices of goods, Smith and early classical economics like David Ricardo tried to find an absolute measure of value such as gold, corn, or labor.

For Ricardo, the common measure should be explained by his labor theory of value. Ricardo, like Smith was familiar with the general ideas of classical physics. Thus, he believed that some of the conclusions of economics were "as certain as the principles of gravitation" [6.15]. As history went on and economic and political problems changed, Ricardo's theory of growth, rent, and labor was influenced by the historical conditions of his own time at the beginning of the 19th century. Prominently, there were the economic problems of feeding a growing population which had been already considered by Malthus.

John Stuart Mill (1806–1873), the British philosopher and economist, had a great interest in the methodology of economics. He defined "political economy" as an axiomatic system of deductive analysis resting on assumed psychological premises and abstracting from all noneconomic aspects of human behavior. These abstractions were compared with disturbing causes like friction in mechanics:

> The disturbing causes have their laws, as the causes which are thereby disturbed have theirs; and from the laws of the disturbing causes, the nature and amount of the disturbance may be predicted a priori ... . The effect of the special causes is then to be added to, or subtracted from, the effect of the general ones. [6.16]

In this quotation, Mill obviously described the principle of causality in classical physics, which basically made forecasting in the long-run possible: similiar causes effect similiar results. Thus, Mill's methodology of economics accords with the Laplacean spirit of classical physics, assuming an approximately correct calculation of forecasts by economic laws if initial conditions are approximately known. Furthermore, Mill's axiomatic hypotheses define a simplified model of economic behavior, and not the complex economic reality.

Thus, Mill was the first economic theorist who explicitly operated with the fictional "economic man" (*homo oeconomicus*), not with the real man in his whole complexity which originally was the subject of Smith's inquiry. The general hypothesis of an economic man maximizing a certain economic utility function is empirically based on a kind of experience, namely, introspection and the observation of Mill's fellow men, but it is not derived from specific observations or concrete events. Analogously, Newton's general law of gravitation was empirically justified by some special observations of falling bodies or moving celestial bodies, but not derived from these events. Mill's methodology corresponds to new insights into the role of formal systems and models in 19th century physics.

Predecessors of modern mathematical economics like Walras and Pareto propagated the use of the mathematical methods of physics in economics. Both thinkers were representatives of the so-called Lausanne School. Classical theorists were already implicitly guided by physico-mathematical concepts. They spoke of a more or less rough correspondence between the play of economic forces and mechanical equilibrium. Actually, the predecessors of mathematical economics borrowed much of their vocabulary from mechanics and thermodynamics, for instance, equilibrium, stability, elasticity, expansion, inflation, contraction, flow, force, pressure, resistance, reaction, movement, friction, and so on [6.17].

In 1874, Walras followed Smith's idea that the maximizing behavior of consumers and producers will result in an equilibrium between amounts demanded and supplied in every product and factor market of the economy [6.18]. Since Walras, general equilibrium theory has become a leading concept which demands existence proofs of equilibrium in mathematical models of an economy. Mathematical economists tried to isolate the elements of a complex system from its environment, which was specified by exogenous parameters. If the exogenous parameters themselves depend on influences from the whole system, however, then separating system and environment and neglecting the actually existing feedback makes an adequate economic model impossible.

In general, classical economists tried to reduce the complexity of economic reality by some typical assumptions characterizing their linear and mechanistic models. First of all, they believed in the fiction of a rational economic agent (*homo oeconomicus*) for each human action. The individual behavior of agents, for instance, in a market should be isolated as a whole. Human behavior is described by general behavioral patterns abstracted from individual behavior. Thus, individual human behavior is assumed to be regular and predicable like elements of a mechanical system with certain mathematical laws of motion. If the initial conditions and the environment are known and precisely measurable, then individual behavior within the environment is believed to be deterministic like that of molecules in a gas.

The linearity of the economic models follows from the superposition principle, assuming that a society consists of the additive actions of its members. The principle of superposition implies that society as a whole does not differ from the sum of the individual actions. Obviously, linear models abstract from unpredictable and irrational individual behavior, from restrictions in the environment, and from non-additive ("nonlinear") and synergetic interdependencies between individuals and their actions.

These linear principles of methodology completely correspond to the Laplacean world view of physics. They still have a strong influence on mainstream economics today, although physics itself has undergone some dramatic revolutions in this century, such as quantum mechanics with its characteristic uncertainty relation. But, Heisenberg's uncertainty relation is a consequence of particular relations between quantum mechanical operators depending on Planck's constant, which seems to have no relevance for the dimensions of the economical world. Nevertheless, the quantum formalism à la Schrödinger and Heisenberg remains linear (compare Sect. 2.3). The fact that classical linear dynamical systems behave in a very regular fashion allows exact forecasts. A nonlinear model demonstrating chaotic behavior as well as the impossibility of predictions in the long run was considered as a bad economic instrument.

In this century, mathematical economists have more and more given up the physicalism of the Lausanne School, for instance, which tried to determine an economy like a classical physical system. Economists have tried to found their own basic mathematical instruments. The assumptions of linear dynamical models have been justified by technical reasons. This formal attitude is expressed in the following statement by John Maynard Keynes in a letter to Roy Harrod from 1938:

> It seems to me that economics is a branch of logic, a way of thinking; and that you do not repel sufficiently firmly attempts . . . to turn it into a pseudo-natural science . . . . Economics is a science of thinking in terms of models joined to the art of choosing models which are relevant to the contemporary world. [6.19]

Under the influence of characteristic economic crashes (for instance at the end of the 1920s), Keynes and others underlined that the economic system is not automatically capable of self-regulation. The "instability of capitalism" has become the popular version of the so-called Keynesian doctrine. So, it was suggested that the economic system be stabilized from outside by particular instruments of policy like fiscalism. Linear models have been employed especially by neoclassical theorists who, again, have concentrated on the investigation of equilibrium economics.

Nonlinear approaches have been preferred mainly by economists who felt uncomfortable with the classical ideal of equilibrium economics. Thus, Keynesian authors have often criticized the linear framework of equilibrium theories without being familiar with the mathematical methods of nonlinearity.

A new epoch of linear mathematical economics was introduced by John von Neumann and Oskar Morgenstern's *Theory of Games and Economic Behavior* (1943) [6.20]. Linear programming, operational research, and even mathematical sociology have been influenced by this famous book. In the

Theory of Games, von Neumann and Morgenstern assumed rationally acting persons who try to maximize their profit in terms of certain utilities. In general, a class of possible actions $a_1, \ldots, a_m$ and a class of possible states $s_1, \ldots, s_n$ are used to form pairs (a_i, s_j) with $1 \leq i \leq m$ and $1 \leq j \leq n$ which are mapped onto utilities u_{ij}. The possible utilities u_{ij} are ordered in a $(m \times n)$-matrix.

Several criteria of rationality have been suggested for decisions under uncertainty, for instance. Uncertainty means that no probabilities of possible utilities are known. The so-called maximin utility criterion is mainly used. In this case each possible action a_i is associated with an index which is the minimum of the utility values $u_{i1}, \ldots, u_{in}$, the smallest value in the i-th line of the matrix (u_{ij}) of utilities. Then, the rule demands: choose the action the index of which is a maximum. In short, the maximin rule chooses the action maximizing the utility of the most unfavorable case. The rule can be applied very easily and mechanically to the matrix of utilities.

Imagine the following example from the philosopher Carl Gustav Hempel [6.21]. There are two urns with balls of equal size which cannot be distinguished by the sense of touch. In the first urn there are balls of lead and platinum. In the second urn there are balls of gold and silver. A player is allowed to take a ball from one of the two urns as a free gift. A probability distribution of the balls in the urns is not known to the player. The utilities of the balls are estimated at 1000 (platinum), 100 (gold), 10 (silver), 1 (lead).

The maximin rule suggests choosing the second urn. In this urn the most unfavorable case is a silver ball, but in the first urn the most unfavorable case is a lead ball. Obviously, the maximin rule corresponds to a pessimistic view of the world. In a game the player assumes a very hostile opponent. Then, the maximin rule suggests the most useful actions.

An optimistic attitude corresponds to the so-called maximax utility criterion. The player is convinced that each possible action has a best possible result. Thus, it seems to be rational to choose an action the best possible result of which is at least as good as the most favorable result of an alternative action. In our urns example the maximax rule suggests choosing the first urn.

A cautious player would not like to choose that rule. But on the other hand the maximin rule is only rational if the opponent is known to be hostile. Some numerical examples may confirm these interpretations. For two possible states s_1, s_2 and two possible actions a_1, a_2 the matrix of utilities is given by Fig. 6.5a.

The maximin rule suggests action a_2. Even if the number 1 is diminished to a tiny value such as 0,000001 and the number 100 is enlarged to a very great value such as 10^{15} (Fig. 6.5b), the maximin rule would suggest action a_2. This decision is actually rational for a player who must assume an absolutely hostile opponent. In any case the opponent would try to prevent a state corresponding to maximum utility for the player. But otherwise the maximin

	s_1	s_2
a_1	0	100
a_2	1	1

	s_1	s_2
a_1	0	10^{15}
a_2	0.000001	0.000001

Fig. 6.5a,b. Utility matrices

rule would be irrational, because a_1 would be the better action. If state s_1 is realized, then the player has to give up an increment of utility which is very small. In the case of state s_2, he would receive a very large growth of profit with action a_1.

In order to justify such a decision, Savage introduced the so-called minimax risk criterion. He suggested replacing the matrix of utilities u_{ij} (Fig. 6.5a) by a matrix of risk values r_{ij} (Fig. 6.5c). A risk value r_{ij} must be added to a utility value u_{ij} in order to get the maximum value of utility in the j-th column.

In matrix 6.5a the largest value of utility in the first column is 1, and in the second column it is 100. Thus, the risk matrix is given by Fig. 6.5c.

	s_1	s_2
a_1	1	0
a_2	0	99

Fig. 6.5c. Risk matrix

The minimax risk rule demands: choose an action minimizing the maximum risk. As the maximum risk of a_2 has the value 99 and that of a_1 has the value 1, it seems to be rational to choose action a_1. But, of course, this rule is only rational under certain particular conditions. There are many other criteria of rationality.

A further example is the so-called pessimism optimism criterion which suggests a solution between the pessimistic maximin rule and the optimistic maximax rule. For an action a_i let m_i be the minimum and M_i the maximum of utilities $u_{i1}, \ldots, u_{in}$. Let α be a constant such that $0 \leq \alpha \leq 1$ called the optimism pessimism index. Then, the action a_i is associated with the so-called α-index $\alpha m_i + (1-\alpha) M_i$. The pessimism optimism rule prefers actions with a larger α-index. But, of course, a special criterion is only defined if a particular number α is given. These examples demonstrate that there is no absolute criterion of rationality, but a class of criteria corresponding to different degrees of optimism and pessimism under certain circumstances.

Von Neumann and Morgenstern's book *Theory of Games* considers the stability of a society or market as the result of interactions of competing or

cooperating persons or groups. In many cases, they accept an oversimplifi-
cation of the actual economic, social, and psychological complexity. Each
player can exactly determine his possible actions to aim at certain states and
possible utilities. In general, the theory of games assumes the principle of
linearity (superposition) that complex interactions of many persons in a society ("game") can be reduced to the sum of many simple interactions of a
few persons.

Thus, the investigation of 2-person games plays a crucial role in game
theory. The event that player 1 chooses action a_1 and player 2 chooses action
a_2 is designated by the pair (a_1, a_2). In the case of this event, the utilities are
defined as $u_1(a_1, a_2)$ for player 1 and $u_2(a_1, a_2)$ for player 2. An important
class of games is characterized by the fact that the utilities of both players
in each event are absolutely opposite with $u_1(a_1, a_2) + u_2(a_1, a_2) = 0$ ("zero-
sum" games). Any cooperation is excluded. Then the maximin rule seems to
be rational if there is no special information about the irrationality of the
opponent. In other cases a cooperation may sometimes be rational.

The mathematically essential problem is the existence of equilibrium
points in games [6.22]. If there is no cooperation, an equilibrium point can
be defined for the possible actions of the two players in the following way. An
event $(\overline{a}_1, \overline{a}_2)$ is an equilibrium point of the game if the utility value $u_1(\overline{a}_1, \overline{a}_2)$
of player 1 is greater than or equal to $u_1(a_1, \overline{a}_2)$ for all his actions a_1 and if
the utility value $u_2(\overline{a}_1, \overline{a}_2)$ of player 2 is greater than or equal to $u_2(\overline{a}_1, a_2)$
for all his actions a_2.

If player 1, assuming that player 2 chooses his action $\overline{a}_2$, tries to maximize his utility, then he can choose action $\overline{a}_2$ and vice versa. The equilibrium
is stable in the sense that a player knowing that his or her opponent is at the
equilibrium point has no reason to alter his or her behavior. Obviously, this
definition of equilibrium does not consider any dynamical aspect. But the
actual behavior of societies and markets is determined by a complex dynamics in time. The trade cycles are well known examples of economic dynamics.
The question arises of whether these dynamical processes are attracted by
equilibria and whether these equilibria are stable. In general, the theory of
games does not consider the "butterfly effect" that a small failure of behavior
may sometimes cause a global crisis and even chaos.

The game theory of von Neumann and Morgenstern is not only in the
tradition of linear mathematical economics. It developes the ideas of economic welfare theory. A rational society is assumed to choose a distribution
of profits which is Pareto-optimal. A distribution of profits is called Pareto-
optimal if the welfare of no individual can be increased by a deviation without
diminishing the welfare of at least some other individual. It is not sufficient
that this weak condition of Pareto-optimal welfare is satisfied. The power
of potential coalitions must be considered, too. The theory of cooperative
solutions in game theory mainly stems back to the ideas of welfare eco-

nomics, diplomacy, and self-interested politicians who are sometimes social. Mathematically, the concepts of justice, impartiality, or fair play, which determine the political and social framework of welfare economics, are reduced to certain principles of symmetry.

Game theory is an exact mathematical theory whose application to economics has sometimes been overestimated. Its limitations are mainly given by its typical linear assumptions about society. Nevertheless, game theory is a brilliant mathematical invention – which was mainly initiated by John von Neumann. It is noteworthy that John von Neumann was a central figure in nearly all the scientific developments of this century which are considered in this book. He was engaged in the development of program-controlled computers, the theory of automata, quantum mechanics, and game theory. Furthermore, he was interested in interdisciplinary mathematical models in the natural and social sciences. All these brilliant developments are mainly ruled by principles of linearity. But John von Neumann was also one of the first scientists to recognize the importance of self-reproduction and self-organization. His theory of cellular automata is a famous example.

6.3 Complex Economical Systems, Chaos, and Self-organization

From a methodological point of view, mainstream economics has often been inspired by models from linear mathematics, classical mechanics, thermodynamics of thermal equilibrium, and sometimes the Darwinian theory of evolution. Classical economic models have postulated a rational economical man (*homo oeconomicus*) who seeks to maximize his utility by minimizing his costs and maximizing his profits. These rational agents are assumed to interact by exchanging commodities in markets that achieve an economic equilibrium of demand and supply by certain price mechanisms.

To describe the dynamics of an economy, it is necessary to have evolution equations for many economic quantities from perhaps thousands of sectors and millions of agents. Since everything depends, in economics like elsewhere, on everything else, the equations will be coupled and nonlinear, in order to model economic complexity as well as possible. But then, even completely deterministic models may produce highly irregular behavior which is unpredictable in the long run. Economics seems to suffer from the same shortcomings as metereology [6.23].

Before the discovery of mathematical chaos and the butterfly effect in meteorology, people believed in the possibility of precise long-term forecasts of the weather. John von Neumann as a computer pioneer maintained that, given sufficient data about global weather and a supercomputer, he could provide accurate predictions of weather over time and space. He did not fail for mathematical reasons, because in the framework of linear mathematics he was as right as a classical astronomer. But the actual behavior of fluids and weather differ dramatically from these models in the long run.

How does one deal with the complexity of weather and the economy? In meteorology, Edward Lorenz has proposed a nonlinear dynamical model which can produce chaotic behavior by its intrinsic ("exogenous") disturbance (compare Sect. 2.4). Analogously, there are two possible ways of explaining the complexity of economic evolution. The mainstream approach has assumed linear models with somewhat ad hoc and unexplained exogenous shocks. The nonlinear approach gives up oversimplified hypotheses of linearity and ad hoc hypotheses of exogenous shocks and tries to explain the complexity of actual economies by their endogenous nonlinear dynamics. In a few cases nonlinearities may be so weak that linear approximations do not constitute an essential error.

In the history of economics, the Great Depression of the 1930s gave rise to theories which tried to explain the economic irregularities. But the models (for instance, the Kalecki and the Hansen-Samuelson models) were linear and could not explain the emergence of oscillations [6.24]. Thus, exogenous shocks were assumed by economists to produce the observed oscillations. If the economists had been more familiar with mathematical developments, they would have been acquainted with nonlinear mathematical models leading to limit cycles as solutions.

Originally, economists only knew about stable equilibrium as a fixed-point attractor. Poincaré generalized the conception of equilibrium to include equilibrium motion in the form of limit cycles. But for chaotic attractors like the Lorenz model (Fig. 2.21) there is neither a fixed point nor a fixed motion, but a never-repeating motion. Nevertheless, it is a bounded motion and a non-wandering set attracting certain dynamical systems to a final state of dynamical equilibrium.

Historically, the economy in this century has been characterized by growth that was interrupted by spectacular collapses, for instance, in the 1930s (the Great Depression) and in the 1970s (the Oil Crisis). Concerning the structure of growth, particular attention must be given to innovations and technical progress. The expansion of a successful innovation is empirically well represented by the logistic function which was introduced in Sect. 2.4. A recursive version can be referred to integers t as time indices and a growth factor $\alpha > 0$. Initially, an innovation is rather unfamiliar. Then as it gains acceptance it reaches its maximum rate of expansion. After that there is a slow deceleration of the process of absorption as the innovation approaches complete integration into the economy.

The resulting curve is shown in Fig. 2.22. For $\alpha \leq 3$ we get a fixed point attractor which is visualized in Fig. 2.22a. For larger α, the result is a periodic oscillation (Fig. 2.22b and Fig. 2.24b) and then a chaotic movement (Fig. 2.22c and Fig. 2.24c). For $\alpha > 3$ the number of periods doubles successively as α increases (Fig. 2.23a), until it is transformed into complete chaos (Fig. 2.20b).

The connection between innovation and economic output is shown in the following model of Fig. 6.6 [6.25]. Initial output q is taken to be in equilibrium so that as the growth rate Δk rises, the output also rises, gradually. But then as innovation reaches its saturation and Δk drops to towards zero, output q falls back to its initial level. So innovation stimulates a boom but also a subsequent collapse. An innovation may be labor-saving. If inputs of labor per unit of output falls by 20 %, employment will fall also.

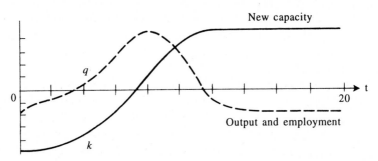

Fig. 6.6. Logistic curve of innovation capacity with output and employment curve [6.25]

As the growth of new ideas is assumed to be exponential, economists like Schumpeter have maintained that at the end of an innovatory burst a new one will start. Then a second boom and collapse is initiated, and so on, as the economic system continues to function and technological conceptions continue to grow at around 4% per year. Innovation is crucial to economic cycle theory, because in a depression there is no other basis for the new investment which is necessary to generate a new expansion.

New ideas arise steadily. When enough ideas have accumulated, a group of innovations are introduced. They develop slowly at first, then accelerate as the methods are improved. A logistic development characterizes the typical trajectory of an innovation. Some investment must precede the introduction of an innovation. Investment stimulates demand. Increased demand facilitates the spread of the innovation. Then, as all the innovations are fully exploited, the process decelerates towards zero.

Schumpeter called this phenomenon the "swarming" of innovations. In his three-cycle model, the first short cycle relates to the stocks cycle and innovations play no role. The following longer cycle is related to innovations. Schumpeter recognized the significance of historical statistics and related the evidence of long waves to the fact that the most important innovations like steam, steel, railways, steamships, and electricity required 30 to 100 years to become completely integrated into the economy.

In general, he described economic evolution as technical progress in the form of "swarms" which were explained in a logistic framework. A technological swarm is assumed to shift the equilibrium to a new fixed point in a cyclical way. The resulting new equilibrium is characterized by a higher real wage and higher consumption and output. But Schumpeter's analysis ignores the essential issue of effective demand as determining output.

Historically, the Great Depression of the 1930s inspired economic models of business cycles. However, the first models (for instance Hansen–Samuelson and Lundberg–Metzler) were linear and hence required exogenous shocks to explain their irregularity. The standard econometric methodology has argued in this tradition, although an intrinsic analysis of cycles has been possible since the mathematical discovery of strange attractors. The traditional linear models of the 1930s can easily be reformulated in the framework of nonlinear systems.

The Metzler model is determined by two evolution equations. In the first equation, the rate of change of output $\dot{q}$ is proportional to the difference between actual stocks s and desired stocks s^*. The desired stocks are proportional to the output. The next equation concerns the rate of change of stocks $\dot{s}$ which is output q less demand. The demand is proportional to the output. The complex system whose dynamics is determined by these two evolution equations will produce simple harmonic motion with increasing amplitude.

If the system is expanded by certain nonlinearities, a different kind of behavior results. A third equation may be associated with perverse behavior of net public surpluses and deficits. The aim is to produce a cycle of a certain duration of years. A mathematical model is suggested by the so-called Rössler band [6.26]. One uses a Möbius strip, top to bottom, sides reversed, giving a one-sided surface (Fig. 6.7a). Following a trajectory, an outer loop expands to the right and upward. Then it folds over and as it descends, contracts to an inner loop and so on. Figure 6.7a gives a two-dimensional projection which shows the two cycles. The lines tend to get grouped with empty spaces in between. If the simulation is continued these bands become more and more dense.

Figure 6.7a is a simple but remarkable example of a chaotic ("strange") attractor. Although each trajectory is exactly determined by the evolution equations, it cannot be calculated and predicted in the long run. Tiny deviations of the initial conditions may dramatically change the path of trajectories in the sense of the butterfly effect. Figure 6.7b shows a trajectory of output in the state space plotted over 15 years, which was simulated in a computer experiment with chosen parameters. Figure 6.7c shows the corresponding development as a time series [6.27].

Fig. 6.7a. Rössler attractor

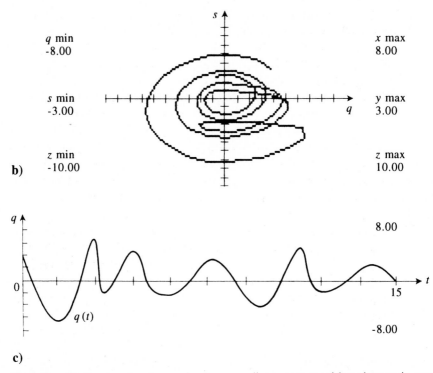

b)

q min -8.00	x max 8.00
s min -3.00	y max 3.00
z min -10.00	z max 10.00

c)

Fig. 6.7b,c. Business cycle of an endogenous nonlinear system with trajectory in state space (**b**) and time series of output (**c**) [6.27]

This highly erratic behavior is produced by a completely endogenous system without any exogenous shocks. In economics the irregularities of time series are usually explained by exogenous shocks. But they are only arbitrary ad hoc hypotheses and hence can explain anything. From a methodological point of view, chaotic endogenous models with strange attractors seem to be more satisfactory. Nevertheless, endogenous nonlinear models and linear models with exogenous shocks must be taken in earnest and tested in economics.

Obviously, an economy consists of many connected and separated parts with endogenous dynamics and exogenous forces. There are individual national economies which are increasingly acted on by movements of the world economy. Within an economy there are several markets with particular dynamics. They are influenced by cycles, for instance, the annual solar cycle determining the agricultural, tourist, or fuel market. The pig cycle and building cycle are further well known economic examples. Thus endogenous nonlinear systems with impressed waves of exogenous forces are realistic models of economies. The impression given is of a disturbed chaotic attractor or a kind of super-chaos. It is this erratic character of economic events which causes great problems for economic agents who have to make decisions that depend on an unpredictable future.

In Sect. 2.3 we saw that self-organizing complex systems may be conservative or dissipative. Their different types of attractors were illustrated in Fig. 2.14a,b. Some well known examples of conservative and dissipative models from the natural sciences have been applied to economies, too. In 1967, Goodwin proposed a conservative dynamical model to make the 19th-century idea of class struggle precise. He considered an economy consisting of workers and capitalists. Workers spend all their income on consumption, while capitalists save all their income. Goodwin used a somewhat modified predator-prey model of Lotka and Volterra which was described in Sect. 3.4 [6.28].

Goodwin's conservative model supports the idea that a capitalist economy is permanently oscillating. Thus, the trajectories describe closed orbits as illustrated in Fig. 3.11b. Goodwin's model was criticized as superficial, because it does not refer directly to the functional income shares of capitalists and workers or to their population size. But it is mainly its conservative character that makes Goodwin's model seem economically unrealistic. The model is put together as an isolated set of assumptions which do not reflect other influences.

Thus, the model has been made more realistic by the assumption of "economic friction". In the biological case, a dissipative Lotka–Volterra model with a point attractor was illustrated in Fig. 3.11c. A dissipative system always possesses attractors or repellers, in the form either of fixed points, limit cycles, or strange attractors. As dissipative systems have an irreversible time evolution, any kind of backward prediction is excluded [6.29].

In reality, a dynamical system cannot be considered as isolated from other dynamical systems. Thus, in Sect. 2.2, we studied coupled oscillating systems like, for instance, two clocks (Fig. 2.11a,b). The state space of the combined systems was represented by a torus (Fig. 2.11c,d). The dynamics of the whole system was illustrated by the phase portrait of trajectories and the vector field on the torus.

An economic model of coupled oscillatory systems is provided by international trade. Imagine a simplified macroeconomic model of a single economy with gross investment and savings, depending both on income and interest rate. The dynamics of the system is determined by an evolution equation for income which adjusts it according to excess demand in the market for goods, and a second evolution equation for the interest rate. These equations constitute a nonlinear oscillator in such a way that the model generates endogenous fluctuations.

The interactions of three economies, for instance, can be described by a system of three independent, two-dimensional limit cycle oscillators. If all three economies are oscillating, the overall motion of the system constitutes a motion on a three-dimensional torus. The coupling of nonlinear oscillators can be understood as a perturbation of the motion of the autonomous economies on a three-dimensional torus. The coupling procedure has been applied to several economic examples, such as international trade models, business cycle models, and interdependent markets.

A crucial problem of practical policy occurs when self-organizing economic systems are allowed to be influenced by political interventions. In some cases the market is not able to develop according to welfare criteria. If the economy is left to itself, it may be characterized by fluctuations. But policy measures can cause effects opposite to those intended if the complexity and nonlinearity of economic growth is not considered.

With respect to the dramatic social and political consequences of economic catastrophes, several policy measures have been discussed in the framework of Keynesianism and Neo-Keynesianism. Compensatory fiscal policy, for instance, may be regarded as a type of dynamical control. It should serve to diminish the amplitude of economic fluctuations. But post-war experiences have shown that there is no hope of reducing fluctuations to zero, or holding the degree of employment constant. Furthermore, a good policy needs considerable time for the collection of data, analyzing the results, and formulating the legislative or executive measures. The result is that any policy may be outdated by the time it is effective. Thus, a policy measure in the complex and nonlinear world of economics may be absolutely counterproductive [6.30].

For example, a Keynesian income policy can be ineffective when the assumed economic dynamics and the proposed time path of policy interventions are too simple. In the framework of complex systems, economic policy measures can be interpreted as impressing exogenous forces on an oscillating system. Thus, it cannot be excluded that the economy becomes chaotic. Forced oscillator systems are well known in physics. For instance, if a dynamical system like a pendulum (Fig. 2.7) is oscillating, and if it is periodically influenced by an exogenous force, the outcome may be unpredictable because of increasing amplitudes, total damping of the oscillation, and complete irregularity.

Since the time of the economic classics until today, it has been the aim of business cycle theory to model the dynamics of an economy with its typical regular fluctuations. According to the linear mechanical view, the actual business cycles were modeled by regular systems with superimposed stochastic exogenous shocks that had to be explained by more or less appropriate economic assumptions. It is, of course, highly unsatisfactory when these exogenous forces determine the main qualitative properties of a model without their having a reasonable economic interpretation. If an actual system is nonlinear and chaotic, further information on possible exogenous forces that may influence the economic dynamics may be superfluous. From a methodological point of view, Ockham's famous razor should be applied to cut out these superfluous ad hoc hypotheses of economics, according to his motto *entia non sunt multiplicanda sine necessitate* ((theoretical) entities should not be multiplied without necessity).

From the practical point of view of an agent, it may be rather irrelevant whether he is confronted with a stochastic linear process or a chaotic nonlinear process. Both kinds of systems seem to prevent him from making precise predictions. As a chaotic model has a sensitive dependence on initial conditions, arbitrarily precise digital computers are in principle unable to calculate the future evolution of the system in the long run. The trajectories diverge exponentially. On the other hand, an agent believing in stochastic exogenous shocks may resign in face of the too-complex behavior of the system.

Nevertheless, nonlinear systems with chaotic time series do not exclude local predictions. If the attractor of a nonlinear system can be reconstructed, then numerical techniques promise to predict the short-run evolution with sufficient accuracy. Short-term economic forecasting may be an interesting application of complex system theory in economics, which is still in its infancy [6.31].

Since the beginning of economics, empirical tests and confirmations have been a severe methodological problem for economic models. In contrast to the natural sciences with their arbitrarily many measurements and experiments in laboratories, economic time series must refer to time units of a day, or annual, quarterly, or monthly data. The length of a standard time series typically consists of a few hundred points. Thus, there are already empirical reasons for the limited reliability of economic models. Empirical experiments are, of course, largely excluded.

Therefore an adequate knowledge of endogenous economic dynamics will at least help to achieve mathematical models whose future developments may be simulated by computer experiments. A politician and manager will at least get the "phase portrait" of alternative economic scenarios if the assumptions of their economic and political surroundings are realized. The qualitative insight into the high sensitivity of nonlinear systems may at least prevent overacting agents from pushing the system from an unstable point to further and perhaps even greater chaos.

The main arguments for nonlinear models in economics are given by the recent structural change of economic growth caused by the development of new dominant technologies. The traditional theory of economics assumes decreasing income. The more goods of a certain kind are produced and put on the market, the more difficult their production and sale will become and the less profit can be gained. The interaction of agents is determined by a negative feedback stabilizing the economy by counteracting each reaction caused by an economic change.

In an economy with negative feedback an equilibrium of prices and market shares may arise and may be forecast. A famous example was the Oil Crisis in the 1970s. The sudden increase of prices for crude oil in the 1970s induced savings and the search for alternative energy which caused a decrease of oil prices in the 1980s. In traditional economies the equilibrium corresponds to an optimal result with respect to the particular surroundings. The theorem of decreasing income implies the existence of a single equilibrium point. Economies with the negative feedback of decreasing income are typical for traditional sectors like agriculture, the mining industry, and mass products.

But economic sectors depending on high technical knowledge gain increasing income. The development and production of high-tech goods like computers, software, airplanes, chemicals, and electronics need complex processes of research, experiment, planning, and design, demanding high investments. But if the high-tech products are put on the market, the enlargement of production capacity is relatively cheap and income is going to increase. Thus, modern high-tech industries must be described in dynamical models as generating the positive feedback of increasing income.

Systems with positive feedback have not only a single, but several states of equilibrium which are not necessarily optimal [6.32]. If a product has some accidental competitive advantages on the market, then the market leader will dominate in the long run and even enlarge its advantage without being necessarily the better product. Many examples of modern high-tech industries show that competing products may have approximately equal market shares in the beginning. But tiny fluctuations which increase the market share of a particular product decide its final success. Sometimes the final market leader may even have less quality from a technical point of view.

These effects cannot be explained in the framework of traditional linear dynamics. But they are well known in nonlinear systems. Figure 6.8 illustrates the competition of two technologies with positive feedback. Some trajectories of market shares are drawn on a convex surface. The more a technology can dominate the market, the more easily it can gain a yet greater market

Only A   only B

Fig. 6.8. Nonlinear dynamics of two competing technologies [6.33]

share. As a leading market position is initiated by random fluctuations, it cannot be forecast. In Fig. 6.8, the left curve illustrates the final dominant of technology A. In the two other cases, the technology B will finally dominate the market after initial fluctuations [6.33].

The nonlinear dynamics of these economic models is determined by initial random fluctuations and positive feedback. Obviously, the bifurcation of the possible paths is a kind of symmetry breaking by initial random fluctuations, which is well known in complex physical systems. The reader need only recall the stationary convection rolls that appear in a heated fluid (Fig. 2.20b), where the rolling direction left or right depends on initial random fluctuations.

Besides dissipative systems there are also conservative ones with symmetry breaking. Consider the self-organization of the dipoles in a spin-glass when the temperature is reduced (Fig. 4.9a). In thermal equilibrium, the spins become aligned in the same direction, depending on initial random fluctuations. The dynamics of market shares seem to develop in the same manner. There are many examples of technologies which have made their way because of initial random fluctuations. In the last century neighboring railway companies accept the same gauge in a larger region. The standard gauge was only the result of historical random events and not distinguished by technical reasons.

The behavior of these complex systems is determined by similar evolution equations, like the symmetry breaking of spin-glass systems or ferromagnets. Figure 6.9a shows the evolution of magnetic dipoles in a ferromagnet. Each dipole or elementary magnet can be either up (North) or down (South). A dipole can interact with its nearest neighbors. At a high temperature the

a) b)

Fig. 6.9a,b. Nonlinear dynamics of magnetic dipoles in a ferromagnet (**a**) and nonlinear dynamics of competing technologies (gauges of railway companies) in an economy (**b**) [6.33]

directions of the dipoles are random. If the temperature is reduced, then the elementary dipoles become aligned in the same direction. As the evolution is a kind of symmetry breaking, it cannot be forecast which direction is distinguished in the final state of equilibrium. Figure 6.9b illustrates the analogous process of self-organization for the gauges of railway companies.

Self-reinforcing mechanisms with positive feedback are typical features of nonlinear complex systems in economies and societies. For example, we may speculate why Santa Clara County in California has become the famous Silicon Valley. In the 1940s and 1950s some key individuals (like the employers Hewlett, Packard, and Shockley) founded electronics firms in the neighborhood of Stanford University. These pioneers created a concentration of high-tech engineers and production which has become an attractor for finally more than 900 firms. At the beginning, there were some random fluctuations favoring Santa Clara County. Thus, it is no miracle, but lawfulness in a nonlinear sense that explains *how* Silicon Valley arose. But it is a miracle in the sense of randomness *that* it arose in Santa Clara County.

Nowadays, self-reinforcing mechanics determine the international trade of high-tech technologies [6.34]. The car industry competition between the USA and Japan can be explained in this framework. Initially, the Japanese industry offered small cars on the American market without any resistance from the American car industry, which was traditionally specialized for production of big cars. The Japanese industry could gain market share, lower costs, and raise quality. Thus, the positive feedback enabled the Japanese industry to invade the American market.

The insight into these nonlinear market effects may have major consequences for political decisions. From the traditional point of view, assuming a constant or decreasing income, governments trust in open markets, try to prevent monopolies and hope that industries will support research and technical developments. They believe in a fixed equilibrium of world market prices and refuse to intervene with subventions or tariffs. Their politics is justified in an economy with decreasing income, but it may be dangerous in sectors depending on high technology with increasing income.

Mechanisms operating with increasing income change the equilibria of competition among nations. Even strong national economies may miss out on the development of important technologies. The technology gap between Western Europe and the USA in the 1960s (for instance in the computer industry) is a famous example. Technical standards or conventions have often been established by positive feedback. Examples like the gauge of railways, the English language as the standard language of air navigation, FORTRAN as a computer language, a particular screw thread, and so on often cannot be changed even if alternative techniques or conventions may be better. They have gained too much market share. But superiority at the beginning is no guarantee for survival in the long run.

Nonlinear systems have several possible equilibria and no final stable state. Nonlinear economies which may be identical in the beginning will not choose the same paths of development, due to their high sensitivity and tiny deviations in their initial conditions. Thus, a nonlinear economy with positive feedback cannot be programmed and directed like a computer, because its evolution cannot be forecast in the long run. The theory of complex systems can help to design a global phase portrait of an economic dynamics. But experience and intuition are sometimes more helpful than scientific knowledge for finding the local equilibria of economic welfare. Politicians themselves must have high sensitivity to deal with highly sensitive complex systems.

6.4 Complex Cultural Systems and the Networks of Communication

In the social sciences and humanities one usually distinguishes strictly between biological evolution and the history of human cultures. The main reason is that the development of nations and cultures is obviously guided by the intentional behavior of humans with their attitudes, emotions, plans, and ideals, while systems in biological evolution are assumed to be driven by unintended self-organization. From a microscopic view, we observe single human individuals with their intentions and desires. Even in biological systems like animal ecologies there are individuals with intentional behavior of some degree.

The crucial point of the complex system approach is that from a macroscopic point of view the development of political, social, or cultural order is not only the sum of single intentions, but the collective result of nonlinear interactions. Adam Smith already knew that the distribution of economic wealth and welfare does not result from the social good will of single bakers and butchers. Individual self-interested and egoistic intentions may be counteracting collective interests. Nevertheless, their (nonlinear) interactions achieve collective equilibria by an "invisible hand" (Smith) or the "cunning of reason" (*List der Vernunft*) (Hegel).

Nonlinear systems of individuals with intentional behavior may be more complex than, for instance, a physical system of atoms or a chemical mixture of molecules. In Sect. 4.3–4, intentional behavior and consciousness were modeled as self-referential global states of complex neural systems which are caused by nonlinear interactions of neurons. The emergence of collective order phenomena with more or less complex degree is an intrinsic common feature of all nonlinear systems which is not necessarily linked with consciousness. A political state as a collective order of a human society obviously has no kind of consciousness or intelligence, as Hegel mistakenly believed, although its emergence may be modeled by a kind of phase transition caused by the nonlinear interaction of conscious humans with intentional behavior.

Thus, in the mathematical framework of complex systems the concept of "evolution" does not in general refer to the particular mechanisms of biological evolution. In a complex system, the so-called evolution equations describe the dynamics of its elements, which may be elementary particles, atoms, molecules, organisms, humans, firms, and so on. Another concept with a broad meaning is complexity itself. In the context of the social sciences there are many aspects of complexity, some of which are designed in Fig. 6.10 [6.35].

In the mathematical framework of complex systems in this book, complexity is at first defined as nonlinearity, which is a necessary but not sufficient

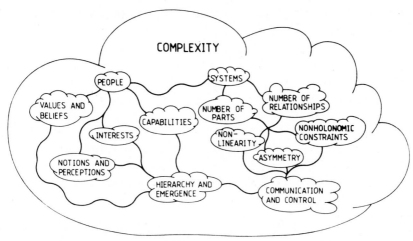

Fig. 6.10. Meanings of complexity [6.35]

condition of chaos and self-organization. On the other hand, linearity implies the superposition principle, which says in a popular slogan "the whole is only the sum of its parts". A second important aspect of complexity is defined by the structure of algorithms which was discussed in Sect. 5.2. Complexity theory in computer science provides a hierarchy of complexity degrees, depending on, for instance, the computational time of computer programs or algorithms. As nonlinear complex systems are sometimes modeled by computer graphics, their degree of algorithmic complexity may be described as their capacity for self-organization. This relationship has been explored in the theory of cellular automata (compare Sect. 5.3), in which different classes of self-organizing complex systems are modeled.

In the social sciences, the complexity of a highly industrialized society mainly consists in the great number of its citizens and their relationships, its organizational substructures, and their dependencies [6.36]. We should remember that in a complex system it is not the great number of its elements that is essential for the emergence of collective (synergetic) order phenomena, but their nonlinear interactions. The reader may recall the astronomical three-body problem with chaotic trajectories as possible solutions.

In the mathematical framework of complex systems, a physical or biological reductionism of human history and sociocultural development is not justified in any case. Models of social and cultural developments must be discussed with their particular constraints and limitations. A severe methodological problem is that of providing empirical tests and confirmation of these models. Thus, the computer-assisted simulation of complex cultural systems has become a crucial instrument for providing new insights into their dynamics which may be helpful for our decisions and actions.

Historically, the interest of social sciences in nonlinear problems dates back to Thomas Malthus [6.37]. He argued that a population would outstrip its food supplies, since the population grows exponentially whereas the food supplies grow only linearly. In 1844, Verhulst modified the exponential equation, arguing that the rate of population growth is proportional to the product of the population and the difference between the total amount of resources and the amount of those resources consumed by the existing population. His famous logistic curve with its characteristical attractor of equilibrium has been used in demography, economics, and many other examples of social sciences. It provides a cascade of possible bifurcations and phase transitions including chaos.

The evolution of a predator–prey ecosystem described by Volterra and Lotka is another model which has been applied to social sciences. For instance, the Lotka–Volterra model helps us to understand the appearance of an agricultural society. Because of their ability to learn, humans can change their coefficients of interaction with the environment faster than nature can genetically evolve countermeasures. A human society which only tries to improve its hunting abilities for its survival diminishes the prey population. Then the society will be diminished, too. Consequently, both predator and prey population will become extinct. But agriculture involves developing the ability to increase the prey birthrate. Thus, human population increases and can be stabilized at a certain equilibrium.

The evolution of biological systems is governed by their genes. In Darwinian evolution a new type of individual emerges by the natural selection of mutants which appear spontaneously. In populations of higher animals the new possibility of behavioral change and adaption by imitation arises. In human societies, there are still more sophisticated strategies of learning, which dominate the behavioral patterns. Societies have developed particular institutions like the legal system, the state, religion, trade, and so on to stabilize the behavioral change for the following generations.

The complex system approach provides the essential insight that neither genetic evolution nor the evolution of behavior needs a global program like a supervising divine will, a vital force, or a global strategy of evolutionary optimization. The survival of genes or the emergence of global patterns of behavior are explained by local interactions of individuals composing the system. To make this point clear, it may be a question of religious or political *Weltanschauung* whether there is a "global program" like God, History or Evolution. In the methodological framework of complex systems, these hypotheses are not necessary for explanations and superfluous in the sense of Ockham's razor and his economy of theoretical concepts.

Obviously, nonlinear systems like biological organisms, animal populations, or human societies have evolved to become more and more complex. Our present society, when compared to Aristotle's *polis* or the political system of the physiocrats, is characterized by a high degree of institutional complexity and information networking. In the last century, Herbert Spencer already maintained that increasing complexity is the hallmark of evolution in general: "Evolution is an increase in complexity of structure and function ... incidental to the ... process of equilibration ... " [6.38]. Spencer still argued in the thermodynamic framework of thermal equilibrium.

In the framework of thermodynamics far from thermal equilibrium, there is not just a single fixed point of equilibrium, but a hierarchy of more or less complex attractors, beginning with fixed points and ending with the fractal structures of strange attractors. Thus, there is no fixed limit of complexity, either in biological or in sociocultural evolution, but there are more or less complex attractors representing metastable equilibria of certain phase transitions which may be overcome if certain threshold parameters are actualized. The structural stability of a society is related to these more or less complex attractors.

The traditional functionalist view of homeostasis and self-regulating systems stems from the concept of a technical thermostat. It may perhaps help us to understand why societies stay the same, but not why they change and why equilibria have been overthrown. In the framework of complex systems, the dynamics of a society is understood in terms of phase transitions of a dissipative system exchanging material, energy, and information with its environment. The institutions of a society are dissipative structures which may emerge and stay the same in a particular interval of threshold conditions. For instance, in neolithic villages, the institution of farming changed from dry farming to irrigation when the food supply was no longer secured by the established social structures.

In the history of industrialized societies, we can distinguish more or less strong economic fluctuations which may initiate the crash of social institutions and the emergence of new ones. For instance, the economic depression of the USA in 1922 was relatively mild and short-lived and did not produce a structural change of the society. In contrast to that phase of American history, the stock market crash of 1929 had a genuine butterfly effect in initiating the Great Depression of 1933 [6.39]. This crisis caused the financial ruin of many firms and a huge number of unemployed, and could not be managed by the established social institutions. The threshold parameters of the established structures were exceeded. New social institutions like the Securities and Exchange Commission, the Federal Deposit Insurance Corporation, and the Public Works Administration came into being to overcome the effects of depression and to prevent excessive fluctuations in the future business cycle. This Keynesian reaction by American society became famous as the New Deal legislation of President Roosevelt.

But as we have learnt from the neoclassic economists and the experience of social development after the Second World War, an optimizing strategy of public welfare may initiate a self-dynamics of administrative bureaucracy which paralyses economic initiatives and counteracts the originally good intentions. Overreactions may be as dangerous as no reactions for the structural stability of a system. On the other hand, the history of political revolutions shows that societies can totally lose their structural stability and achieve new constitutions, institutions, and social structures, but, of course, without any guarantee of durability.

From a methodological point of view, the question arises of how to represent the sociocultural evolution of societies in the mathematical framework of complex systems. The recognition of attractors and equilibria needs a phase portrait of the sociocultural dynamics which presumes the definition of a "sociocultural state" and a "sociocultural state space". But what is the sociocultural state space of Victorian England or the Weimar Republic? These questions demonstrate obvious limitations. What are the possibilities of the complex system approach in the historical and social sciences?

It cannot be the aim of research to represent a historical period in a complete mathematical state space. The relevant data are often unknown, contingent, and not quantified. In the last section, complex systems with state spaces and dynamical phase portraits were used to model the economic evolution of human societies. Economists do not claim to represent the complete economic development of the Weimar Republic, for instance. But nonlinear endogenous or linear exogenous models of business cycles can describe typical economic scenarios that influenced or depended on situations in political and cultural history.

Economic models are not studied for their own sake. Economists want to understand economic dynamics in order to support practical decisions by better structural insights. The economic dynamics of a society is embedded in its global sociocultural development. In view of its complexity, attempts have been made to model the sociocultural development only of such subsystems as urban centers. These models may grasp the typical features of evolving urban systems, which can help politicians and citizens to make better decisions in appropriate stituations.

In modern industrialized societies, there is a great variety of centers of all sizes, forms, and characters, from very large cities with high densities to small villages with few inhabitants. We may ask what is the reason for the spatial distribution of these different centers and what will be their evolution in time. To answer this, we need to know the global spatio-temporal state of an urban system resulting from local nonlinear interactions of its agents – individuals, families, firms, administrators, and so on – who may pursue different cooperative or conflicting interests. The structure of an urban center depends on commercial and industrial interests, the flows of goods and services, traffic connections, cultural attractiveness, and ecological demands. These factors are to be made precise as measurable quantities. The urban system has several exchanges with the outside world. Thus, it can be interpreted as a dissipative structure modeled by a complex dynamical system.

Peter Allen has suggested a system with evolution equations expressing the nonlinear interactions of its different actors. The spatio-temporal structure of the urban system with its changing centers and concentrations of inhabitants is not the trival sum of its actors. It is not the result of some global optimizer or some collective utility function, but the result of the instability of successive equilibria caused by nonlinear phase transitions. In this sense, the evolution of an urban system is not ruled by a Platonic king (or dictator), not constructed by a Cartesian architect or forecast by a Laplacean spirit. In the mathematical framework of complex systems the form of an urban system seems to grow like a living organism.

In Allen's analysis [6.40], the geographic space of the urban system is simulated by a triangular lattice of 50 local points (Fig. 6.11). The growth of the urban system is determined by two equations describing the change of the population on the local points and the evolution of the employment offered at the points. The local population and the local capacity for employment are linked by the urban multiplier as a positive feedback. The employment concentration offers externalities and common infrastructure which in turn gives rise to a positive feedback, while the residents and entrepreneurs together compete for space in a center that provides negative feedback.

The computer-drawn pictures Fig. 6.11a–e show the evolution of the population distribution of a region that starts off initially an area with no interaction between local centers. The urbanization process is revealed in phase transitions with changing local attractors. In Fig. 6.11b at time $t = 12$ units, the structure is beginning to solidify around five main centers. In Fig. 6.11c, the central core of the largest center is going through a maximum. In Fig. 6.11d at time $t = 34$, the basic structure is essentially stable. Two centers have undergone central core decay. In Fig. 6.11e, the basic pattern is stable. The decline, centralization, and decentralization result from complex nonlinear dynamics.

Figure 6.11a–e form a speeded-up motion picture of the urban system's global evolution. Each picture is a phase portrait of a global dynamical state at a particular time. The model is simplified, of course. But it can be enlarged by integrating more functional aspects and by studying their more complicated nonlinear interactions. Nevertheless, the potential of the model for the exploration of decision alternatives can be analyzed in computer-simulated case studies. Either local or global changes can be imposed on the system. These simulations are extremely interesting for a government.

Fig. 6.11a–e. Computer-assisted model of urban evolution at time (**a**) $t = 4$, (**b**) $t = 12$, (**c**) $t = 20$, (**d**) $t = 34$, (**e**) $t = 46$ [6.40]

e)

A possible strategy may be to interfere in the urban structure by placing a specific investment at a particular point. This decision strategy corresponds to the development of a hitherto undeveloped region in an urban system. Investments mean not only economic measures but also cultural attractions and traffic connections, for instance. Sometimes, an investment may intiate a local butterfly effect with global consequences that counteract the good intentions of the urban planners. This may occur because of the model's nonlinearity, which limits the possibilities of forecasting in the long run.

An urban dynamics is a practical example of a complex system demonstrating that the good will of single individuals is not enough and may even be dangerous if the nonlinear consequences are neglected. The collective effects of individual actions characterize our society. Decisions should be made, as far as possible, with an awareness of these collective effects. The importance of these results lies not only in computer simulations of concrete decisions and their nonlinear effects. Even citizens who are not engaged in concrete planning activities must be aware of the complex interdependencies in a society.

The demagogic demand for a strong political leader who can solve all problems is not only dangerous from a democratic point of view. It is false for mathematical reasons due to the complexity characterizing modern highly industrialized societies. On the other hand, we should not hold out high hopes for single politicians or parties and then react with total political frustration when our exaggerated expectations are disappointed. Human societies are characterized by the intentionality of their members. Nevertheless, they are governed by the nonlinear laws of complexity like atomic groups, molecular mixtures, cellular organisms, or ecological populations.

Sociological theories considering the epistemological consequences of complexity and nonlinearity are still in their infancy. The development of an appropriate statistical mathematics capable of handling the complexity of social problems could serve as a bridge to traditional concepts of sociology. In the complex system approach, social phenomena are described by nonlinear equations. When, for instance, Emil Durkheim [6.41] speaks of solidarity in society, then we may ascribe the functional aspects of this concept to nonlinear and collective effects of complex systems. We can distinguish political decisions as "linear", corresponding to "individual" choices, and "nonlinear", corresponding to the institutional environment of administrations, mass media, political parties, for instance. The actions and reactions of many citizens and institutions may be understood as fluctuations inherent to the statistical description of the society. The deterministic character of the society reflects only the averages of the distribution functions, which develop in time according to nonlinear laws like a master equation.

This approach to socio-economic dynamics has been realized by the Stuttgart school of Wolfgang Weidlich. The mathematical modeling methods are derived from synergetics and statistical physics, and allow a quantitative description of collective developments in the society. Synergetics suggests a relationship between the micro-level of individual decisions and the macro-level of dynamical collective processes in a society. The distinction between micro- and macro-economics or micro- and macro-sociology are well known traditional concepts in the social sciences. The synergetic approach of Weidlich aims at a probabilistic description of macro-processes including stochastic fluctuations and derivation of a quasi-deterministic description neglecting the fluctuations.

Models of solutions can be investigated either by analytical methods, for instance, the exact or approximate solution of master equations or average equations, or by numerical and computer-assisted simulations of characteristic scenarios. Empirical systems can be analyzed and evaluated by determining the model parameters with field research on the micro-level or by estimating future developments with model simulation. The methodological framework of the synergetic approach to modeling social dynamics is illustrated in Fig. 6.12 [6.42].

The synergetic concept of modeling has been applied to several questions in the social sciences, for instance, collective formation of political opinions, demography, migration of populations, and regional geography. The synergetic concept is especially appropriate for integrating the interactions of several sectors of a society like, such as the relationship between the economy

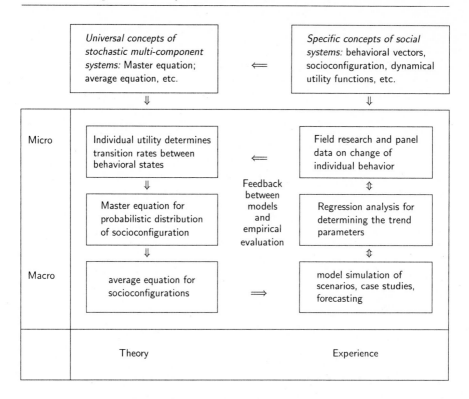

Fig. 6.12. Synergetic methodology for modeling social dynamics [6.42]

and collective formation of political opinions, or the interaction between the economy and processes of migration. Migration is a very dramatic topic nowadays, and demonstrates how dangerous linear and mono-causal thinking may be. It is not sufficient to have good individual intentions without considering the nonlinear effects of single decisions. Linear thinking and acting may provoke global chaos, although we act locally with the best intentions.

According to the synergetic approach, a socio-economic system is characterized on two levels, distinguishing the micro-aspect of individual decisions and the macro-aspect of collective dynamical processes in a society. The probabilistic macro-processes with stochastic fluctuations are described by the master equation of human socioconfigurations. Each components of a socioconfiguration refers to a subpopulation with a characteristic vector of behavior. Concerning the migration of populations, the behavior and the decisions to rest or to leave a region can be identified with the spatial distribution of populations and their change. Thus, the dynamics of the model allows us to describe the phase transitions between different global macro-states of the populations.

The empirical administrative data can be used to test the theory. The models may concern regional migration within a country, motivated by different economic and urban developments, or even the dramatic worldwide migration between the "South" and the "North", between the poor countries and the highly industrialized states of Western Europe and the USA, driven by political and economic depression. Physical transport or migration of animal populations are often uncontrolled, random, and linear, without interaction between the elements and collective aggregations. But human migration is intentional (driven by considerations of utilities) and nonlinear, because the transition rates depend not linearly on the whole socioconfiguration.

The migration interaction of two human populations may cause several synergetic macro-phenomena, such as the emergence of a stable mixture, the emergence of two separated, but stable ghettos, or the emergence of a restless migration process. In numerical simulations and phase portraits of the migration dynamics, the synergetic macro-phenomena can be identified with corresponding attractors. Figure 6.13a,b shows the homogenous mixture of both populations with weak trends of agglomeration and segregation in each population. Figure 6.13a is the phase portrait of the average equations with a stable point of equilibrium. Figure 6.13b shows the stationary solution of the master equation and a stationary probabilistic distribution with a maximum at the origin. Figure 6.14a,b shows the emergence of two stable ghettos with a weak trend of agglomeration and a strong trend of segregation between the populations. Figure 6.14a is a phase portrait with two stationary fixed points, while Fig. 6.14b describes the probabilistic distribution with maxima at the stationary points.

Figure 6.15a,b shows a moderate trend of agglomeration in each population and a strong asymmetric interaction between the populations. Figure 6.15a illustrates a vortex picture and Fig. 6.15b the corresponding probabilistic distribution with a maximum at the origin. Figure 6.16a,b corresponds to a restless migration process with a strong trend of agglomeration in each population and a strong asymmetric interaction between the populations. The phase portrait 6.16a shows a limit cycle with unstable origin. The stationary probabilistic distribution of Fig. 6.16b has four maximum values with connecting ridges along the limit cycle. Sociologically, this case may be interpreted as sequential erosion of regions by asymmetric invasion and emigration of the populations.

If we consider three instead of two populations in three regions, then the phenomenon of deterministic chaos emerges in the nonlinear model of migration. Some numerical simulations have a strange attractor as the final

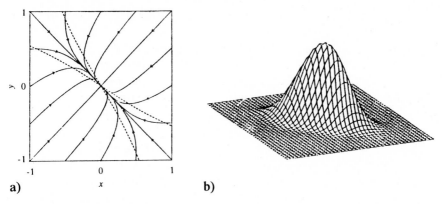

Fig. 6.13a,b. Migration dynamics with a stable point of equilibrium: (a) phase portrait, (b) probabilistic distribution [6.42]

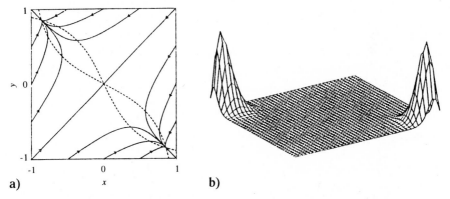

Fig. 6.14a,b. Migration dynamics with two stationary fixed points [6.42]

state of trajectories. In other cases, a successive bifurcation becomes more and more complex with a final transition to chaos.

Another practical application of the complex system approach is management and organizational sociology [6.43]. Actually, modern firms have begun to reorganize and decentralize their large organizations, in order to enable successful strategies in spite of the increasing complexity of their problems. They have started to support organizational fluidity, for instance, with new pathways that allow project-centered groups to form rapidly and reconfigure as circumstances demand. Fluid organizations display higher levels of cooperation than groups with a fixed social structure. Faced with a social dilemma, fluid organizations show a huge variety of complex cooperative behaviors caused by the nonlinear interplay between individual strategies and structural changes.

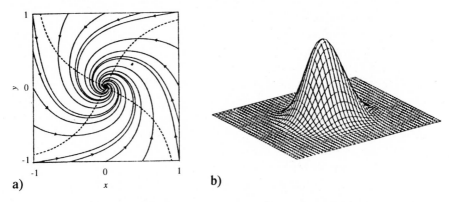

Fig. 6.15a,b. Migration dynamics with a vortex [6.42]

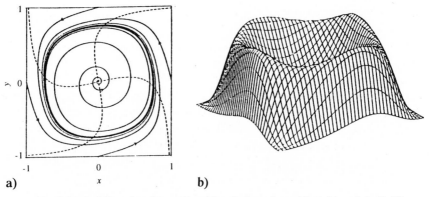

Fig. 6.16a,b. Migration dynamics with a limit cycle and instable origin [6.42]

Thus, the dynamics of these social groups may by modeled by complex systems. Computer simulations can deliver global insights into the development of possible behavioral patterns which may help managers to achieve appropriate conditions of development. If models of complex systems are applicable, then forecasting in the long run and overall control by a centralized supervisor is, of course, excluded.

The model consists of intentional agents making choices that depend on their individual preferences, expectations, and beliefs as well as upon incomplete knowledge of the past. Patterns of cooperation merge from individual choices at certain threshold values. An agent will cooperate when the fraction of the group perceived as cooperating exceeds a critical threshold. Critical thresholds depend on the group size and the social organizational structure emerging from the pattern of interdependencies among individuals. The potential for cooperative solutions of social dilemmas increases if groups are allowed to change their social structure. The advantages of organizational fluidity must be balanced against possible loss of effectiveness. The effectiveness of an organization may be measured by its capability to obtain an overall utility over time.

In firms, there is generally an informal structure emerging from the pattern of affective ties among the agents as well as a formal one dictated by the hierarchy. The informal structure is achieved by a kind of self-organization and can be represented by the sociometric structure of a social network. This approach dates back to early sociological explorations on social networks of urban families in the 1950s and has developed to a sophisticated computer-assisted instrument of sociology [6.44]. From the micro-view of individual interdependencies a global perspective on social structures emerges.

In Figs. 6.17 and 6.18, these structures are visualized as trees. Each branch represents a subdivision of the higher level in the hierarchy. The modes at the lowest level represent individuals, with filled circles marking cooperators and open circles marking defectors. The number of organizational layers which separate two individuals is determined by the number of modes backwards up the tree from each individual until a common ancestor is reached. The distance between two agents in the organization is measured by the number of separating layers. The larger the distance between two agents, the less their actions are assumed to affect each other. Thus the tree illustrates the amount and extent of clustering in a group.

The crucial question is how the structure and fluidity of a group will affect the dynamics of cooperation. Fluidity depends on how easily individuals can move within the social structure, and how easily they can break away on their own, extending the structure. In the framework of complex systems, macroscopic properties of the system are derived from the underlying interactions among the constituents, which are mathematically modeled by nonlinear evolution equations [6.45].

Figure 6.17a–d, due to Glance and Huberman, shows a computer simulation of some phase transitions in a fixed social structure, a three-level hierarchy consisting of three large clusters, each subsuming three clusters of three agents [6.46]. The final overall cooperation of Fig. 6.17d has been initiated by the actions of a few agents clustered together in Fig. 6.17a. These agents reinforce each other and at the same time can spur agents one level further removed from them to begin cooperating. This increase of cooperation can affect cooperation in agents even further removed in the structure.

A tiny act of cooperation within a hierarchy can initiate a widespread transition to cooperation within the entire organization. This cascade of increasing cooperation leads to a fixed point of equilibrium. But groups with fixed structures easily grow beyond the bounds within which cooperation is sustainable. In this case, the group rapidly evolves into its equilibrium state of overall defection. But even in these bounds, cooperation patterns may be metastable in the sense that the agents remain cooperating for a long time,

Fig. 6.17. Social dynamics in a fixed three-level hierarchy: the nodes at the lowest level of each tree represent individuals, with filled circles marking cooperators and open circles marking defectors [6.46]

until suddenly a symmetry breaking takes place with a transition to overall defection.

In fluid structures, individual agents are able to move within the organization. Individuals make their decision to cooperate or defect according to the long-term benefit they expect to obtain. In order to evaluate his or her position in the structure, an individual compares the long-term payoff he or she expects if he or she stays with the long-term payoff expected if he or she moves to another location, chosen randomly. When evaluating his or her position the individual also considers the possibility of breaking away to form a new cluster if he or she feels there is nothing to lose. How easily agents are tempted to break away determines the breakaway threshold, which is given as a fraction of the maximum possible payoff over time.

Figure 6.18a–e shows snapshots of a phase transition in a fluid organization [6.47]. Initially (Fig. 6.19a) all members of the group, divided into four clusters of four agents each, are defecting. Figure 6.18b shows that nearly all the agents have broken away on their own. In this situation, agents are much more likely to switch to a cooperative strategy, as realized in Fig. 6.18c. Because of uncertainty, agents will occasionally switch between clusters (Fig. 6.18d). When a cluster becomes too large, a transition to defection may start within that cluster. At this phase of transition (Fig. 6.18e), more and more

Fig. 6.18. Social dynamics in a hierarchy with fluid structure [6.47]

agents will break away on their own, and a similar development is repeated. Cycles of this type have been observed to appear frequently in simulated organizations.

As in the case of urban growth (Figs. 6.11) or migration dynamics (Figs. 6.13–6.16), computer experiments simulating social organizations cannot deliver deterministic forecasts of individual behavior, but they help one to understand the sensitivity and complexity of social dynamics. Thereafter, appropriate circumstances and conditions may be achieved that allow the living conditions of humans in the corresponding social systems to be improved.

Models of sociocultural evolution must consider several interacting sections of a society. If a society is composed of overlapping layers and sectors of dissipative structures, we must find an appropriate picture of how these come about. In complex developments involving interacting dissipative systems, the emergence of new macro-structures is an uncentralized and unplanned event in the long run. Moreover, each section includes many more or less unclear human ideas, feelings, and projections, motivating and inspiring human actions. They cannot be directly identified as measurable quantities, because they are dissolved in a kind of mainstream which is called "lifestyle". Nevertheless, the lifestyle of a society is a typical sociocultural macro-phenomenon depending on several interacting factors which can be identified, like conditions relating to the economy, technology, work, travel, ecology, and the mass media.

In modern times, the evolution of technologies has been a driving force for change, influencing numerous lifestyle components. With respect to self-organizing processes, it is a significant feature of technological developments that they are autocatalytic, each innovation catalyzing the generation of the next. If technological and social evolution is interpreted as a consequence of sequential replacements of a technique, a mainstream idea ("paradigm"), or an artifact by another, then the development from growth to saturation can be mathematically modeled by a pattern of interdependent logistic curves. The assertion that a technology advances through a series of phase transitions and substitutions implies that it might be considered as a succession of logistic curves. Each curve reaches a saturation level. With each level of evolutionary innovation a transition to a new logistic curve develops.

In Sect. 6.3, we have already discussed that these phase transitions in technology are connected with the growth and decline of economies. The development of computer and information technologies has influenced the lifestyle of mankind in nearly all sectors as never before in history. It seems to be comparable with a quasi-evolutionary process producing computers and information systems of increasing complexity. Computer scientists use the terminology of Herbert Spencer when they speak of computer generations overwhelming the complexity of earlier generations. Actually, the complexity of the systems' functionality has increased. But on the other hand, the complexity of problems, which is measured by their computational time, for instance, has been reduced. The reduction of problem complexity is an order parameter of the quasi-evolutionary process of these technologies.

Computers and informational systems have become crucial techniques in sociocultural development, evolving in a quasi-evolutionary process. The replicators of this process are any of the information patterns that make up a culture and spread with variation from human to human. As humans, unlike molecules or primitive organisms, have their own intentionality, the spreading process of information patterns is realized not via mechanical imitation but via communication. By analogy with genes, these replicators are sometimes called "memes" [6.48]. They include ideas, beliefs, habits, morals, fashions, techniques, and so on.

Any pattern which can spread via communication of information is a meme, even if its human host cannot articulate it or is unaware of its existence. It is important to recognize that the replicators of human culture are memes, not people. Our ability to change our minds allows cultural evolution to proceed not by selection of humans, but by "letting our theories die in our stead", as Karl Popper has proclaimed [6.49].

In the framework of complex systems, we may, of course, speak of the system's "evolution" in the abstract sense of mathematical evolution equations. Biological evolution with its special biochemical mechanisms is only a special model of the general mathematical scheme characterizing complex systems. Thus, the evolutionary character of human culture cannot be reduced to the biochemical mechanisms of biological evolution. But concepts like "memes" should not be misunderstood as a merely social-Darwinistic jargon. They may illustrate basic features of complex systems which can be mathematically defined and empirically tested.

In this sense, the development of a worldwide communication network can be interpreted as the evolution of complex systems for aiding the spread of memes among humans and for establishing a memetic ecosystem [6.50]. The memes making up human culture are diverse, as are their variation and selection mechanisms. In Sect. 6.3, we have already discussed "economic memes" with their selection mechanism of markets. Economic markets are more or less open to their own environment of human society. They operate under a wide range of more or less rigorously enforced rules, imposed by a variety of legal and regulatory institutions.

In human societies, legal systems and governmental activities provide a framework for the market. In the framework of complex systems, they are not protected from evolutionary forces. They evolve within a political ecosystem with its own mechanisms for the variation and selection of laws [6.51]. Some political memes, like political desires, slogans, or programs, may become attractors in the dynamical phase portrait of a society. In an open democratic society, they may emerge, but also decline, if their attractiveness decreases due to the selection pressure of competing alternatives.

The capability to manage the complexity of modern societies depends decisively on an effective communication network. Like the neural nets of biological brains, this network determines the learning capability that can help mankind to survive. In the framework of complex systems, we have to model the dynamics of information technologies spreading in their economic and cultural environment. Thus, we speak of informational and computational ecologies. There are actually realized examples, like those used in airline reservation, bank connections, or research laboratories, which include networks containing many different kind of computers.

Incomplete knowledge and delayed information are the typical features of open computational systems which have no central controls. These large networks, emerging from the increasing connectivity between diverse computer-assisted information centers, are becoming self-organizing systems which are different from their individual program-controlled components. Their unplanned growth leads to an immense diversity of technical composition and use, with increasing difficulties of interoperability. The horrifying vision of a separated robotic world enslaving human culture may be the ultimate consequence of many possible uncomfortable scenarios.

As the worldwide growth of local information and computation centers cannot be planned by a centralized processor, it seems to have a nonlinear dynamics which should be studied in the framework of complex systems. Even simplified case studies would provide crucial insights in the complex dynamics of modern sociocultural evolution. The complex interdependencies of computational ecologies violate the traditional requirements for a hierarchical decomposition into technical, industrial, or administrative modules as used in traditional management. Modern technical communication networks are growing open systems which must be used without central control, synchronicity, or consistent data from other agents like machines or humans. Thus, the dynamical theory of informational and computational ecologies which incorporates the features of incomplete knowledge and delayed information will provide well known evolutionary patterns like fixed points, oscillations, or chaos.

Imperfect knowledge leads to an optimality gap, while delays in information access induce oscillations in the number of the agents involved. Synergetic effects may arise by cooperation and competition for finite resources. Chaos prevents any stable strategy of problem solving. Marvin Minsky has studied a simplified model of collective problem solving by nearly independent agents working on a set of related problems and interacting with each other. This example can be used as a design model for a distributed system of computer-assisted information systems [6.52].

Bernardo A. Huberman and Tad Hogg analyzed a model with a number of centers or agents capable of choosing among various possible strategies in order to complete various assigned tasks (Fig. 6.19) [6.53]. The strategies correspond to their perceived payoffs. Agents repeatedly evaluate the perceived payoff for different strategies and switch to the one with the highest payoff. The evolution equation of the model determines the rate at which processes change their strategies. As only a fraction of the agents decide to switch during a time interval, the dynamics of the system must be modeled by probabilistic equations.

The variety of behaviors which the system can exhibit range from fixed points to oscillations and chaos. Unavoidable oscillations in the system can be induced by delayed information. Imperfect knowledge and delays can even effect chaotic behavior. As the payoffs depend on which alternative strategies are preferred, the dynamics and the initial conditions will prevent predictions of the behavior in the long run.

The world is going to be populated with a huge number of computing systems with increasing complexity [6.54]. Among them will be traditional von-Neumann machines, vectorizing super-computers, shared-memory multiprocessors, connection machines, neural-net simulators, millions of personal

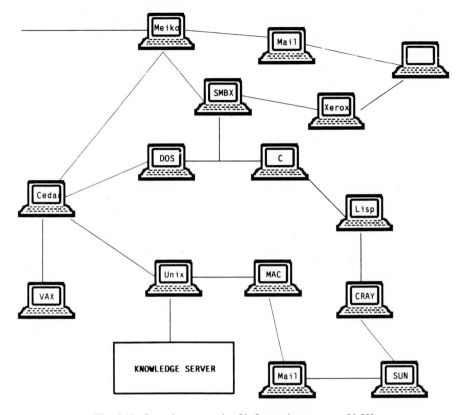

Fig. 6.19. Complex network of information systems [6.53]

computers overcrowding the world like amebas, and future molecular machines. These computing systems are becoming more and more linked with
such information systems as satellites, phones, and optical fiber. The idea of
a self-organizing worldwide network of software and hardware systems has
become reality.

For example, in 1969, a computer-assisted information network was established in the U.S. Department of Defense. It has become the nucleus of
a pullulating network with 175 000 computers, 936 subnets, and innumer-
able humans. The growth of INTERNET was not planned or controlled by
any central processor unit, but a more or less anarchic process. Nevertheless,
patterns of organization both arise from chaos or and decline in a kind of
global self-organization.

In complex information networks, knowledge and information is distributed among several centers and individual programmers. Their complexity excludes central planning. Like all systems involving goals, resources,

and actions, computation has been described in economic terms. Obviously, there is already a computational market for software and hardware using market mechanisms. As we have seen in Sect. 6.2–3, markets are a form of self-organizing complex ecosystem. According to Smith's fundamental insight, the force of consumer choice can make computational market ecosystems serve human purposes better than any programmer or central processor could plan or understand. The reason is the enormous complexity and diversity of computational ecosystems linked to the human market.

The reader may be reminded of the complexity of a biological ecosystem with several levels like cells, organs, and organisms. The elements of a computational ecosystem are likewise grouped on different levels according to the increasing complexity of its computational systems. Figure 6.20 shows the global network of USENET, which has grown through many local initiatives [6.55]. The network in its present state of development consists of 37 000 nodes, which is a fraction of INTERNET. If one wants to send or receive electronic post or use other information services, one only has to find a local USENET station to set up the desired global connections.

However, the analogy between computational ecosystems and biological ecosystems does not imply any reductionism or biologism. In the framework of complex systems, computational ecosystems and biological ecosystems are only models of mathematical evolution equations characterizing the nonlinear dynamics of complex systems. According to Minsky, the computer-

Fig. 6.20. Global network of information systems [6.55]

assisted networks all over the world can be interpreted as a "marketplace" or "society of mind" [6.56].

Philosophically, this new kind of worldwide "knowledge medium" or computer-assisted "intelligence" may remind us of Hegelian ideas concerning the emergence of an "objective mind" embodied in human societies and their legal systems, economies, and bureaucracies, that overcomes the "subjective mind" of individual humans [6.57]. But these computational ecosystems have neither the consciousness nor the intentionality possessed by the cerebral neural networks of human individuals (compare Chap. 4).

Nevertheless, the question arises of whether computational ecosystems may be called "intelligent" to some degree. Individuals taking intelligence tests are judged by their ability to achieve goals set by a test-giver using time provided for the purpose. In this context, intelligence is no metaphysical universal concept; instead, there are several more or less well defined and testable standards of behavior. Some authors have suggested judging the "intelligence" of a society likewise by its ability to achieve goals set by certain legitimated subgroups (for instance, parliaments), using resources provided for the purpose. The degree of "intelligence" would depend on the range of goals that can be achieved, the speed with which they can be achieved, and the efficiency of the means used. The details of these definitions may differ, but "intelligence" based on these concepts would actually refer to a macro-feature of the whole system. The practical purpose of these definitions would be to compare computational ecosystems and their degrees of successful problem solving. In contrast, terms like the collective "intelligence" of a nation are ideologically dangerous. Furthermore, we should be aware that technical standards of intelligence are to be distinguished from concepts like consciousness which have actually been realized by the self-referentiality of certain brain networks in biological evolution.

The growth of informational and computational ecosystems is connected with a fundamental change of society characterized by a switch from traditional industries handling goods to knowledge industries for information and an economy of information services. The production, distribution, and administration of information have become the main acitivites of modern knowledge based societies. Thus, the interface between humans and information systems must be continously improved in order to realize the ideal of worldwide human communication. Human means of expression like speech, gestures, or handwriting should be understood immediately by computation and information systems. The "whole-person paradigm" and the "human-machine network" are highlights in the future world of communication.

Human communication involves not only bit strings of information, but also intuition, feeling, and emotion. The future world of communication is sometimes called a "global village" to emphasize the degree of familiarization created by the high-tech environment. But its acceptance decisively depends on the realization of human-friendly interfaces. We have to consider a new kind of complexity, referring to human intuition and emotion. The old ideals of rationality abstracting from these essentials of human life are absolutely ignorant of the human world. Even the process of scientific research is inspired by human intuition and driven by human emotions that must be considered in the future world of communication.

Some people are afraid that the final attractor of sociocultural evolution will not be a global village or worldwide *polis*, but a gigantic "Leviathan" dominating mankind with the effectiveness of modern high-tech procedures. In the computerized world of virtual realities, all human means of expressions would be digitized without any niche for personal intimacy. But the complexity of sociocultural evolution could allow several attractors. They cannot be forecast or determined by human decisions, but they may be influenced by conditions and constraints humans are able to achieve. What is the chance of human freedom in a world of high complexity? What is the degree of individual responsibility in a complex world of collective effects with high nonlinearity? These questions lead to an epilogue on ethics in a complex and nonlinear world.

7. Epilogue on Future, Science, and Ethics

The principles of complex systems suggest that the physical, social, and mental world is nonlinear and complex. This essential result of epistemology has important consequences for our present and future behavior. Science and technology will have a crucial impact on future developments. Thus this book finishes with an outlook on future, science, and ethics in a complex and nonlinear world. What can we know about its future? What should we do?

7.1 Complexity, Forecasts, and Future

In ancient times the ability to predict the future seemed to be a mysterious power of prophets, priests and astrologists. In the oracle of Delphi, for example, the seer Pythia (6th century B.C.) revealed the destiny of kings and heroes in a state of trance (Fig. 7.1) In modern times people came to believe in the unbounded capabilities of Laplace's demon: Forecasting in a linear and conservative world without friction and irreversibility would be perfect. We only need to know the exact initial conditions and equations of motion of a process in order to predict the future events by solving the equations for future times. Philosophers of science have tried to analyze the logical conditions of forecasting in the natural and social sciences [7.1]. Belief in man's forecasting power has been shaken over the course of this century by several scientific developments. Quantum theory teaches us that, in general, we can only make predictions in terms of probabilities (cf. Sect. 2.3). A wide class of phenomena is governed by deterministic chaos: Although their motions obey the laws of Newtonian physics, their trajectories depend sensitively on their initial conditions and thereby exclude predictions in the long run. In dissipative systems, such as the fluid layer of a Bénard experiment (Fig. 2.20), the emergence of order cannot be predicted, but depends on microscopically small initial fluctuations. A random event such as the stroke of a butterfly's wing, can, in principle, influence the global dynamics of weather. In Chap.

Fig. 7.1. Aigeas, king of Athens, asking the Oracle at Delphi about his future (Greek bowl: 440–430 B.C.)

6 we have learnt that patterns and relationships in economics, business, and society sometimes change dramatically. Going beyond the natural sciences, people's actions, which are observed in the social sciences, can and do influence future events. A forecast can, therefore, become a self-fulfilling or self-defeating prophecy that itself changes established patterns or relationships of the past. Is forecasting nothing more than staring into a crystal ball?

But nearly all our decisions are related to future events and require forecasts of circumstances surrounding that future environment. This is true for personal decisions, such as when and whom to marry or when and how to invest savings, and for complex decisions affecting an entire organization, firm, society, or the global state of the earth. In recent years increased emphasis has been placed on improving forecasting and decision making in economy and ecology, management and politics. Economic shocks, ecological catastrophes, political disasters, but also chances such as new markets, new technological trends, and new social structures, should no longer be random and fateful events sent by the gods. People want to be prepared and have thus developed a variety of quantitative forecasting methods for different situations, e.g., in business and management. From a methodological point of view, every quantitative forecasting instrument can be characterized by a particular predictability horizon which limits its reliable application. Let us have a look at the strengths and weakness of some forecasting instruments.

The most common quantitative methods of forecasting are the time-series procedures [7.2]. They assume that some pattern in a data series is recurring over time and can be extrapolated to future periods. Thus, a time-series procedure may be appropriate for forecasting environmental factors such as the level of employment or the pattern of weekly supermarket sales where individual decisions have little impact. But time-series methods cannot explain the causes behind the data patterns. In historical times, the method was used by the Babylonian astronomers who extrapolated the data pattern of moonrise into the future without any explanation based on models of planetary motion. In the 18th century physicists knew little about the causes of sunspots. But in the observations of sunspots a pattern of frequency and magnitude was found and predictions were possible by its continuation through time-series analysis. In business and economics, there are various underlying patterns in data series. A horizontal pattern exists where there is no trend in the data (e.g., products with stable sales). A seasonal pattern exists when a series fluctuates according to some seasonal factor such as products whose sale depends on the weather. A cyclic pattern may not repeat itself at constant intervals of time, e.g., the price of metals or the gross national product. A trend pattern exists when there is a general increase or decrease in the value of the variable over time. When an underlying pattern exists in a data series, that pattern must be distinguished from randomness by averaging and weighting ("smoothing") the past data values. Mathematically, a linear smoothing method can be used effectively with data that exhibit a trend pattern. But smoothing methods make no attempt to identify individual components of the basic underlying patterns. There may be subpatterns of trend, cycle, and seasonal factors, which must be separated and decomposed in analyzing the overall pattern of the data series.

While in time-series procedures some data pattern from the past is merely extrapolated to the future, an explanatory model assumes a relationship between the ("dependent") variable y that we want to forecast and another ("independent") variable x. For example, the dependent variable y is the cost of production per unit, and the independent variable x determining the cost of production is the number of units produced. In this case, we can model the relationship in a two-dimensional coordinate system of y and x and draw a straight line that in some sense will give the best linear approximation of the relationship. Regression analysis uses the method of least squares in order to minimize the distance between the actual observations y and the corresponding points $\hat{y}$ on the straight line of linear approximation. Obviously, there are many situations in which this is not a valid approach. An example is the forecast of monthly sales varying nonlinearily according to the seasons of the year. Furthermore, every manager knows that sales are not influenced by time alone, but by a variety of other factors such as the gross national product, prices, competitors, production costs, taxes, etc. The linear interaction of two factors only is a simplification in economy similar to the two-body problems in the linear and conservative world of classical physics.

But, of course, a complex model that is more accurate requires a larger amount of effort, greater expertise and more computational time. In many decision-making situations more than one variable can be used to explain or forecast a certain dependent variable. An ordinary example is a marketing manager who wants to forecast corporate sales for the coming year and to better understand the factors that influence them. Since he has more than one independent variable, his analysis is known as multiple regression analysis. Nevertheless, the dependent variable he wishes to forecast is expressed as a linear function of the independent variables. The computation of the coefficients in the regression equation is based on the use of a sample of past observations. Consequently the reliability of forecasts based on that regression equation depends largely on the specific sample of observations that were used. Therefore degrees of reliability must be measured by tests of statistical significance. While multiple regression involves a single equation, econometric models can include any number of simultaneous multiple regression equations [7.3]. In the case of linear equations, the mathematical methods of solution are based on linear algebra and linear optimization methods (e.g., simplex method). In spite of their linearity, the econometric models may be highly complicated with many variables which can only be mastered by computer programs and machines. The solution strategy of nonlinear programming in economics often decomposes complex problems into subproblems which can be approximately treated as linear.

An implicit assumption in using these methods is that the model best fitting the available historical data will also be the best model to predict the future beyond these data. But this assumption does not hold true for the great majority of real-world situations. Furthermore, most data series used in economics and business are short, measurement errors abound, and controlled experimentation is not possible. It is therefore necessary to understand how various forecasting methods succeed when changes in the established patterns of the past take place. The predictions are different at the various forecasting horizons characterizing each method. Obviously, there is no unique method that can forecast best for all series and forecasting horizons. Sometimes there is nothing in the past data to indicate that a change will be forthcoming. Thus, it may be impossible to anticipate a pattern change without inside knowledge. Pattern shifts or the "change of paradigms" is an everyday experience of business people and managers and by no means an extraordinary insight of some philosophers of science in the tradition of Kuhn et al.

Are there quantitative procedures for determining when a pattern or relationship in a data series has changed? Such methods indeed exist and use a tracking signal to identify when changes in the forecasting errors indicate that a nonrandom shift has occurred. In a quality control chart of, e.g., a production series of cars, the output of the equipment is sampled periodically. As long as that sample mean is within the control limits, the equipment is operating correctly. When this is not the case, the production is stopped and an appropriate action is taken to return it to correct operation. In general, automatic monitoring of quantitative forecasting methods follows the concept of a quality control chart. Every time a forecast is made, its error (i.e., actual minus predicted value) is checked against the upper and lower control limits. If it is within an acceptable range, the extrapolated pattern has not changed. If the forecasting error is outside the control limits, there has probably been some systematic change in the established pattern. Automatic monitoring through tracking signals may be appropriate when large numbers of forecasts are involved. But in the case of one or only a few series, one must still play a waiting game to discover whether changes in the trends of business data are occurring.

Forecasting the future of technological trends and markets, the profitability of new products or services, and the associated trends in employment and unemployment is one of the most difficult, but also most necessary tasks of managers and politicians. Their decisions depend on a large number of technological, economic, competitive, social, and political factors. Since the emergence of commercial computers in the 1950s there has been hope that one might master these complex problems by increasing computational speed and data memory. Indeed, any quantitative forecasting method can be programmed to run on a computer. As no single forecasting method is appropriate for all situations, computer-based multiple forecasting systems have been developed in order to provide a menu of alternative methods for a manager. An example is the forecasting system SIBYL which is named after the ancient seer Sibyl. The story goes that Sibyl of Cumae sold the famous Sibylian books to the Roman king Tarquinius Superbus.

Indeed, SIBYL is a knowledge-based system (cf. Sect. 5.2) for a computerized package of forecasting methods [7.4]. It provides programs for data preparation and data handling, screening of available forecasting methods, application of selected methods, and comparing, selecting, and combining of forecasts. In screening alternative forecasting techniques, the inference component of the knowledge-based system suggests those methods that most closely match the specific situation and its characteristics based on a broad sample of forecasting applications and decision rules. The final function of SIBYL is that of testing and comparing which method provides the best results. The interface of user and system is as friendly and efficient as possible, in order to suit a forecasting expert as well as a novice. Nevertheless, we must not forget that SIBYL can only optimize the application of stored forecasting methods. In principle, the predictability horizon of forecasting methods cannot be enlarged by the application of computers. Contrary to the learning ability of a human expert, forecasting systems such as SIBYL are still program-controlled with the typical limitations of knowledge-based systems.

In general, the computer-based automation of forecasting followed along the lines of linear thinking. On the other hand, the increasing capability of modern computers encouraged researches to analyze nonlinear problems. In the mid-1950s meteorologists preferred statistical methods of forecasting based on the concept of linear regression. This development was supported by Norbert Wiener's successful predicting of stationary random processes. Edward Lorenz was sceptical about the idea of statistical forecasting and decided to test its validity experimentally against a nonlinear dynamical model (cf. Sect. 2.4). Weather and climate is an example of an open system with energy dissipation. The state of such a system is modeled by a point in a phase space, the behavior of the system by a phase trajectory. After some transient process a trajectory reaches an attracting set ("attractor") which may be a stable singular point of the system (Fig. 2.14a or 3.11c), a periodic oscillation called a limit cycle (Fig. 3.11d) or a strange attractor (Fig. 2.21). If one wants to predict the behavior of a system containing a stable singular point or a limit cycle, one may observe that the divergence of nearby trajectories appears not to be growing and may even diminish (Fig. 7.2). In this case, a whole class of initial conditions will be able to reach the steady state and the corresponding systems are predictable. An example is an ecological system with periodic trajectorties of prey and predator populations modeled by nonlinear Lotka–Volterra equations. The divergence or convergence of nearby trajectories can be measured numerically by the so-called Lyapunov exponent:

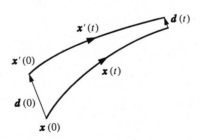

Fig. 7.2. Predictable system with stable point attractor or limit cycle and convergence of nearby trajectories [7.5]

Let us consider two nearby trajectories $x(t)$ and $x'(t)$ with the initial states $x(0)$ and $x'(0)$ at time $t = 0$ and the length $d(t) = |x'(t) - x(t)|$ of the vector $d(t)$. If the trajectories converge, then $d(t) \approx e^{\Lambda t}$ and $\Lambda < 0$. The quantity Λ is called Lyapunov exponent and defined as $\Lambda(x(0), d(0)) = \lim_{t \to \infty} \lim_{d(0) \to 0} [(1/t) \ln(d(t)/d(0))]$. If it is positive, the Lyapunov exponent gives the rate of convergence. In Fig. 7.2, the model process $x'(t)$ delivers reliable predictions of the real process $x(t)$, because the system is assumed to have converging trajectories independent of their initial conditions.

A phase portrait of a nonlinear system may have a number of attractors with different regions ("separatrices") of approaching trajectories (cf. Fig. 2.10). For forecasting the future of the evolving system it is not sufficient to know all possible attractors and the initial state $x(0)$. What we need to know in addition are the separatrices for attraction basins of the different attractors. If the initial state of a system happens to be far away from the basin of a certain attractor, the final state of the corresponding attractor cannot be predicted.

In Fig. 2.22a-c, the nonlinear logistic map describes a transition from order to chaos depending on an increasing control parameter. Figure 2.23a,b illustrates the corresponding sequence of bifurcations with the chaotic regime occurring beyond a critical threshold. If the corresponding Lyapunov exponent is positive, the behavior of the system is chaotic. If it is zero, the system has a tendency to bifurcate. If it is negative, the system is in a stable state or branch of the bifurcation tree. In this case the system is predictable. In the other cases the sensitivity to initial conditions comes into play. It is remarkable that a nonlinear system in the chaotic regime is nonetheless not completely unpredictable. The white stripes or "windows" in the grey veil of a chaotic future (Fig. 2.23b) indicate local states of order with negative Lyapunov exponents. Thus, in a sea of chaos we may find predictable islands of order. In this case the system is at least predictable for characteristic short intervals of time.

In general, the degree of predictability is measured by a statistical correlation between the observed process and the model at the particular time since the start of the observation. Values close to unity correspond to a satisfactory forecast, while small values indicate a discrepancy between observation and prediction. Every forecasting model has a certain time of predictable behavior after which the degree of predictability decreases more or less rapidly to zero. With improvement of the model the time of predictable behavior may be enlarged to some extent. But the predictability range depends upon fluctuational parameters. Weak microscopic perturbations of locally unstable chaotic systems can reach a macroscopic scale in a short time. Thus, local instabilities reduce the improvement of predictable behavior drastically. The predictability horizon of a forecasting system means a finite timespan of predictable behavior that cannot be surpassed by either improved measuring instruments or a refined prediction model. When we remember that the atmosphere is modeled, following Lorenz, by nonlinear systems with local and global instabilities, we realize the difficulties encountered by meteorologists in obtaining efficient long- or even medium-term forecasting. The belief in a linear progress of weather forecasting by increasing computational capacities was an illusion of the 1950s.

As nonlinear models are applied in different fields of research, we gain general insights into the predictable horizons of oscillatory chemical reactions, fluctuations of species, populations, fluid turbulence, and economic processes. The emergence of sunspots, for instance, which was formerly analyzed by statistical methods of time-series is by no means a random activity. It can be modeled by a nonlinear chaotic system with several characteristic periods and a strange attractor only allowing bounded forecasts of the variations. In nonlinear models of public opinion formation, for instance, we may distinguish a predictable stable state before the public voting ("bifurcation") when neither of two possible opinions is preferred, the short interval of bifurcation when tiny unpredictable fluctuations may induce abrupt changes, and the transition to a stable majority. The situation reminds us of growing air bubbles in turbulently boiling water: When a bubble has become big enough, its steady growth on its way upward is predictable. But its origin and early growth is a question of random fluctuation. Obviously, nonlinear modeling explains the difficulties of the modern Pythias and Sibyls of demoscopy.

Today, nonlinear forecasting models do not always deliver better and more efficient predictions than the standard linear procedures. Their main advantage is the explanation of the actual nonlinear dynamics in real processes, the identification and improvement of local horizons with short-term predictions. But first of all an appropriate dynamical equation governing an observation at time t must be reconstructed, in order to predict future behavior by solving that equation. Even in the natural sciences, it is still unclear whether appropriate equations for complex fields such as earthquakes can be derived. We may hope to set up a list in a computer memory with typical nonlinear equations whose coefficients can be automatically adjusted for the observed process. Instead, to make an exhaustive search for all possible relevant parameters, a learning strategy may start with a crude model operating over relatively short times and then specify a smaller number of parameters in a relatively narrow range of values. An improvement of short-term forecasting has been realized by the learning strategies of neural networks. On the basis of learned data, neural nets can weight the input data and minimize the forecasting errors of short-term stock quotations by self-organizing procedures (Fig. 5.22 a, b). So long as only some stock market advisors use this technical support, they may do well. But if all agents in a market use the same learning strategy, the forecasting will become a self-defeating prophecy.

The reason is that human societies are not complex systems of molecules or ants, but the result of highly intentional acting beings with a greater or lesser amount of free will [7.6]. A particular kind of self-fulfilling prophecy is the Oedipus effect in which people like the legendary Greek king try, in vain, to change their future as forecasted to them. From a macroscopic viewpoint we may, of course, observe single individuals contributing with their activities to the collective macrostate of society representing cultural, political, and economic order ("order parameters"). Yet, macrostates of a society, of course, do not simply average over its parts. Its order parameters strongly influence the individuals of the society by orientating ("enslaving") their activities and by activating or deactivating their attitudes and capabilities. This kind of feedback is typical for complex dynamical systems. If the control parameters of the environmental conditions attain certain critical values due to internal or external interactions, the macrovariables may move into an unstable domain out of which highly divergent alternative paths are possible. Tiny unpredictable microfluctuations (e.g., actions of very few influential people, scientific dicoveries, new technologies) may decide which of the diverging paths in an unstable state of bifurcation society will follow.

7.2 Complexity, Science, and Technology

Despite the difficulties referred to above, we need reliable support for short-, medium-, and long-term forecasts of our local and global future. A recent demand from politics is the modeling of future developments in science and technology which have become a crucial factor of modern civilization. Actually, this kind of development seems to be governed by the complex dynamics of scientific ideas and research groups which are embedded in the complex network of human society. Common topics of research groups attract the interest and capacity of researchers for longer or shorter periods of time. These research "attractors" seem to dominate the activities of scientists like the attractors and vortices in fluid dynamics. When states of research become unstable, research groups may split up into subgroups following particular paths of research which may end with solutions or may bifurcate again, and so forth. The dynamics of science seems to be realized by phase transitions in a bifurcation tree with increasing complexity. Sometimes scientific problems are well-defined and lead to clear solutions. But there are also "strange" and "diffuse" states like the strange attractors of chaos theory.

Historically, quantitative inquiries into scientific growth started with statistical approaches such as Rainoff's work on "Wave-like fluctuations of creativity in the development of West-European physics in the 18th and 19th century" (1929). From a sociological point of view Robert Merton discussed "Changing foci of interest in the sciences and technology", while Pitirim

Sorokin analyzed the exponential increase of scientific discoveries and technological inventions since the 15th century. He argued that the importance of an invention or discovery does not depend on subjective weighting, but on the amount of subsequent scientific work inspired by the basic innovation. As early as 1912 Alfred Lotka had the idea of describing true epidemic processes like the spread of malaria and chemical oscillations with the help of differential equations. Later on, the information scientist William Goffman applied the epidemic model to the spread of scientific ideas. There is an initial focus of "infectious ideas" infecting more and more people in quasi-epidemic waves. Thus, from the viewpoint of epidemiology, the cumulation and concentration in a scientific field is modeled by so-called Lotka- and Bradford-distributions, starting with a few articles of some individual authors which are the nuclei of publication clusters [7.7]. The epidemic model was also applied to the spread of technical innovations. In all these examples we find the well-known S-curve of a logistic map (Fig. 2.22a) with a slow start followed by an exponential increase and then a final slow growth towards saturation. Obviously a learning process is also described in the three phases of an S-curve with slow learning success of an individual in the beginning, then a rapid exponential increase and finally a slow final phase approaching saturation.

The transition from statistical analysis to dynamical models has the great methodological advantage that incomprehensible phenomena such as strange fluctuations or statistical correlations of scientific activities can be illustrated in computer-assisted simulation experiments with varying dynamical scenarios. The epidemic model and Lotka–Volterra equation were only a first attempt to simulate coupled growth processes of scientific communities. However, essential properties of evolutionary processes like creation of new structural elements (mutation, innovation, etc.) cannot be reflected. Evolutionary processes in social systems have to be pictured through unstable transitions by which new ideas, research fields, and technologies (like new products in economic models) replace already existing ones and thereby change the structure of the scientific system. In a generalization of Eigen's equation of prebiotic evolution (cf. Sect. 3.3), the scientific system is described by an enumerable set of fields (i.e., subdisciplines of a scientific research field), each of which is characterized by a number of occupying elements (i.e., scientists working in the particular subdiscipline). Elementary processes of self-reproduction, decline, exchange, and input from external sources or spontaneous generation have to be modeled. Each self-replication or death process changes only the occupation of a single field. For simple linear self-reproduction processes without exchange, the selection value of a field is given by the difference between the "birth" and "death" rates of the field. When a new field is first populated, it is its selection value that decides whether the system is stable or unstable with respect to the innovation. If its selection value is larger than any other selection value of existing fields, the new field will outgrow the others, and the system may become unstable. The evolution of new fields with higher selection values characterizes a simple selection process according to Darwinian "survival of the fittest".

But we must not forget that such mathematical models do not imply the reduction of scientific activities to biological mechanisms. The variables and constants of the evolution equation do not refer to biochemical quantities and measurements, but to the statistical tables of scientometrics. Self-reproduction corresponds to young scientists joining the field of research they want to start working in. Their choice is influenced by education processes, social needs, individual interest, scientific schools, etc. Decline means that scientists are active in science for a limited number of years. The scientists may leave the scientific system for different reasons (e.g., age). Field mobility means the process of exchange of scientists between research fields according to the model of migration. Scientists might prefer the direction of higher attractiveness of a field expressed by a higher self-reproduction rate. When processes include exchange between fields with nonlinear growth functions of self-reproduction and decline, then the calculation of selection values of an innovation is a rather complicated mathematical task. In general, a new field with higher selection value is indicated by the instability of the system with respect to a corresponding perturbation.

Actually, scientific growth is a stochastic process. When, for example, only a few pioneers are working in the initial phase of a new field, stochastic fluctuations are typical. The stochastic dynamics of the probable occupation density in the scientific subfields is modeled by a master equation with a transition operator which is defined by transition probabilities of self-reproduction, decline, and field mobility. The stochastic model provides the basis for several computer-assisted simulations of scientific growth processes. The corresponding deterministic curves, as average over a large number of identical stochastic systems, are considered for trend analysis, too. As a result, the general S-shaped growth law for scientific communities in subdisciplines with a delayed initial phase, a rapid growth phase, and a saturation phase has been established in several simulations. In a series of simulations (Fig. 7.3), a research field was assumed to comprise about 120–160 members. For five fields, 100 scientists were chosen as initial condition with the saturation domain near the initial conditions. A sixth field is not yet set up (with the initial condition of zero members). In a first example, the influence of the self-reproduction process on the growth curve of the new field was simulated for several cases. With increasing self-reproduction rates the new field grows ever more rapidly at the expense of neighboring fields.

The emergence of a new field may have a tendency to more coexistence or selection. The growth of the initial phase may be more or less rapid or can also be delayed. A famous example of delayed growth in the history of science

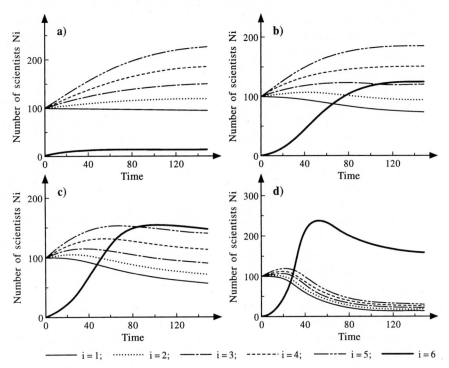

Fig. 7.3. Influence of the self-reproduction rate of a new scientific field on the growth curves of neighboring fields [7.8]

is chaos theory itself, which was treated by only very few scientists (e.g., Poincaré) in its initial phase. Although the mathematical principles of the new field were quite clear, its exponential growth began only some years ago when computational technology could handle nonlinear equations. Sometimes an emerging field cannot expand to a real domain of science, because it has only a weak selection advantage in comparison with mighty surrounding fields. It is a pity that some technological fields such as alternative energies (e.g., wind, solar) are still in such a poor state, surrounded by the powerful industries of traditional or nuclear energy. If a new attractive field emerges, a strong influx of scientists from the surrounding fields can be observed. These people are adapting to the style and problem solving pattern of the new field. This kind of directed field mobility sometimes leads to the phenomena of fashion in science.

It is well-known that the S-shaped nonlinear logistic map gives rise to a variety of complex dynamical behaviors such as fixed points, oscillations, and deterministic chaos, if the appropriate control parameters increase beyond certain critical values (Fig. 2.22). Obviously, both the stochastic and the deterministic models reflect some typical properties of scientific growth. Such effects are structural differentiation, deletion, creation, extension of new fields with delay, disappearance, rapid growth, overshooting fashions, and regression. The computer-assisted graphic simulations of these dynamical effects allow characterization by appropriate order parameters which are testable on the basis of scientometric data. Possible scenarios under varying conditions can be simulated, in order to predict the landmarks and the scope of future developments.

But so far, the evolution of scientific research fields has been considered in the model only in terms of changes of the scientific manpower in the selected fields. A more adequate representation of scientific growth must take account of the problem-solving processes of scientific endeavors. But it is a difficult methodological problem to find an adequate state space representing the development of problem solving in a scientific field. In the mathematical theory of biological evolution, the species can be represented by points in a high-dimensional space of biological characters (Fig. 3.4). The evolution of a species corresponds to the movement of a point through the phenotypic character space. Analogously, in the science system, a high-dimensional character space of scientific problems has to be established. Configurations of scientific articles which are analyzed by the technique of multidimensional scaling in co-citation clusters can be represented by points in a space of two or three dimensions. Sometimes research problems are indicated by sequences of keywords ("macro-terms") which are registered according to the frequency of their occurrence or co-occurrence in a scientific text.

In a continuous evolution model each point of the problem space is described by a vector corresponding to a research problem (Fig. 7.4a). The problem space consists of all scientific problems of a scientific field, of which some are perhaps still unknown and not under investigation. This space is metric, because the distance between two points corresponds to the degree of thematic connection between the problems represented. The scientists working on problem q at time t distribute themselves over the problem space with the density $x(q, t)$. In the continuous model $x(q, t)\,dq$ means the number of scientists working at time t in the "problem element" dq (Fig. 7.4b).

Thus the research fields may correspond to more or less closely connected point clouds in the problem space. Single points between these areas of greater density correspond to scientists working on isolated research problems which may represent possible nuclei of new research fields. History of science shows that it may take decades before a cluster of research problems grows up into a research field. In the continuous model, field mobility processes are reflected by density change: If a scientist changes from problem q to problem q', then the density $x(q, t)$ will get smaller and $x(q', t)$ will increase. The movement of scientists in the problem space is modeled by a cer-

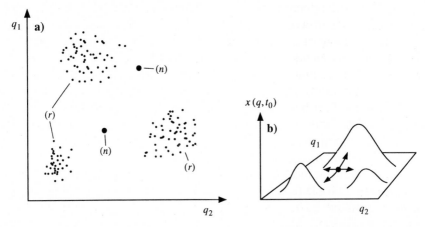

Fig. 7.4a,b. A two-dimensional problem space (**a**) with research fields (r) as clouds of related problems, possible nuclei (n) of new research fields, and a potential landscape (**b**) of research activities $x(q, t)$ in problem $q = (q_1, q_2)$ of the problem space at time t [7.9]

tain reproduction–transport equation. A function $a(q)$ expresses the rate at which the number of scientists in field q is growing through self-reproduction and decline. Thus, it is a function with many maxima and minima over the problem space, expressing the increasing or decreasing attractiveness of the problems in a scientific field. In analogy to physical potentials (e.g., Fig. 4.10), one may interpret $a(q)$ as a potential landscape of attractiveness with hills and valleys, representing the attractors and deadlocked areas of a research field (Fig. 7.4b).

Dynamical models of the growth of knowledge become testable by scientometrics. Thus, they may open a bridge between philosophy of science with its conceptual ideas of scientific growth and history of science with its evaluation of scientific documents. In cognitive scientometrics an attempt has recently been made to quantify the concept of research problems and to represent them in appropriate problem spaces by bibliometric, cognitive, and social characteristics. The simplified schemes of the history of science which have been suggested by Popper, Kuhn, and others, could perhaps be replaced by testable hypotheses. Kuhn's discontinuous sequence with phases of "normal" and "revolutionary" science is obviously not able to tackle the growth of knowledge. On the other hand, the naive belief of some historians that the growth of science is a continuous cumulation of eternal truths is not appropriate to the complex dynamics of research in any way. Even Popper's sophisticated late philosophy that science does not grow through a monotonic increase of the number of indubitably established laws but through learning strategies of hypotheses and criticism needs more precision and clarification with reference to the changing historical standards of methodology, institutionalization, and organization. The increasing computational capacities of modern computers enable a new quantitative approach with simulation experiments in social sciences. The great advantage of dynamical models is their computer-assisted graphic illustration of several scenarios with varying parameters. These scenarios may confirm, restrict, or refute the chosen model. Last but not least, we need reliable support for decisions in science policy. Different scenarios of future developments may help us to decide where to invest our limited resources of research budget and how to realize desirable future states of society.

Thus nonlinear modeling and computer-assisted simulation may enable us to derive multiple futures, but provides no algorithm to decide between them. Normative goals must be involved in order to realize desirable future states of society. Since the 1960s the reports of the Club of Rome have tried to initiate international debates about the goals and alternative futures of mankind, supported by quantitative long-term forecasts. In Sect. 7.1, we saw the limitations of quantitative long-term forecasting in a nonlinear world. Consequently, scientific ideas and technological innovations cannot be forced into being by political decisions. But they must no longer be fateful random events which may or may not happen. We need instruments to evaluate desirable goals and their chance of realization.

A nonquantitative approach is the so-called Delphi method, used to prepare decisions and forecasts of scientific and technological trends by a panel of experts. The name "Delphi" is a reference to the legendary Pythia (Fig. 7.1) who was said to prepare her prophecies by gathering information about her clients. The Delphi method of today uses the estimates of scientific experts. The individual experts are kept apart so that their judgement is not influenced by social pressure or group behavior. The experts were asked in a letter to name and to weight inventions and scientific breakthroughs that are possible and/or desirable in a certain period of time. Sometimes they are not only asked for the probability of each development: Additionally, they are asked to estimate the probability that the occurrence of any one of the potential developments will influence the likelihood of occurrence of each of the others. Thus one gets a correlated network of future developments which can be represented by a matrix of subjective conditional probabilities. In the next phase the experts are informed about the items with general consensus. When they are asked to state the reasons for their disagreement with the majority, several of the experts re-evaluate their time estimates, and a narrower range for each breakthrough may be arranged.

The Delphi method, of course, cannot deliver a single answer. But the spread of expert opinions gathers considerable information about potential major breakthroughs. The average deviations from the majority should be narrowed down without pressuring the experts with extreme responses. But the Delphi method therefore cannot predict the unexpected. Sometimes the Delphi method is supported by the relevance-tree method, in order to select the best actions from alternatives by constructing decision trees. The relevance-tree method uses the ideas of decision theory to assess the desirability of a certain future and to select those areas of science and technology whose development is necessary for the achievment of those goals.

Obviously there is no single method of forecasting and deciding in a complex nonlinear world. We need an integrative ("hybrid") network of quantitative and qualitative methods. Finally, we need ethical landmarks to guide us in applying these instruments and in mastering the future.

7.3 Complexity, Responsibility, and Freedom

In recent years, ethics has become a major topic attracting increasing interest from a wide variety of professionals including engineers, physicians, scientists, managers, and politicians. The reasons for this interest are the growing problems with the environment, the economy, and modern technologies, questions of responsibility, increasing alarm, and decreasing acceptance of critical consequences in a highly industrialized world. But we must be aware that our standards of ethical behavior have not fallen to Earth from Heaven and have not been revealed by some mysterious higher authority. They have changed and they will continue to change, because they are involved in the evolution of our sociocultural world.

In modeling human society we must not forget the highly nonlinear self-referentiality of a complex system with intentionally acting beings. There is a particular measurement problem in social sciences arising from the fact that scientists observing and recording behavior of society are themselves members of the social system they observe. Well-known examples are the effects of demoscopic opinion-polling during political elections. Furthermore, theoretical models of society may have a normative function influencing the future behavior of its agents. A well-known example was the social Darwinism of the 19th century which tried to explain the social development of mankind as a linear continuation of biological evolution. Actually, that social theory initiated a brutal ideology legitimating the ruthless selection of the social, economic, and racial victors of history. Today, it is sometimes fashionable to legitimate political ideas of basic democracy and ecological economy by biological models of self-organization [7.10]. But nature is neither good nor bad, neither peaceful nor militant. These are human evaluations. Biological strategies over millions of years have operated at the expense of myriads of populations and species with gene defects, cancer, etc., and have, from a human point of view, perpetrated many other cruelties. They cannot deliver the ethical standards for our political, economic, and social developments.

In this book we have seen that the historical models of life, mind, and society often depend on historical concepts of nature and historical standards of technology. Especially the linear and mechanistic view of causality was a dominant paradigm in the history of natural, social, and technical sciences. It also influenced ethical norms and values which cannot be understood without the epistemic concepts of the historical epochs in which they arose. The historical interdependence of epistemology and ethics does not mean any kind of relativism or naturalism. As in the case of scientific theories and hypotheses we have to distinguish their context of historical and psychological invention and discovery from their context of justification and validity. Even human rights have a historical development with changing meanings [7.11]. Hegel once mentioned that the history of mankind can be understood as a "development to freedom". Thus, before we discuss possible ethical consequences in a complex and nonlinear world, we should take a short glance at the historical development of ethical standards.

Ethics is a discipline of philosophy like logic, epistemology, philosophy of science, language, law, religion, and so on [7.12]. Historically, the word "ethics" stems back to the Greek word $\vec{\eta}\vartheta o\varsigma$, which means custom and practice. Originally, ethics was understood as a doctrine of moral customs and institutions teaching people how to live. The central problem of ethics has become that of finding a good moral code with advice on how to live well, how to act justly, and how to decide rationally. Some essential concepts of ethics were already discussed in Greek philosophy following Socrates. His student Plato generalized the Socratic quest for a good life to the universal idea of the greatest good, which is eternal and independent of the historical life behind the transitory and continously changing world of matter [7.13].

Aristotle criticized his teacher's doctrine of eternal values as ignorant of real human life. For Aristotle the validity of the good, the just, and the rational is referred to the political society (*polis*), the family, and the interaction of single persons [7.14]. Justice in the polis is realized by the proportionality or the equilibrium of natural interest of free men. The greatest good of man is happiness, which is realized by a successful life according to the natural customs and practice in the polis and the family. Obviously, Aristotle's concept of ethics corresponds to his organic view of a nature filled with growing and maturing organisms like plants, animals, and humans.

After the dissolution of the Greek polis, ethics needed a new framework of standards. In Epicurean ethics, the internal equality of individual life, action, and feeling was emphasized, while the ethics of the Stoics underlined the external equality of all people achieved by nature. In the Christian Middle Ages a hierarchy of eternal values was guaranteed by the divine order of the world. At the beginning of modern times the theological framework as a universally accepted foundation of ethics was ripe for dissolution.

Descartes not only suggested a mechanistic model of nature but also demanded a moral system founded upon scientific reason. Baruch Spinoza derived an axiomatized system of rationalist morality corresponding to the deterministic and mechanistic model of nature. As the laws of nature are believed to be identical with the laws of rationality, human freedom only could mean acting according to deterministic laws which were recognized as rational. The greatest good meant the dominance of rationality over the affects of the material human body. Hobbes defended a mechanistic view of nature and society, but he doubted human rationality. Political laws and customs can only be guaranteed by the centralized power of "Leviathan". The greatest good is peace as a fixed and final equilibrium in an absolutist state.

The liberal society of Locke, Hume, and Smith was understood by analogy with Newton's model of separable forces and interacting celestial bodies. In the American and French revolutions, individual freedom was proclaimed as a natural right [7.15]. But how to justify individual freedom in a mechanistic world with deterministic causality? Every natural event is the effect of a linear chain of causes which in principle can be derived by mechanistic equations of motion. Only humans are assumed to be capable of spontaneous and free decisions initiating causal chains of actions without being influenced by outer circumstances. Kant called this human feature the "causality of freedom".

As nobody's opinions and desires are privileged, only advice which is acceptable for everybody is justified as reasonable. In the words of Kant, only those generally accepted "maxims" can be justified as universal moral laws. This formal principle of moral universality is Kant's famous categorical imperative of reason: we should act according to imperatives which can be justified as general laws of morality. The freedom of a person is limited by the freedom of his or her neighbor. In another famous formulation, Kant said that humans as free beings should not be misused as instruments for the interests of other humans. Thus, besides the mechanistic world of nature, which is ruled by deterministic laws, there is the internal world of reason with the laws of freedom and morality. Kant's ethic of freedom has been incorporated in the formal principles of every modern constitutional state [7.16].

But how can the laws of freedom be realized in the real world of politics and economics? During the initial process of industrialization the ethics of Anglo-American utilitarianism (due to Bentham and Mill) required an estimation of personal happiness. The happiness of the majority of people was declared to be the greatest good of ethics. While Kant suggested a formal principle of individual freedom, the utilitarian principle of happiness can be interpreted as its material completion. It was explicitly demanded as a natural human right in the American constitution. The utilitarian philosophers and economists defined the demand for happiness as an utility function that must be optimized with as few costs as possible in order to realize the greatest welfare for the majority of people. The principles of utilitarianism have become the ethical framework of welfare economics [7.17].

Modern philosophers like John Rawls have argued that the utilitarian principle in combination with Kant's demand for ethical universality can help to realize the demand for a just distribution of goods in modern welfare politics [7.18]. From a methodological point of view, the ethical, political, and economic model of utilitarianism corresponds to a self-organizing complex system with a single fixed point of equilibrium which is realized by an optimization of the utility function of a society and which is connected with a just distribution of goods to the majority of people.

Obviously, Kant's ethics as well as Anglo-American utilitarianism are normative demands to judge our actions. They may be accepted or not by individuals. Hegel argued that the subjective ethical standards of individuals were products of objective historical processes in history which were realized by the institutions of a society. Thus, he distinguished between the subjective morality and subjective reason of individuals and the objective morality and objective reason of institutions in a society. Historically, Hegel's foundation of ethics in the actual customs and morality of a civil society reminds the reader of Aristotle's realistic ethics of the Greek polis. But Aristotle's order of society was static, while Hegel assumed a historical evolution of states and their institutions.

From a methodological point of view, it is remarkable that Hegel already distinguished between the micro-level of individuals and a macro-level of societies and their institutions which is not only the sum of their citizens. Furthermore, he described an evolution of society which is not determined by the intentions and the subjective reason of single individuals, but by the self-organizing process of a collective reason. Nevertheless, Hegel believed in a rather simplified model of evolution with sequential states of equilibrium leading to a final fixed point which is realized by the attractor of a just civil society. Actual history after Hegel showed that his belief in the rational forces of history driving human society to a final state of justice by self-organization was a dangerous illusion. It is well known that his model was modified and misused by totalitarian politicians of the political right and left.

Friedrich Nietzsche attacted the belief in objective reason as well as in eternal ethical values as idealistic ideologies which were contrary to the real forces of life. Nietzsche's philosophy of life was influenced by Darwin's biology of evolution, which had become a popular philosophy in the late 19th century. Although Nietzsche had criticized nationalism and racism in his writings, his glorification of life and the victors in the struggle of life was terribly misused in the politics of our century. Nevertheless, he is another example to show that concepts from the natural sciences have influenced political and ethical ideas [7.19].

Nietzsche's nihilism and his critique of modern civilization were continued by Martin Heidegger in our century. In Heidegger's view, the technical evolution of mankind is an automatism without orientation which has forgotten the essential foundation of man and humanity. A philosopher like Heidegger cannot and will not change or influence this evolution. He only has the freedom to bear this fate with composure. But in what way is Heidegger's attitude against technology and civilization more than resignation, fatalism, and an escape into an idyllic utopia without technology which has never existed in history? It seems to be the extreme counterposition to the Laplacean belief in an omnipotent planning and controlling capacity in nature and society [7.20].

What are the ethical consequences of the complex system approach which has been discussed in this book? First, we must be aware that the theory of complex systems is not a metaphysical process ontology. It is not an epistemic doctrine in the traditional sense of philosophy. The principles of this methodology deliver a heuristic scheme for constructing models of nonlinear complex systems in the natural and social sciences. If these models can be mathematized and their properties quantified, then we get empirical models which may or may not fit the data. Moreover, it tries to use a minimum of hypotheses in the sense of Ockham's razor. Thus it is a *mathematical, empirical, testable*, and *heuristically economical methodology*. Furthermore, it is an *interdisplinary research program* in which several natural and social sciences are engaged. However, it is not an ethical doctrine in the traditional sense of philosophy.

Nevertheless, our models of complex and nonlinear processes in nature and society have important consequences for our behavior. In general, linear thinking may be dangerous in a nonlinear complex reality. We have seen that traditional concepts of freedom were based on linear models of behavior. In this framework every event is the effect of a well defined initial cause. Thus, if we assume a linear model of behavior, the responsibility for an event or effect seems to be uniquely decidable. But what about the global ecological damage which is caused by the local nonlinear interactions of billions of self-interested people? Recall, as one example, the demand for a well balanced complex system of ecology and economics. As ecological chaos can be global and uncontrollable, some philosophers like Hans Jonas have proposed that we stop any activity which could perhaps have some unknown consequences [7.21]. But we can never forecast all developments of a complex system in the long run. Must we therefore retire into a Heidegger-like attitude of resignation? The problem is that doing nothing does not necessarily stabilize the equilibrium of a complex system and can drive it to another metastable state. Nevertheless, short-term forecasting is possible in complex systems and attempts are being made to improve it in economics, for instance.

In a linear model, the extent of an effect is believed to be similar to the extent of its cause. Thus, a legal punishment of a punishable action can be proportional to the degree of damage effected. But what about the butterfly effect of tiny fluctuations which are initiated by some persons, groups, or firms, and which may result in a global crisis in politics and economics? For instance, consider the responsibility of managers and politicians whose failure can cause the misery of thousands or millions of people [7.22].

As the ecological, economic, and political problems of mankind have become global, complex, and nonlinear, the traditional concept of individual responsibility is questionable. We need new models of collective behavior depending on the different degrees of our individual faculties and insights. Individual freedom of decision is not abolished, but restricted by collective effects of complex systems in nature and society which cannot be forecast or controlled in the long run. Thus, it is not enough to have good individual intentions. We have to consider their nonlinear effects. Global dynamical phase portraits deliver possible scenarios under certain circumstances. They may help to achieve the appropriate conditions for fostering desired developments and preventing evil ones.

The electronic vision of a global village, for instance, may imply a severe threat to personal freedom. If information about citizens can easily be gained and evaluated in large communication networks, then the danger of misuse by interested institutions is to be taken in earnest. As in the traditional economy of goods, there may arise information monopolies prejudicing other people, classes, and countries. For instance, consider the former "Third World" or the "South" with its less developed systems for information services which could have no fair chance against the "North" in a global communications village.

Our physicians and psychologists must learn to consider humans as complex nonlinear entities of mind and body. Linear thinking may fail to yield a successful diagnosis. Local, isolated, and "linear" therapies of medical treatment may cause negative synergetic effects. Thus, it is noteworthy that mathematical modeling of complex medical and psychological situations advises high sensitivity and cautiousness, in order to heal and help ill people. The complex system approach cannot explain to us *what* life is. But it can show us *how* complex and sensitive life is. Thus, it can help us to become aware of the value of our life. Obviously, the theory of complex systems has consequences for the ethics of politics, economics, ecology, medicine, and biological, computational, and information sciences. These ethical consequences strongly depend on our knowledge about complex nonlinear dynamics in nature and society, but they are not derived from the principles of complex systems. Thus, we do not defend any kind of ethical naturalism or reductionism. Dynamical models of urban developments, global ecologies, human organs, or information networks only deliver possible scenarios with different attractors. It is a question for us to evaluate which attractor we should prefer ethically and help to realize by achievement of the appropriate conditions. Immanuel Kant summarized the problems of philosophy in the three famous questions [7.23]:

What can I know?
What must I do?
What may I hope?

The first question concerns epistemology with the possibilities and limitations of our recognition. The theory of complex systems explains what we can know and what we cannot know about nonlinear dynamics in nature and society. In general, the question invites a demand for scientific research, in order to improve our knowledge about complexity and evolution.

The second question concerns ethics and the evaluation of our actions. In general, it invites a demand for sensitivity in dealing with highly sensitive complex systems in nature and society. We should neither overact nor retire, because overaction as well as retirement can push the system from one chaotic state to another. We should be both cautious and courageous, according to the conditions of nonlinearity and complexity in evolution. In politics we should be aware that any kind of mono-causality may lead to dogmatism, intolerance, and fanaticism.

Kant's last question "What may we hope?" concerns the Greatest Good, which has traditionally been discussed as *summum bonum* in the philosophy of religion. At first glance, it seems to be beyond the theory of complex systems, which only allows us to derive global scenarios in the long run and short-term forecasts under particular conditions. But when we consider the long sociocultural evolution of mankind, the greatest good that people have struggled for has been the dignity of their personal life. This does not depend on individual abilities, the degree of intelligence, or social advantages acquired by the contingencies of birth. It has been a free act of human self-determination in the struggles of history. We have to protect the Greatest Good in an ongoing evolution with increasing complexity.

References

Chapter 1

1.1   Mainzer, K./Schirmacher, W. (eds.): Quanten, Chaos und Dämonen. Erkenntnis-theoretische Aspekte der modernen Physik. B.I. Wissenschaftsverlag: Mannheim (1994)

1.2   Stein, D.L. (ed.): Lectures in the Sciences of Complexity, Santa Fe Institute Stud-ies in the Sciences of Complexity vol. 1. Addison-Wesley: Redwood City, CA, (1989); Jen, E. (ed.): Lectures in Complex Systems, Santa Fe Institute Studies in the Sciences of Complexity vol. 2 (1990); Stein, D.L. (ed.): Lectures in Complex Systems, Santa Fe Institute Studies in the Sciences of Complexity vol. 3 (1991); Kurdyumov, S.P.: Evolution and self-organization laws in complex systems, In-tern. J. Modern Physics C vol. 1, no. **4** (1990) 299–327

1.3   Nicolis, G./Prigogine, I.: Exploring Complexity. An Introduction. W.H. Freeman: New York (1989)

1.4   Haken, H.: Synergetics. An Introduction, 3rd Edn. Springer: Berlin (1983)

1.5   Mainzer, K.: Symmetries in Nature. De Gruyter: New York (1995) (German orig-inal: Symmetrien der Natur 1988)

Chapter 2

2.1   For historical sources of Sect. 2.1 compare Mainzer, K.: Symmetries in Nature. De Gruyter: New York (1994) (German original 1988) Chapter 1

2.2   Diels, H.: Die Fragmente der Vorsokratiker, 6th ed., revised by W. Kranz, 3 vol. Berlin (1960/1961) (abbrev.: Diels-Kranz), 12 A 10 (Pseudo-Plutarch)

2.3   Diels-Kranz 13 A 5, B 1

2.4   Diels-Kranz 22 B 64, B 30

2.5   Heisenberg, W.: Physik und Philosophie. Ullstein: Frankfurt (1970) 44

2.6   Diels-Kranz 22 B8

2.7   Diels-Kranz 31 B8

2.8   Heisenberg, W.: Die Plancksche Entdeckung und die philosophischen Grund-lagen der Atomlehre, in: Heisenberg, W.: Wandlungen in den Grundlagen der Naturwissenschaften. S. Hirzel: Stuttgart (1959) 163

2.9   Cf. also Hanson, N.R.: Constellations and Conjectures. Boston (1973) 101

2.10   Hanson, N.R. (see Note 9, 113) carried out corresponding calculations.

2.11   Bohr, H.: Fastperiodische Funktionen. Berlin (1932)

2.12   Forke, A.: Geschichte der alten chinesischen Philosophie. Hamburg (1927) 486; Fêng Yu-Lan: A History of Chinese Philosophy vol. 2: The Period of Classical Learning. Princeton NJ (1953) 120

2.13   Mainzer, K.: Geschichte der Geometrie. B. I. Wissenschaftsverlag: Mannheim/ Wien/Zürich (1980) 83; Edwards, C.H.: The Historical Development of the Calculus. Springer: Berlin (1979) 89

2.14   Mainzer, K.: Geschichte der Geometrie (see Note 13) 100; Abraham, R.H./Shaw, C.D.: Dynamics – The Geometry of Behavior Part 1. Aerial Press: Santa Cruz (1984) 20

2.15   Audretsch, J./Mainzer, K. (eds.): Philosophie und Physik der Raum-Zeit. B.I. Wissenschaftsverlag: Mannheim (1988) 30

2.16   Audretsch, J./Mainzer, K. (eds.): Philosophie und Physik der Raum-Zeit (see Note 15) 40; Weyl, H.: Raum, Zeit, Materie. Vorlesung über Allgemeine Relativitätstheorie. Wissenschaftliche Buchgesellschaft: Darmstadt (1961) (Reprint of the 5th Edition (1923)) 141

2.17   Mach, E.: Die Mechanik. Historisch-kritisch dargestellt. Wissenschaftliche Buchgesellschaft: Darmstadt (1976) (Reprint of the 9th Edition (1933)) 149; Abraham, R.H./Shaw, C.D.: Dynamics – The Geometry of Behavior (see Note 14) 57

2.18   Ruelle, D.: Small random pertubations of dynamical systems and the definition of attractors. Commun. Math. Phys. **82** (1981) 137–151; Abraham, R.H./ Shaw, C.D.: Dynamics – The Geometry of Behavior (see Note 14) 45

2.19   For an analytical elaboration cf. Stauffer, D./ Stanley, H.E.: From Newton to Mandelbrot. A Primer in Theoretical Physics. Springer: Berlin (1990) 26

2.20   Nicolis, G./Prigogine, I.: Die Erforschung des Komplexen (see Chapter 1, Note 3) 132; Abraham, R.H./Shaw, C.D.: Dynamics – The Geometry of Behavior (see Note 14) 168, 174

2.21   For an analytical elaboration cf. Mainzer, K.: Symmetries in Nature (see Note 1) Chapter 3.31; Stauffer, D./Stanley, H.E.: From Newton to Mandelbrot (see Note 19) 24

2.22   Arnold, V.I.: Mathematical Methods of Classical Mechanics. Springer: Berlin (1978); Davies, P.C.W.: The Physics of Time Asymmetry. Surrey University Press: London (1974); Penrose, R.: The Emperor's New Mind. Oxford University Press: Oxford (1989) 181

2.23   Lichtenberg, A.J./Liebermann, M.A.: Regular and Stochastic Motion. Springer: Berlin (1982); Schuster, H.G.: Deterministic Chaos. An Introduction. Physik-Verlag: Weinheim (1984) 137

2.24   Poincaré, H.: Les Méthodes Nouvelles de la Méchanique Céleste. Gauthier-Villars: Paris (1892)

2.25   Arnold, V.I.: Small Denominators II, Proof of a theorem of A.N. Kolmogorov on the preservation of conditionally-periodic motions under a small perturbation of the Hamiltonian, Russ. Math. Surveys **18** (1963) 5; Kolmogorov, A.N.: On Conservation of Conditionally-Periodic Motions for a Small Change in Hamilton's Function, Dokl. Akad. Nauk. USSR **98** (1954) 525; Moser, J.: Convergent series expansions of quasi-periodic motions, Math. Anm. **169** (1967) 163

2.26   Cf. Arnold, V.I.: Mathematical Methods of Classical Mechanics (see Note 22); Schuster, H.G.: Deterministic Chaos (see Note 23), 141

2.27   Hénon, M./Heiles, C.: The applicability of the third integral of the motion: Some numerical experiments, Astron. J. **69** (1964) pp. 73; Schuster, H.G.: Deterministic Chaos (see Note 23), 150; Figures 2.16a–d from M.V. Berry in S. Jorna (ed.), Topics in nonlinear dynamics, Am. Inst. Phys. Conf. Proc. vol. **46** (1978)

2.28   For mathematical details compare, e.g. Staufer, D./Stanley, H.E.: From Newton to Mandelbrot (see Note 19), 83

2.29   Mainzer, K.: Symmetrien der Natur (see Note 1), 423; Primas, H./Müller-Herold, U.: Elementare Quantenchemie. Teubner: Stuttgart (1984) with an elementary introduction to the Galileo-invariant quantum mechanics (Chapter 3)

2.30   Audretsch, J./Mainzer, K. (eds.): Wieviele Leben hat Schrödingers Katze? B.I. Wissenschaftsverlag: Mannheim (1990)

2.31   Gutzwiller, M.C.: Chaos in Classical and Quantum Mechanics. Springer: Berlin (1990)

2.32   Friedrich, H.: Chaos in Atomen, in: Mainzer, K./Schirmacher, W. (eds.): Quanten, Chaos und Dämonen (see Note 1 of Chapter 1); Friedrich, H./Wintgen, D.: The hydrogen atom in a uniform magnetic field, Physics Reports **183** (1989) 37–79

2.33   Birkhoff, G.D.: Nouvelles recherches sur les systèmes dynamiques, Mem. Pont. Acad. Sci. Novi Lyncaei **1** (1935) 85

2.34   Enz, C.P.: Beschreibung nicht-konservativer nicht-linearer Systeme I–II, Physik in unserer Zeit **4** (1979) 119–126, **5** (1979) 141–144 (II)

2.35   Lorenz, E.N.: Deterministic nonperiodic flow, J. Atoms. Sci. **20** (1963) 130; Schuster, H.G.: Deterministic Chaos (see Note 23) 9

2.36   Eckmann, J.P.: Roads to turbulence in dissipative dynamical systems, Rev. Mod. Phys. **53** (1981) 643; Computer simulation of Fig. 2.21 from Lanford, O.E., Turbulence Seminar, in: Bernard, P./Rativ, T. (eds.): Lecture Notes in Mathematics 615, Springer: Berlin (1977) 114

2.37   Mandelbrot, B.B.: The Fractal Geometry of Nature, Freeman: San Fransisco (1982); Grassberger, P.: On the Hausdorff dimension of fractal attractors, J. Stat. Phys. **19** (1981) 25; Lichtenberg, A.J./Liebermann, M.A.: Regular and Stochastic Motions (see Note 23)

2.38   Collet, P./Eckmann, J.P.: Iterated Maps of the Interval as Dynamical Systems, Birkhäuser: Boston (1980) (see Figures 2.22–24)

2.39   Großmann, S./Thomae, E.: Invariant distributions and stationary correlation functions of one-dimensional discrete processes, Z. Naturforsch. **32 A** (1977) 353; Feigenbaum, M.J.: Quantitative universality for a class of nonlinear transformations, J. Stat. Phys. **19** (1978) 25

2.40   Mainzer, K.: Symmetrien der Natur (see Note 1)

2.41   Cf. Nicolis, G./Prigogine, I.: Die Erforschung des Komplexen (see Note 3, Chapter 1) 205

2.42   Cf. Prigogine, I.: From Being to Becoming – Time and Complexity in Physical Sciences, Freemann: San Fransisco (1980); Introduction to Thermodynamics of Irreversible Processes, Wiley: New York (1961)

2.43   Fig. 2.27 from Feynman, R.P./Leighton, R.B./Sands, M.: The Feynman Lectures of Physics vol. II., Addison-Wesley (1965)

2.44   Haken, H.: Synergetics (see Note 4, Chapter 1) 5

2.45   Haken, H.: Synergetics (see Note 4, Chapter 1) 202; Haken, H.: Advanced Synergetics. Instability Hierarchies of Self-Organizing Systems and Devices. Springer: Berlin (1983) 187; Weinberg, S.: Gravitation and Cosmology. Principles and Applications of the General Theory of Relativity. Wiley: New York (1972)

2.46   Cf. Mainzer, K.: Symmetrien der Natur (see Note 1) Chapter 4

2.47   Curie, P.: Sur la Symétrie dans les Phénomènes Physiques, Journal de Physique **3** (1894) 3

2.48   Audretsch, J./Mainzer, K. (eds.): Vom Anfang der Welt. C.H. Beck: München (^{2}1990); Mainzer, K.: Symmetrien der Natur (see Note 1) 515; Fritzsch, H.: Vom

Urknall zum Zerfall. Die Welt zwischen Anfang und Ende. Piper: München (1983) 278

2.49   Hawking, S.: A Brief History of Time. From the Big Bang to Black Holes. Bantam Press: London (1988); Hoyle, F./Burbridge, G./Narlikar, J.V.: A quasi-steady state cosmological model with creation of matter. Astrophys. Journal **410** (1993) 437–457

2.50   Audretsch, J./Mainzer, K. (eds.): Vom Anfang der Welt (see Note 48) 165

2.51   Whitesides, G.M./Mathias, J.P./Seto, C.T.: Molecular self-assembly and nanochemistry: A chemical strategy for the synthesis of nanostructures. Science **254** (1991) 1312–1319

2.52   Newkome, G.R. (ed.): Advances in Dendritic Macromolecules. JAI Press: Greenwich, Conn. (1994)

2.53   Curl, R.F./Smalley, R.E.: Probing C_{60}. Science **242** (1988) 1017–1022; Smalley, R.W.: Great balls of carbon: The story of Buckminsterfullerene. The Sciences **31** (1991) 22–28

2.54   Müller, A.: Supramolecular inorganic species: An expedition into a fascinating rather unknown land mesoscopia with interdisciplinary expectations and discoveries, J. Molecular Structure **325** (1994) 24; Angewandte Chemie (International Edition in English) **34** (1995) 2122–2124; Müller, A./Mainzer, K.: From molecular systems to more complex ones. In: Müller, A./Dress, A./Vögtle, F. (Eds.): From Simplicity to Complexity in Chemistry – and Beyond. Vieweg: Wiesbaden (1995) 1–11

2.55   Dry, C.M.: Passive smart materials for sensing and actuation. Journal of Intelligent Materials Systems and Structures **4** (1993) 415

2.56   Amato, I.: Animating the material world. Science **255** (1992) 284–286

Chapter 3

3.1   For historical sources of Sect. 3.1 compare Mainzer, K.: Die Philosophen und das Leben. In: Fischer, E.P./Mainzer, K. (eds.): Die Frage nach dem Leben. Piper: München (1990) 11–44

3.2   Diels-Kranz (see Note 2, Chapter 2) 12 A 30

3.3   Aristotle: Historia animalium 588 b 4

3.4   Aristotle: De generatione animalium II 736 b 12-15. a 35-b2

3.5   Descartes, R.: Discours de la méthode. Leipzig (1919/20) 39

3.6   Borelli, G.A.: De motu animalium. Leipzig (1927) 1

3.7   Leibniz, G.W.: Monadology §64

3.8   Bonnet, C.: Contemplation de la nature (1764). Oeuvres VII, 45

3.9   Kant, I.: Kritik der Urteilskraft. Ed. G. Lehmann, Reclam: Stuttgart (1971) 340

3.10   Goethe, J.W.: Dichtung und Wahrheit. In: Werke (Hamburger Ausgabe) Bd. IX 490

3.11   Schelling, F.W.J.: Sämtliche Werke Bd.II (ed. Schröter, M.), München (1927) 206

3.12   Darwin, C.: On the Origin of Species by Means of Natural Selection, or the Preservation of Favoured Races in the Struggle for Life. London (1859); The Descent of Man, and Selection in Relation to Sex. London (1871); Spencer, H.: Structure, Function and Evolution (ed. Andrenski, S.), London (1971); For a modern evaluation of Darwin's position compare Richards, R.: The Meaning of Evolution. University of Chicago Press: Chicago (1992)

3.13   Boltzmann, L.: Der zweite Hauptsatz der mechanischen Wärmetheorie. In: Boltzmann, L. (ed.): Populäre Schriften. Leipzig (1905) 24–46

3.14   Cf. Schneider, I.: Rudolph Clausius' Beitrag zur Einführung wahrscheinlichkeits-
        theoretischer Methoden in die Physik der Gase nach 1856. Archive for the History
        of Exact Sciences **14** (1974/75) 237–261

3.15   Prigogine, I.: Introduction to Thermodynamics of Irreversible Processes (see Note
        43, Chapter 2)

3.16   Cf. Boltzmann, L.: Über die mechanische Bedeutung des zweiten Hauptsatzes der
        Wärmetheorie (1866). In: Boltzmann, L.: Wissenschaftliche Abhandlungen (ed.
        Hasenöhrl, F.) vol. **1** Leipzig (1909), repr. New York (1968) 9–33; Analytischer
        Beweis des zweiten Hauptsatzes der mechanischen Wärmetheorie aus den Sätzen
        über das Gleichgewicht der lebendigen Kraft (1871) 288–308

3.17   Cf., e.g., Einstein's famous article 'Über die von der molekularkinetischen Theorie
        der Wärme geforderte Bewegung von in ruhenden Flüssigkeiten suspendierten
        Teilchen'. Annalen der Physik **17** (1905) 549–560

3.18   Poincaré, H.: Sur les tentatives d'explication méchanique des principes de la ther-
        modynamique. Comptes rendus de l'Académie des Sciences **108** (1889) 550–553;
        Zermelo, E.: Über einen Satz der Dynamik und die mechanische Wärmetheorie.
        Annalen der Physik **57** (1896) 485

3.19   Cf. Popper, K.R.: Irreversible processes in physical theory. Nature **181** (1958)
        402–403; Reichenbach, H.: The Direction of Time. Berkeley (1956); Grünbaum,
        A.: Philosophical Problems of Space and Time. Dordrecht (1973); Hintikka,
        J./Gruender, D./Agazzi, E. (eds.): Probabilistic Thinking, Thermodynamics and
        the Interaction of the History and Philosophy of Science II. Dordrecht/Boston/
        London (1978)

3.20   Boltzmann, L.: Der zweite Hauptsatz der mechanischen Wärmetheorie. In: Boltz-
        mann, L.: Populäre Schriften (see Note 13) 26–46

3.21   Boltzmann, L.: Über die Frage nach der objektiven Existenz der Vorgänge in der
        unbelebten Natur. In: Boltzmann, L.: Populäre Schriften (see Note 13) 94–119

3.22   Monod, J.: Le Hasard et la Nécessité. Editions du Seuil: Paris (1970)

3.23   Primas, H.: Kann Chemie auf Physik reduziert werden? Chemie in unserer Zeit
        19 (1985) 109–119, 160–166

3.24   Bergson, H.L.: L'évolution créative. Paris (1907); Heitler, W.H.: Über die Komple-
        mentarität von lebloser und lebender Materie. Abhandlungen der Math.-Naturw.
        Klasse d. Ak. d. Wiss. u. Lit. Mainz Nr. **1** (1976) 3–21; Driesch, A.: Philosophie
        des Organischen. Leipzig (1909); Whitehead, A.N.: Process and Reality. An Essay
        in Cosmology. New York (1978)

3.25   Schrödinger, E.: Was ist Leben? Piper: München (1987) 133

3.26   Schrödinger, E.: Was ist Leben? (see Note 25) 147

3.27   Thompson, W.: The Sorting Demon of Maxwell (1879). In: Thompson, W.: Phys-
        ical Papers I–VI, Cambridge (1882–1911), V, 21–23

3.28   Prigogine, I.: Time, irreversibility and structure. In: Mehra, J. (ed.): The Physicist's
        Conception of Nature. D. Reidel: Dordrecht/Boston (1973) 589

3.29   Eigen, M.: The origin of biological information. In: Mehra, J. (ed.): The Physi-
        cist's Conception of Nature (see Note 28) 607; Fig. 3.2 shows a so-called coat gene
        obtained by nuclease digestion of phage MS2-RNA (Min Jou, W./Haegemann,
        G./Ysebaert, M./Fiers, W.: Nature **237** (1972) 82). This gene codes for a sequence
        of 129 amino acids. The structure is further spatially folded. Also compare Perel-
        son, A.S./Kauffman, S.A. (eds.): Molecular Evolution on Ragged Landscapes:
        Proteins, RNA, and the Immune System. Santa Fé Institute Studies in the Sci-
        ences of Complexity, Proceedings vol. **9**. Addison-Wesley: Redwood City (1990)

3.30   For a survey cf. Depew, D.J./Weber, B.H.: Evolution at a Crossroads. The New
        Biology and the New Philosophy of Science. MIT Press: Cambridge, MA (1985);

Ebeling, W./Feistel, R.: Physik der Selbstorganisation und Evolution. Akademie-Verlag: Berlin (1982); Haken, H./Haken-Krell, M.: Entstehung von biologischer Information und Ordnung. Wissenschaftliche Buchgesellschaft: Darmstadt (1989); Hofbauer, L.: Evolutionstheorie und dynamische Systeme. Mathematische Aspekte der Selektion. Springer: Berlin (1984)

3.31   Eigen, M.: Homunculus im Zeitalter der Biotechnologie – Physikochemische Grundlagen der Lebensvorgänge. In: Gross, R. (ed.): Geistige Grundlagen der Medizin. Springer: Berlin (1985) 26, 36 for Fig. 3.4a–d; Maynard Smith, J.: Optimization theory in evolution. Annual Review of Ecological Systems **9** (1978) 31–56; Mainzer, K.: Metaphysics of nature and mathematics in the philosophy of Leibniz. In: Rescher, N. (ed.): Leibnizian Inquiries. University Press of America: Lanham/New York/London (1989) 105–130

3.32   Dyson, F.: Origins of Life. Cambridge University Press: Cambridge (1985)

3.33   Kauffman, S.: Autocatalytic sets of proteins. Journal of Theoretical Biology **119** (1986) 1–24

3.34   For a survey cf. Kauffmann, A.S.: Origins of Order: Self-Organization and Selection in Evolution. Oxford University Press: Oxford (1992)

3.35   Haken, H.: Synergetics (see Note 4, Chapter 1) 310

3.36   Hess, B./ Mikhailov, A.: Self-organization in living cells. In: Science **264** (1994) 223–224; Ber. Bunsenges. Phys. Chem. **98** (1994) 1198–1201 (extended version)

3.37   Susman, M.: Growth and Development. Prentice-Hall: Englewood Cliffs, NJ (1964); Prigogine, I.: Order through fluctuation. In: Jantsch, E./Waddington, C.H. (eds.): Evolution and Consciousness. Human Systems in Transition. Addison-Wesley: London (1976) 108

3.38   Gerisch, G./Hess, B.: Cyclic-AMP-controlled oscillations in suspended dictyostelium cells: Their relation to morphogenetic cell interactions. Proc. Natl. Acad. Sci. **71** (1974) 2118

3.39   Rosen, R.: Dynamical System Theory in Biology. Wiley-Interscience: New York (1970); Abraham, R.H./Shaw, C.D.: Dynamics – The Geometry of Bahavior (see Note 14, Chapter 2) 110 for Figs. 3.5

3.40   Meinhardt, H./Gierer, A.: Applications of a theory of biological pattern formation based on lateral inhibition. J. Cell. Sci. **15** (1974) 321 (Figs. 3.7–8); Meinhardt, M.: Models of Biological Pattern Formation. Academic Press: London (1982)

3.41   For a survey compare Gerok, W. (ed.): Ordnung und Chaos in der belebten und unbelebten Natur. Verhandlungen der Gesellschaft Deutscher Naturforscher und Ärzte. 115. Versammlung (1988), Stuttgart (1989); Mainzer, K.: Chaos und Selbstorganisation als medizinische Paradigmen. In: Deppert, W./Kliemt, H./Lohff, B./Schaefer, J. (eds.): Wissenschaftstheorien in der Medizin. De Gruyter: Berlin/New York (1992) 225–258

3.42   Bassingthwaighte, J.B./van Beek, J.H.G.M: Lightning and the heart: Fractal behavior in cardiac function. Proceedings of the IEEE **76** (1988) 696

3.43   Goldberger, A.L./Bhargava, V./West, B.J.: Nonlinear dynamics of the heartbeat. Physica **17D** (1985) 207–214; Nonlinear dynamics in heart failure: Implications of long-wavelength cardiopulmonary oscillations. American Heart Journal **107** (1984) 612–615; Ree Chay, T./Rinzel, J.: Bursting, beating, and chaos in an excitable membrane model. Biophysical Journal **47** (1985) 357–366; Winfree, A.T.: When Time Breaks Down: The Three-Dimensional Dynamics of Electrochemical Waves and Cardiac Arrhythmias. Princeton (1987); Guevara, M.R./Glass, L./Schrier, A.: Phase locking, period-doubling bifurcations, and irregular dynamics in periodically stimulated cardiac cells. Science **214** (1981) 1350

3.44   Cf. Johnson, L.: The thermodynamic origin of ecosystems: a tale of broken sym-
        metry. In: Weber, B.H./Depew, D.J./Smith, J.D. (eds.): Entropy, Information, and
        Evolution. New Perspectives on Physical and Biological Evolution. MIT Press.
        Cambridge, MA (1988) 75–105; Schneider, E.D.: Thermodynamics, ecological
        succession, and natural selection: a common thread. In: Weber, B.H./Depew,
        D.J./Smith, J.D. (eds.): Entropy, Information, and Evolution (see Note 42) 107–
        138
3.45   Odum, E.P.: The strategy of ecosystem development. Science 164 (1969) 262–270;
        Margalef, R.: Perspectives in Ecological Theory. University of Chicago Press:
        Chicago (1968)
3.46   Lovelock, J.E.: The Ages of Gaia. Bantam (1990); Schneider, S.H./Boston, P.J.
        (eds.): Scientists on Gaia. MIT Press: Cambridge, MA (1991); Pimm, S.: The
        Balance of Nature, University of Chicago Press: Chicago (1991)
3.47   Cf. Rosen, R.: Dynamical System Theory in Biology (see Note 37); Freedmann,
        H.I.: Deterministic Mathematical Models in Population Ecology. Decker: New
        York (1980); Abraham, R.H./Shaw, C.D.: Dynamics – The Geometry of Behavior
        (see Note 14, Chapter 2) 85
3.48   Lotka, A.J.: Elements of Mathematical Biology. Dover: New York (1925); Volter-
        ra, V.: Leçons sur la théorie mathématique de la lutte pour la vie. Paris (1931);
        Haken, H.: Synergetics (see Note 4, Chapter 1) 130, 308
3.49   Rettenmeyer, C.W.: Behavioral studies of army ants. Kansas Univ. Bull. 44 (1963)
        281; Prigogine, I.: Order through Fluctuation: Self-Organization and Social Sys-
        tem. In: Jantsch, E./Waddington, C.H. (eds.): Evolution and Consciousness (see
        Note 35) 111
3.50   Prigogine, I./Allen, P.M.: The challenge of complexity. In: Schieve, W.C./Allen,
        P.M.: Self-Organization and Dissipative Structures. Applications in the Physical
        and Social Sciences. University of Texas Press: Austin (1982) 28; Wicken, J.S.:
        Thermodynamics, evolution, and emergence: Ingredients of a new synthesis. In:
        Weber, B.H./Depew, D.J./Smith, J.D. (eds.): Entropy, Information, and Evolution
        (see Note 42) 139–169

Chapter 4

4.1   Diels-Kranz: B 36
4.2   Cf. Guthrie, W.K.C.: A History of Greek Philosophy vol. I: The Earlier Preso-
        cratics and the Pythagoreans. Cambridge University Press: Cambridge (1962) 349;
        Popper, K.R./Eccles, J.C.: The Self and its Brain. Springer: Berlin (1977) 161
4.3   Aristotle: De anima 403 b 31
4.4   Plato: Menon
4.5   Cf. Galen: Galen on Anatomical Procedures. Translation of the Surviving Books
        with Introduction and Notes. Oxford University Press: London (1956)
4.6   Wickens, G.M: Avicenna. Scientist and Philosopher. A Millenary Symposium:
        London (1952)
4.7   Descartes, R.: Meditations (1641). Eds. E. Haldane, G. Ross. Cambridge Univer-
        sity Press: Cambridge (1968) 153
4.8   Descartes, R.: Treatise on Man (1664). Harvard University Press: Cambridge,
        Mass. (1972)
4.9   Spinoza, B.: Ethics
4.10   Leibniz, G.W.: Monadology; Rescher, N.: Leibniz: An Introduction to his Philos-
        ophy. Basil Blackwell: Oxford (1979)

4.11   Hume, D.: A Treatise of Human Nature (1739). Penguin: Harmondsworth (1969) 82

4.12   Mainzer, K.: Kants Begründung der Mathematik und die Entwicklung von Gaußbis Hilbert. In: Akten des V. Intern. Kant-Kongresses in Mainz 1981 (ed. Funke, G.). Bouvier: Bonn (1981) 120–129

4.13   Brazier, M.A.B.: A History of Neurophysiology in the 17th and 18th Centuries. Raven: New York (1984); Cf. Clarke, E./O'Malley, C.D.: The Human Brain and Spinal Cord: A Historical Study illustrated by Writings from Antiquity to the Twentieth Century. University of California Press: Berkeley (1968)

4.14   Helmholtz, H.v.: Schriften zur Erkenntnistheorie (eds. Hertz, P./Schlick, M.). Berlin (1921); Mainzer, K.: Geschichte der Geometrie (see Note 13 Chapter 2) 172

4.15   Müller, J.: Handbuch der Physiologie des Menschen. Koblenz (1835)

4.16   Helmholtz, H.v.: Vorläufiger Bericht über die Fortpflanzungsgeschwindigkeit der Nervenreizung. Archiv für Anatomie, Physiologie und wissenschaftliche Medizin (1850) 71–73

4.17   James, W.: Psychology (Briefer Course). Holt: New York (1890) 3

4.18   James, W.: Psychology (see Note 17) 254

4.19   James, W.: Psychology (see Note 17) Fig. 57

4.20   Cf. Baron, R.J.: The Cerebral Computer. An Introduction to the Computational Structure of the Human Brain. Lawrence Erlbaum: Hillsdale N.J. (1987); Braitenberg, V.: Gehirngespinste. Neuroanatomie für kybernetisch Interessierte. Springer: Berlin (1973)

4.21   Churchland, P.S./Sejnowski, T.J.: Perspectives in cognitive neuroscience. Science **242** (1988) 741–745. The subset of visual cortex is adapted from van Essen, D./Maunsell, J.H.R.: Two-dimensional maps of the cerebral cortex. Journal of Comparative Neurology **191** (1980) 255–281. The network model of ganglion cells is given in Hubel, D.H./Wiesel, T.N.: Receptive fields, binocular interaction and functional architecture in the cat's visual cortex. Journal of Physiology **160** (1962) 106–154. An example of chemical synapses is shown in Kand, E.R./Schwartz J.: Principles of Neural Science. Elsevier: New York (1985)

4.22   Cf. Churchland, P.M.: A Neurocomputational Perspective: The Nature of Mind and the Structure of Science. MIT Press: Cambridge, Mass./London (1989) 99

4.23   Pellionisz, A.J.: Vistas from tensor network theory: A horizon from reductionalist neurophilosophy to the geometry of multi-unit recordings. In: Cotterill, R.M.J. (ed.): Computer Simulation in Brain Science. Cambridge University Press: Cambridge/New York/Sydney (1988) 44–73; Churchland, P.M.: A Neurocomputational Perspective (see Note 22) 83, 89

4.24   Cf. Schwartz, E.L. (ed.): Computational Neuroscience. MIT Press: Cambridge, Mass. (1990)

4.25   Cf. Churchland, P.S./Sejnowski, T.J.: The Computational Brain. MIT Press: Cambridge, Mass. (1992) 169

4.26   Hebb, D.O.: The Organization of Behavior. Wiley: New York (1949) 50

4.27   Kohonen, T.: Self-Organization and Associative Memory. Springer: Berlin (1989) 105; Churchland, P.S./Sejnowski, T.J.: The Computational Brain (see Note 25) 54; Ritter, H./Martinetz, T./Schulten, K.: Neuronale Netze. Eine Einführung in die Neuroinformatik selbstorganisierender Netzwerke. Addison-Wesley: Reading, Mass. (1991) 35

4.28   Hopfield, J.J.: Neural Network and physical systems with emergent collective computational abilities. Proceedings of the National Academy of Sciences **79** (1982) 2554–2558

4.29   Hertz, J./Krogh, A./Palmer, R.G.: Introduction to the Theory of Neural Compu-
        tation. Addison-Wesley: Redwood City (1991)
4.30   Serra, R./Zanarini, G.: Complex Systems and Cognitive Processes. Springer:
        Berlin (1990) 78
4.31   Hertz, J./Krogh, A./Palmer, R.G.: Introduction to the Theory of Neural Compu-
        tation (see Note 29); Hopfield, J.J./Tank, D.W.: Computing with neural circuits:
        A model. Science 233 (1986) 625–633
4.32   Ackley, D.H./Hinton, G.E./Sejnowski, T.J.: A learning algorithm for Boltzmann
        machines. Cognitive Science 9 (1985) 147–169
4.33   A mathematical elaboration of the learning algorithm for a Boltzmann machine
        is given in Serra, R./Zanarini, G.: Complex Systems and Cognitive Processes (see
        Note 30) 137. An illustration is shown in Churchland, P.S./Sejnowski, T.J.: The
        Computational Brain (see Note 25) 101
4.34   Rumelhart, D.E./Zipser, D.: Feature discovery by competitive learning. In:
        McClelland, J.L./Rumelhart, D.E. (eds.): Parallel Distributed Processing. MIT
        Press: Cambridge, Mass. (1986)
4.35   Kohonen, T.: Self-Organization and Associative Memory (see Note 27) 123
4.36   Kohonen, T.: Self-Organization and Associative Memory (see Note 27) 125
4.37   Ritter, H./Martinetz, T./Schulten, K.: Neuronale Netze (see Note 27) 75
4.38   Suga, N./O'Neill, W.E.: Neural axis representing target range in the auditory
        cortex of the mustache Bat. Science 206 (1979) 351–353; Ritter, H./Martinetz,
        T./Schulten, K.: Neuronale Netze (see Note 27) 88
4.39   Widrow, B./Hoff, M.E.: Adaptive switching circuits. 1960 IRE WESCON Con-
        vention Record. IRE: New York (1960) 36–104
4.40   Cf. Churchland, P.S./Sejnowski, T.J.: The Computational Brain (see Note 25) 106
4.41   Rumelhart, D.E./Hinton, G.E./Williams, R.J.: Learning representations by back-
        propagating errors. Nature 323 (1986) 533–536; Arbib, M.A.: Brains, Machines,
        and Mathematics. Springer: New York (1987) 117
4.42   Köhler, W.: Die physischen Gestalten in Ruhe und im stationären Zustand.
        Vieweg: Braunschweig (1920); Jahresberichte für die ges. Physiol. und exp. Phar-
        makol. 3 (1925) 512–539; Stadler, M./Kruse, P.: The self-organization perspective
        in cognitive research: Historical remarks and new experimental approaches. In:
        Haken, H./Stadler, M. (eds.): Synergetics of Cognition. Springer: Berlin (1990) 33
4.43   Cf. Churchland, P.M.: A Neurocomputational Perspective (see Note 22) 209
4.44   Cf. Churchland, P.M.: A Neurocomputational Perspective (see Note 22) 211
4.45   Cf. Feigl, H./Scriven, M./Maxwell, G. (eds.): Concepts, Theories and the Mind-
        Body Problem. University of Minnesota Press: Minneapolis (1958); Marcel,
        A.J./Bisiach, E. (eds.): Consciousness in Contemporary Science. Clarendon Press:
        Oxford (1988); Bieri, P.: Pain: A case study for the mind-body problem. Acta Neu-
        rochirurgica 38 (1987) 157–164; Lycan, W.G.: Consciousness. MIT Press: Cam-
        bridge, Mass. (1987)
4.46   Flohr, H.: Brain processes and phenomenal consciousness. A new and specific
        hypothesis. Theory & Psychology 1(2) (1991) 248
4.47   von der Malsburg, C.: Self-organization of orientation sensitive cells in the striate
        cortex. Kybernetik 14 (1973) 85–100; Wilshaw, D.J./von der Malsburg, C.: How
        patterned neural connections can be set up by self-organization. Proceedings of
        the Royal Society Series B 194 (1976) 431–445
4.48   Cf. Pöppel, E. (ed.): Gehirn und Bewußtsein. VCH Verlagsgesellschaft: Wein-
        heim (1989); Singer, W. (ed.): Gehirn und Kognition. Spektrum der Wissenschaft:
        Heidelberg (1990)
4.49   Haken, H./Stadler, M. (eds.): Synergetics of Cognition (see Note 42) 206

4.50   Haken, H./Stadler, M. (eds.): Synergetics of Cognition (see Note 42) 204
4.51   Pöppel, E.: Die neurophysiologische Definition des Zustands "bewußt". In: Pöppel, E. (ed.): Gehirn und Bewußtsein (see Note 48) 18
4.52   Searle, J.R.: Intentionality. An Essay in the Philosophy of Mind. Cambridge University Press: Cambridge (1983); Dennett, D.: The Intentional Stance, MIT Press: Cambridge, Mass. (1987)
4.53   Shaw, R.E./Kinsella-Shaw, J.M.: Ecological mechanics: A physical geometry for intentional constraints. Hum. Mov. Sci. 7 (1988) 155
4.54   For Figs. 4.22a–d, 4.23 Kugler, P.N./Shaw, R.E.: Symmetry and symmetry breaking in thermodynamic and epistemic engines: A coupling of first and second laws. In: Haken, H./Stadler, M. (eds.): Synergetics of Cognition (see Note 42) 317, 318, 319, 328
4.55   Kelso, J.A.S./Mandell, A.J./Shlesinger, M.F. (eds.): Dynamic Patterns in Complex Systems. World Scientific: Singapore (1988); For Figs. 4.24a–b, 4.25 compare Haken, H./Haken-Krell, M.: Erfolgsgeheimnisse der Wahrnehmung. Deutsche Verlags-Anstalt: Stuttgart (1992) 36, 38
4.56   Kelso, J.A.S.: Phase transitions: Foundations of behavior. In: Haken, H./Stadler, M. (eds.): Synergetics of Cognition (see Note 42) 260
4.57   Searle, J.R.: Mind, brains and programs. Behavioral and Brain Science 3 (1980) 417–424; Intrinsic intentionality. Behavioral and Brain Science 3 (1980) 450–456; Analytic philosophy and mental phenomena. Midwest Studies in Philosophy 5 (1980) 405–423. For a critique of Searle's position compare Putnam, H.: Representation and Reality. MIT Press: Cambridge, Mass. (1988) 26
4.58   Eccles, J.C.: The Neurophysiological Basis of Mind. Clarendon Press: Oxford (1953); Facing Reality. Springer: New York (1970); Eccles, J.C. (ed.): Mind and Brain, Paragon: Washington, D.C. (1982)
4.59   Palm, G.; Assoziatives Gedächtnis und Gehirn. In: Singer, W. (ed.): Gehirn und Kognition (see Note 48) 172; Palm, G. (ed.): Neural Assemblies: An Alternative Approach to Artificial Intelligence. Springer: Berlin (1984)

Chapter 5

5.1   For this chapter compare Mainzer, K.: Computer – Neue Flügel des Geistes? De Gruyter: Berlin/New York (1993); Mainzer, K.: Die Evolution intelligenter Systeme. Zeitschrift für Semiotik 12 (1990) 81–104
5.2   Feigenbaum, E.A./McCorduck, P.: The Fifth Generation. Artificial Intelligence and Japan's Computer Challenge to the World. Michael Joseph: London (1984)
5.3   Cf. Williams, M.R.: A History of Computing Technology. Prentice-Hall: Englewood Cliffs (1985)
5.4   Cohors-Fresenborg, E.: Mathematik mit Kalkülen und Maschinen. Vieweg: Braunschweig (1977) 7
5.5   Herrn von Leibniz' Rechnung mit Null und Eins. Siemens AG: Berlin (1966); Mackensen, L. von: Leibniz als Ahnherr der Kybernetik – ein bisher unbekannter Leibnizscher Vorschlag einer "Machina arithmetica dyadicae". In: Akten des II. Internationalen Leibniz-Kongresses 1972 Bd. 2. Steiner: Wiesbaden (1974) 255–268
5.6   Scholz, H.: Mathesis Universalis. Schwabe: Basel (1961)
5.7   Babbage, C.: Passages from the Life of a Philosopher. Longman and Co.: London (1864); Bromley, A.G.: Charles Babbage's Analytical Engine 1838. Annals of the History of Computing 4 (1982) 196–219

5.8   Cf. Minsky, M.: Recursive unsolvability of Post's problem of "tag" and other topics in the theory of Turing machines. Annals of Math. **74** 437–454; Sheperdson, J.C./Sturgis, H.E.: Computability of recursive functions. J. Assoc. Comp. Mach. **10** (1963) 217–255. For the following description of register machines compare Rödding, D.: Klassen rekursiver Funktionen. In: Löb, M.H. (ed.): Proceedings of the Summer School in Logic. Springer: Berlin (1968) 159–222; Cohors-Fresenborg, E.: Mathematik mit Kalkülen und Maschinen (see Note 4)

5.9   Turing, A.M.: On computable numbers, with an application to the "Entscheidungsproblem". Proc. London Math. Soc., Ser. **2 42** (1936) 230–265; Post, E.L.: Finite combinatory processes – Formulation I. Symbolic Logic I (1936) 103–105; Davis, M.: Computability & Unsolvability. McGraw-Hill: New York (1958) 3

5.10   Arbib, M.A.: Brains, Machines, and Mathematics. Springer: New York (1987) 131

5.11   Mainzer, K.: Der Konstruktionsbegriff in der Mathematik. Philosophia Naturalis **12** (1970) 367–412

5.12   Cf. Knuth, D.M.: The Art of Computer Programming. Vol. 2. Addison-Wesley: Reading, MA (1981); Börger, E.: Berechenbarkeit, Komplexität, Logik. Vieweg: Braunschweig (1985)

5.13   Grötschel, M./Lovásc, L./Schryver, A.: Geometric Algorithms and Combinatorial Optimization. Springer: Berlin (1988)

5.14   Gardner, A.: Penrose Tiles to Trapdoor Ciphers. W.H. Freeman: New York (1989)

5.15   Cf. Arbib, M.A.: Speed-up theorems and incompleteness theorem. In: Caianiello, E.R. (ed.): Automata Theory. Academic Press (1966) 6–24; Mostowski, A.: Sentences Undecidable in Formalized Arithmetic. North-Holland: Amsterdam (1957); Beth, E.W.: The Foundations of Mathematics, North-Holland: Amsterdam (1959); Kleene, S.C.: Introduction to Metamathematics. North-Holland: Amsterdam (1967)

5.16   Mainzer, K.: Rationale Heuristik und Problem Solving. In: Burrichter, C./Inhetveen, R./Kötter, R. (eds.): Technische Rationalität und rationale Heuristik. Schöningh: Paderborn (1986) 83–97

5.17   Mainzer, K.: Knowledge-based systems. Remarks on the philosophy of technology and Artificial Intelligence. Journal for General Philosophy of Science **21** (1990) 47–74

5.18   Bibel, W./Siekmann, J.: Künstliche Intelligenz. Springer: Berlin (1982); Nilson, N.J.: Principles of Artificial Intelligence. Springer: Berlin (1982); Kredel, L.: Künstliche Intelligenz und Expertensysteme. Droemer Knaur: München (1988)

5.19   Puppe, F.: Expertensysteme. Informatik-Spektrum **9** (1986) 1–13

5.20   For instance, one of the constructors of DENDRAL and MYCIN received a Ph.D. in philosophy for his dissertation "Scientific Discoveries", supervised by Gerald Massey (today Center for Philosophy of Science, Pittsburgh). Also refer to Glymour, C.: AI is philosophy. In: Fetzer, J.H. (ed.): Aspects in Artificial Intelligence. Kluwer Academic Publisher: Boston (1988) 195–207

5.21   Buchanan, B.G./Sutherland, G.L./Feigenbaum, E.A.: Heuristic DENDRAL: A program for generating processes in organic chemistry. In: Meltzer, B./Michie, D. (eds.): Machine Intelligence **4**. Edinburgh (1969); Buchanan, B.G./Feigenbaum, E.A.: DENDRAL and META-DENDRAL: Their applications dimension. In: Artificial Intelligence **11** (1978) 5–24

5.22   McCarthy, J.: LISP Programmer's Manual (Rep. MIT Press). Cambridge, MA (1962); Stoyan, H./Goerz, G.: LISP – Eine Einführung in die Programmierung. Springer: Berlin (1984)

5.23   Randall, D./Buchanan, B.G./Shortliffe, E.H.: Production rules as a representation for a knowledge-based consultation program. Artificial Intelligence **8** (1977); Shortliffe, E.H.: MYCIN: A rule-based computer program for advising physicians regarding antimicrobial therapy selection. AI Laboratory, Memo 251, STAN-CS-74-465, Stanford University; Winston, P.H.: Artificial Intelligence. Addison-Wesley: Reading, MA (1977) Chap. 9

5.24   Doyle, J.: A truth maintenance system. Artificial Intelligence **12** (1979) 231–272

5.25   Lenat, D.B./Harris, G.: Designing a rule system that searches for scientific discoveries. In: Waterman, D.A./Hayes-Roth, F. (eds.): Pattern Directed Inference Systems. New York (1978) 26; Lenat, D.B.: AM: Discovery in mathematics as heuristic search. In: Davis, R./Lenat, D.B. (eds.): Knowledge-Based Systems in Artificial Intelligence. New York (1982) 1–225

5.26   Langley, P.: Data-driven discovery on physical laws. Cognitive Science **5** (1981) 31–54

5.27   Langley, P./Simon, H.A./Bradshaw, G.L./Zytkow, J.M.: Scientific Discovery: Computational Explorations of the Creative Processes. Cambridge, MA (1987)

5.28   Kulkarni, D./Simon, H.A.: The process of scientific discovery: The strategy of experimentation. Cognitive Science **12** (1988) 139–175

5.29   Cf. Winston, P.H.: Artificial Intelligence (see Note 23)

5.30   Audretsch, J./Mainzer, K. (eds.): Wieviele Leben hat Schrödingers Katze? B.I. Wissenschaftsverlag: Mannheim (1990) 273 (Fig. 5.14a), 274 (Fig. 5.14.b), 276 (Fig. 5.14c)

5.31   Penrose, R.: Newton, quantum theory and reality. In: Hawking, S.W./Israel, W.: 300 Years of Gravity. Cambridge University Press: Cambridge (1987)

5.32   Deutsch, D.: Quantum theory, the Church-Turing principle and the universal quantum computer. Proc. Roy. Soc. **A400** (1985) 97–117

5.33   For this chapter compare Mainzer, K.: Computer – Neue Flügel des Geistes? De Gruyter: Berlin (1993); Mainzer, K.: Philosophical concepts of computational neuroscience. In: Eckmiller, R./Hartmann, G./Hauske, G. (eds.): Parallel Processing in Neural Systems and Computers. North-Holland: Amsterdam (1990) 9–12

5.34   McCullock, W.S./Pitts, W.: A logical calculus of the ideas immanent in nervous activity. In: Bulletin of Mathematical Biophysics **5** (1943) 115–133; Arbib, M.A.: Brains, Machines and Mathematics (see Note 10) 18

5.35   Neumann, J. von: The Computer and the Brain. Yale University Press: New Haven (1958)

5.36   Cf. Burks, A.W. (ed.): Essays on Cellular Automata. University of Illinois Press (1970); Myhill, J.: The abstract theory of self-reproduction. In: Mesarovic, M.D. (ed.): Views on General Systems Theory. John Wiley (1964) 106–118

5.37   Wolfram, S.: Universality and complexity in cellular automata. In: Farmer, D./Toffoli, T./Wolfram, S. (eds.): Cellular Automata. Proceedings of an Interdisciplinary Workshop. North-Holland: Amsterdam (1984) 1–35; Wolfram, S.: Theory and Applications of Cellular Automata. World Scientific: Singapore (1986); Demongeot, J./Golés, E./Tchuente, M. (eds.): Dynamical Systems and Cellular Automata. Academic Press: London (1985); Poundstone, W.: The Recursive Universe. Cosmic Complexity and the Limits of Scientific Knowledge. Oxford University Press: Oxford (1985)

5.38   Rosenblatt, F.: The perceptron: A probabilistic model for information storage and organization in the brain. Psychological Review **65** (1958) 386–408

5.39   Minsky, M./Papert, S.A.: Perceptrons. MIT Press: Cambridge MA (1969) (expanded edition 1988)

5.40   Gorman, R.P./Sejnowski, T.J.: Analysis of hidden units in a layered network trained to classify sonar targets. Neural Networks **1** (1988) 75–89; Churchland, P.M.: A Neurocomputational Perspective. The Nature of Mind and the Structure of Science. MIT Press: Cambridge MA (1989)

5.41   Sejnowski, T.J./Rosenberg, C.R.: NETtalk: A parallel network that learns to read aloud. The Johns Hopkins University Electrical Engineering and Computer Science Technical Report IHU/EECS-86/01 (1986) 32; Cf. Kinzel, W./Deker, U.: Der ganz andere Computer: Denken nach Menschenart. Bild der Wissenschaft **1** (1988) 43

5.42   Schöneburg, E./Hansen, N./Gawelczyk: Neuronale Netzwerke. Markt & Technik: München (1990) 176, 177

5.43   Rumelhart, D.E./Smolensky, P./McClelland, J.L./Hinton, G.E.: Schemata and sequential thought processes. In: McClelland, J.L./Rumelhart, D.E. (eds.): Parallel Distributed Processing: Explorations in the Microstructure of Cognition vol. 2: Applications. MIT Press: Cambridge MA (1986); Churchland, P.S.: Neurophilosophy. Toward a Unified Science of the Mind-Brain. MIT Press: Cambridge MA (1988) 465

5.44   Haken, H. (ed.): Neural and Synergetic Computers. Springer: Berlin (1988) 5–7

5.45   Haken, H.: Synergetics as a tool for the conceptualization and mathematization of cognition and behaviour – How far can we go? In: Haken, H./Stadler, M. (eds.): Synergetics of Cognition. Springer: Berlin (1990) 23–24

5.46   For chapter 5.4 compare Mainzer, K.: Philosophical foundations of neurobionics. In: Bothe, H.W./Samii, M./Eckmiller, R. (eds.): Neurobionics. An Interdisciplinary Approach to Substitute Impaired Functions of the Human Nervous System. North-Holland: Amsterdam (1993) 31–47

5.47   Cf. Samii, M. (ed.): Peripheral Nerve Lesions. Springer: Berlin (1990)

5.48   Eckmiller, R. (ed.): Neurotechnologie-Report. Bundesministerium für Bildung, Wissenschaft, Forschung und Technologie: Bonn (1995); Stein, R.B./Peckham, P.H. (eds.): Neuroprostheses: Replacing Motor Function After Disease or Disability. Oxford University Press: New York (1991)

5.49   Cosman, E.R.: Computer-assisted technologies for neurosurgical procedures. In: Bothe, H.W./Samii, M./Eckmiller, R. (eds.): Neurobionics (see Note 46) 365

5.50   Foley, J.D./Dam, A. van: Fundamentals of Interactive Computer Graphics. Addison-Wesley: Amsterdam (1982); Foley, J.D.: Neuartige Schnittstellen zwischen Mensch und Computer. Spektrum des Wissenschaft **12** (1987)

5.51   Gibson, W.: Neuromancer. Grafton: London (1986) 67

5.52   Putnam, H.: Reason, Truth and History. Cambridge University Press: Cambridge (1981) 6

5.53   Langton, C.G. (ed.): Artificial Life, Addison Wesley: Redwood City (1989)

5.54   Lindenmayer, A./Rozenberg, G. (eds.): Automata, Languages, Development. North-Holland: Amsterdam (1976); Lindenmayer, A.: Models for multicellular development: Characterization, inference and complexity of L-systems. In: Kelemenová, A./Kelemen, J. (eds.): Trends, Techniques and Problems in Theoretical Computer Science, Springer: Berlin (1987) 138–168

5.55   Langton, C.G.: Self-reproduction in cellular automata. Physica **D 10** (1984) 135–144; Artificial Life, in: Langton [5.53] Plates 1–8

5.56   Langton, C.G.: Life at the edge of chaos. In: Langton, C.G./ Taylor, C./ Farmer, J. D./Rasmussen, S. (eds.): Artificial Life II, Addison-Wesley: Redwood City (1992) 41–91

5.57   Rechenberg, I.: Evolutionsstrategie: Optimierung technischer Systeme nach Prinzipien der biologischen Evolution, Frommann-Holzboog: Stuttgart (1973)

5.58   Holland, J.H.: Adaption in Natural and Artificial Systems. The MIT Press: Cambridge, MA (1992, 1st ed. 1975)
5.59   Mitchell, M./Forrest, S.: Genetic Algorithms and Artificial Life. Artificial Life 1 (1994) 267–289
5.60   Forrest, S./Javornik, B./Smith, R./Perelson, A.: Using genetic algorithms to explore pattern recognition in the immune system. Evolutionary Computation 1 (1993) 191–211
5.61   Highland, H.J. (ed.): Computer Virus Handbook. Elsevier Advanced Technology: Oxford (1990); Spafford, E.H.: Computer viruses as artificial life. Artificial Life 1 (1994) 249–265
5.62   Taylor, C./Jefferson, D.: Artificial life as a tool for biology inquiry. Artificial Life 1 (1994) 1–13
5.63   Morán, F./Moreno, A./Merelo, J.J./Chacón, P. (eds.): Advances in Artificial Life, Springer: Berlin (1995)
5.64   Mainzer, K.: Gehirn; Computer, Komplexität. Springer: Berlin 1997

Chapter 6

6.1   Ehrenberg, V.: Der Staat der Griechen. Zürich (1965); Meier, C.: Die Entstehung des Politischen bei den Griechen. Frankfurt (1983)
6.2   Platonis Opera, 5 vols. Ed. J. Burnet. Oxford (1899–1906) (especially Politeia, Politikos); Cf. Cross, R.C./Woozley, A.D.: Plato's Republic. A Philosophical Commentary. London (1964)
6.3   Aristotle: Ethica Nicomachae, Politica (Oxford Classical Texts); Atheniesium Res Publica. Ed. F.G. Kenyon. Oxford (1920); Höffe, O.: Praktische Philosophie – Das Model des Aristoteles. München (1971); Barker, L.E.: The Political Thought of Plato and Aristotle. New York (1959)
6.4   Hobbes, T.: The Elements of Law Natural and Politic (1640). Ed. F. Tönnies. London (1889); Leviathan, or the Matter, Form and Power of a Commonwealth, Ecclesiasticial and Civil (1651). Ed. C.B. Macpherson. Harmondsworth (1968); Gough, J.W.: The Social Contract. A Critical Study of its Development. Oxford (1957); Höffe, O. (ed.): Thomas Hobbes: Anthropologie und Staatsphilosophie. Freiburg (1981)
6.5   Quesnay: Oeuvres économiques et philosophiques. Francfort (1988)
6.6   Figure 6.3a shows a "Tableau économique" from the "Elements de la philosophie rurale" (1767) of Marquis of Mirabeau; cf. Meek, R.L.: The interpretation of the "Tableau economique", Economica 27 (1960) 322–347
6.7   Serviere, G. de: Recueil d'ouvrages curieux de mathématique et de méchanique, ou description du cabinet de Monsieur Grollier de Serviere, avec des figures en taille-douce. Seconde édition. Lyon (1733) (Fig. 38); Rieter, H.: Quesnays Tableau Economique als Uhren-Analogie. In: Scherf, H. (ed.): Studien zur Entwicklung der ökonomischen Theorie IX. Duncker & Humblot: Berlin (1990) 73
6.8   Locke, J.: Two Treatises on Government (1689). Ed. P. Laslett. Cambridge (1970). Book II: Essay concerning the true original, and end of civil government
6.9   Hume, D.: Essays and Treatise on Several Subjects. In two Volumes. I. Containing Essays, Moral, Political, and Literary. A New Edition. London (1784)
6.10   Smith, A.: The Theory of Moral Sentiments (1759). Ed. D.D. Raphael/A.L. Macfie. Oxford (1979)

6.11   Cf. Skinner, A.S.: Adam Smith, science and the role of imagination. In: Todd, W.B. (ed.): Hume and the Enlightenment. Edinburgh University Press: Edinburgh (1974) 164–188

6.12   Smith, A.: An Inquiry into the Nature and Causes of the Wealth of Nations. Ed. E. Cannan. The University of Chicago Press: Chicago (1976) vol. 1, 478

6.13   Smith, A.: (see Note 12), vol. 1, 18

6.14   Mayr, O.: Adam Smith und das Konzept der Regelung. In: Troitzsch/Wohlauf, G. (eds.): Technikgeschichte. Frankfurt (1980) 245

6.15   Cf. Marchi, N.B. de: The empirical content and longevity of Ricardian economics. Economica **37** (1970) 258–259; Sowell, T.: Classical Economics Reconsidered. Princeton University Press: Princeton (1974) 118–120; Blaug, M.: The Methodology of Economics. Cambridge University Press: Cambridge (1980)

6.16   Mill, J.S.: On the definition of political economy. In: Collected Works. Essays on Economy and Society. Ed. J.M. Robson. University of Toronto Press: Toronto (1967) vol. 4, 330; Hicks, J.: From classical to post-classical: The work of J.S. Mill. In: Hicks, J.: Classics and Moderns. Collected Essays on Economic Theory. Vol. III. Basil Blackwell: Oxford (1983) 60–70

6.17   Georgescu-Roegen, N.: The Entropy Law and Economic Progress. Harvard University Press: Cambridge, MA (1971)

6.18   Walras, L.: Principe d'une théorie mathématique de l'echange. Journal des Economistes **33** (1874) 1–21; Hicks, J.: Léon Walras. In: Hicks, J.: (see Note 16) 85–95

6.19   Keynes, J.M.: The Collected Writings of John Maynard Keynes. Vol. XIV. The General Theory and After. Ed. D. Moggridge. Macmillan: London (1973) 296

6.20   Neumann, J. von/Morgenstern, O.: Theory of Games and Economic Behavior. Princeton University Press: Princeton, NJ (1943)

6.21   Cf. Stegmüller, W.: Probleme und Resultate der Wissenschaftstheorie und Analytischen Philosophie Bd. 1. Springer: Berlin (1969) 391

6.22   Cf. Shubik, M. (ed.): Game Theory and Related Approaches to Social Behavior. John Wiley: New York/London/Sidney (1964); Nash, J.F.: Equilibrium points in n-person games. Proceedings of National Academy of Science (USA) **36** (1950) 48–49

6.23   For a survey compare Anderson, P.W./Arrow, K.J./Pines, D. (eds.): The Economy as an Evolving Complex System. Santa Fé Institute Studies in the Sciences of Complexity vol. 3. Addison-Wesley: Redwood City (1988); Barnett, W.A./Geweke, J./Shell, K. (eds.): Economic Complexity. Chaos, Sunspots, Bubbles, and Nonlinearity. Cambridge University Press: Cambridge (1989)

6.24   Kalecki, M.: Theory of Economic Dynamics. Unwin University Books: London (1954); Samuelson, P.A.: Foundations of Economic Analysis. Harvard University Press: Cambridge, MA (1947)

6.25   Goodwin, R.M.: Chaotic Economic Dynamics. Clarendon Press: Oxford (1990) 19

6.26   Rössler, O.E.: Chaos und Endophysik. In: Mainzer, K./Schirmacher, W. (eds.): Quanten, Chaos und Dämonen. Erkenntnistheoretische Aspekte der modernen Physik. B.I. Wissenschaftsverlag: Mannheim (1993)

6.27   Goodwin, R.M.: Chaotic Economic Dynamics (see Note 25) 113

6.28   Lorenz, H.-W.: Nonlinear Dynamical Economics and Chaotic Motion. Springer: Berlin (1989); Goodwin, R.W.: A growth cycle. In: Hunt, E.K./Schwartz, J.G. (eds.): A Critique of Economic Theory. Penguin: Harmondsworth (1969) 442–449

6.29   This idea stems back to Samuelson, P.A.: Generalized predator-prey oscillations in ecological and economic equilibrium. In: The Collected Scientific Papers of

P.A. Samuelson. Ed. R.C. Merton. Vol III. MIT Press: Cambridge, MA (1972) 487–490

6.30   Cf. Lorenz, H.-W.: Nonlinear Dynamical Economics and Chaotic Motion. (See Note 27) 157; Goodwin, R.M.: Chaotic Economic Dynamics. (See Note 25) 117

6.31   Cf. Farmer, J.D./Sidorowich, J.J.: Predicting Chaotic Time Series. Mineo: Los Alamos National Laboratory (1987)

6.32   In his "History of Economical Analysis", J.A. Schumpeter had already discussed the possibility of multiple equilibrium points. But like Poincaré in the physical sciences, he was afraid of chaotic scenarios in economics: Schumpeter, J.A.: Geschichte der ökonomischen Analyse. Vandenhoeck & Ruprecht: Göttingen (1965)

6.33   Arthur, W.B.: Positive feedbacks in the economy. Scientific American, February (1990) 92–99

6.34   Arthur, W.B./Ermoliew, J.M./Kaniowski, J.M.: Path-dependent processes and the emergence of macro-structure. European Journal of Operational Research **30** (1987) 294–303; Helpman, E./Krugman, P.: Market Structure and Foreign Trade. MIT Press: Cambridge, MA (1985)

6.35   Flood, R.L.: Complexity: A definition by construction of a conceptual framework. Systems Research **4**(3) (1987) 177–185; Flood, R.L./Carson, E.R.: Dealing with Complexity. An Introduction to the Theory and Application of Systems Science. Plenum Press: New York (1988); An early inquiry of complexity in social sciences was written by Hayek, F.A. von: The Theory of Complex Phenomena. J.C.B. Mohr: Tübingen (1972)

6.36   For a survey compare Cambel, A.B./Fritsch, B./Keller, J.U. (eds.): Dissipative Strukturen in integrierten Systemen. Nomos Verlagsgesellschaft: Baden-Baden (1989); Tainter, J.: The Collapse of Complex Societies. Cambridge University Press (1988); Dumouchel, P./Dupuy, J.-P. (eds.): L'Auto-Organisation de la Physique au Politique. Editions Du Seuil: Paris (1983)

6.37   Malthus, T.R.: An Essay on the Principle of Population. London (1798)

6.38   Spencer, H.: First Principles. D. Appleton & Co.: New York (1896) 527; Carneiro, R.L.: Structure, function, and equilibrium in the evolutionism of Herbert Spencer. Journal of Anthropological Research **29** (1973) 77–95

6.39   For a review compare, for instance, Carneiro, R.L.: Successive reequilibrations as the mechanism of cultural evolution. In: Schieve, W.C./Allen, P.M. (eds.): Self-Organization and Dissipative Structures. Applications in the Physical and Social Sciences. University of Texas Press: Austin (1982) 110–115

6.40   Allen, P.M.: Self-organization in the urban system. In: Schieve, W.C./Allen, P.M. (eds.): Self-Organization and Dissipative Structures. (See Note 39) 142–146

6.41   Durkheim, E.: De la Division du Travail Social. Paris (1893); Les Règles de la Méthode Sociologique. Paris (1895)

6.42   Weidlich, W.: Das Modellierungskonzept der Synergetik für dynamisch sozio-ökonomische Prozesse. In: Mainzer, K./ Schirmacher, W. (eds.): Quanten, Chaos und Dämonen. (See Note 26) Figs. 6.12–16; Weidlich, W.: Stability and cyclicity in social systems. In: Cambel, A.B. Fritsch, B./Keller, J.U. (eds.): Dissipative Strukturen in Integrierten Systemen. (See Note 36) 193–222

6.43   For a survey compare Scott, W.R.: Organizations: Rational, Natural, and Open Systems. Prentice-Hall: Englewood Cliffs, NJ (1992): Taylor, M.: The Possibility of Cooperation. Cambridge University Press: Cambridge (1987)

6.44   Bott, E.: Urban families: Conjugal roles and social networks. Human Relations **8** (1955) 345–383; Knipscheer, C.P.M./Antonucci, T.C. (eds.): Social Network Research: Substantive Issues and Methodological Questions. Swets and Zeitlinger: Amsterdam (1990)

6.45 Schelling. T.C.: Micromotives and Macrobehavior. W.W. Norton & Co.: New York (1978)

6.46 Glance, N.S./Huberman, B.A.; Organizational fluidity and sustainable cooperation, dynamics of computation group: Xerox Palo Alto Research Center: March 12 (1993) (Fig. 3)

6.47 See Note 46, Fig. 4

6.48 Dawkins, R.: The Selfish Gene. Oxford University Press: New York (1976)

6.49 Popper, K.R.: Evolutionary epistemology. In: Miller, D. (ed.): Popper Selections. Princeton University Press: Princeton, NJ (1985) 78–86

6.50 Miller, M.S./Drexler, K.E.: Comparative ecology: A computational perspective. In: Huberman, B.A. (ed.): The Ecology of Computation. North-Holland: Amsterdam (1988) 51–76

6.51 Compare Luhmann, N.: The self-reproduction of the law and its limits. In: Teubner, G. (ed.): Dilemmas of Law in the Welfare State. Berlin (1986) 111–127; Mayntz, R.: Differenzierung und Verselbständigung. Zur Entwicklung gesellschaftlicher Teilsysteme. Frankfurt (1988)

6.52 Minsky, M.: The Society of Mind. Simon & Schuster: New York (1985) 314

6.53 Huberman, B.A. (ed.): The Ecology of Computation. (See Note 50) 81

6.54 For a survey compare Kellermayr, K.H.: Lokale Computernetze LAN. Technologische Grundlagen, Architektur, Übersicht und Anwendungsbereiche. Springer: Wien (1986); Sharp, J.A.: Verteilte und parallele Computernetze. Architektur und Programmierung. VCH-Verlagsgesellschaft: Weinheim (1989); Barz, H.W.: Kommunikation und Computernetze. Konzepte, Protokolle und Standards. Carl Hanser: München (1991)

6.55 Cerf, V.G.: Netztechnik, Spektrum der Wissenschaft Nov. 11 (1991) 76

6.56 Minsky, M.: The Society of Mind. (See Note 52)

6.57 Hegel, G.W.F.: Phänomenologie des Geistes. (1807) In: Werke in zwanzig Bänden, vol. 3. Suhrkamp: Frankfurt (1972)

Chapter 7

7.1 Goodmann, N.: Fact, Fiction and Forecast. London (1954); Lenk, H.: Erklärung, Prognose, Planung. Rombach: Freiburg (1972)

7.2 Montgomery, D.C./Johnson, L.A.: Forecasting and Time Series Data Analysis and Theory. Holt, Rinehart and Winston: New York (1975)

7.3 Abraham, B./Ledolter, J.: Statistical Methods of Forecasting. Wiley: New York (1983); Pindyek, R./Rubinfeld, D.: Econometric Models and Economic Forecasts. McGraw-Hill: New York (1980)

7.4 Makridakis, S./Wheelwright, S.C.: Forecasting Methods for Management. Wiley: New York (1989); Makridakis, S./Wheelwright, S.C./Mc Gree, V.E.: Forecasting: Methods and Applications. Wiley: New York (1989)

7.5 Kravtsov, Y. A. (ed.): Limits of Predictability. Springer: Berlin (1993) 82; Sarder, Z./Ravetz, J.R. (eds.): Complexity: fad or future? In: Futures. J. Forecasting, Planning, and Policy 26 6 (1994)

7.6 Mayntz, R.: The Influence of natural science theories on contemporary social science. In: Meinolf, D./Biervert, B. (eds.): European Social Science in Transition. Assessment and Outlook. Westview Press: Boulder, Colorado (1992) 27–79

7.7   Merton, R.K.: The Sociology of Science. Theoretical and Empirical Investiga-
        tions. Chicago (1973); Sorokin, P.A.: Social and Cultural Dynamics. New York
        (1962); Price, D. de Solla: Science since Babylon. New Haven (1961); Goffman,
        W.: An epidemic process in an open population. In: Nature **205** (1965), 831–832 ;
        Mathematical approach to the spread of scientific ideas – the history of mast cell
        research. In: Nature **212** (1966), 449–452; Berg, J./Wagner-Döbler, R.: A multidi-
        mensional analysis of scientific dynamics. In: Scientometrics **3** (1996), 321–346
7.8   Bruckner, E.W./Ebeling, W./Scharnhorst, A.: The application of evolution models
        in scientometrics. In: Scientometrics **18** (1990), 33
7.9   Small, H./Sweeny/Greenlee, E.: Clustering the science citation index using co-
        citations. II. Mapping science. In: Scientometrics **8** (1985) 321–340; Callon,
        M./Courtial, J.-P./Turner, W.A./Bauin, S.: From translation to problematic net-
        works – An introduction to co-word analyses. In: Social Science Information **22**
        (1983), 191–235
7.10   Mayntz, R.: Soziale Diskontinuitäten: Erscheinungsformen und Ursachen. In:
        Hierholzer, K./Wittmann, H.-G. (eds.): Phasensprünge und Stetigkeit in der
        natürlichen und kulturellen Welt. Wiss. Verlagsgesellschaft: Stuttgart (1988) 15–
        37
7.11   Mainzer, K.: Die Würde des Menschen ist unantastbar. Zur Herausforderung
        von Artikel 1I Grundgesetz durch den technisch-wissenschaftlichen Fortschritt.
        Jahrbuch der Universität 1989. Augsburg (1990) 211–221
7.12   For a historical survey compare Schwemmer, O.: Ethik. In: Mittelstraß, J. (ed.):
        Enzyklopädie Philosophie und Wissenschaftstheorie. Vol. 1. B.I. Wissenschafts-
        verlag: Mannheim/Wien/Zürich (1980) 592–599; Abelson, R./Nielsen, K.: History
        of ethics. In: Edwards (ed.): The Encyclopedia of Philosophy. Vol. III. New York
        (1967) 81–117
7.13   Irwin, T.: Plato's Moral Theory. The Early and Middle Dialogues. Oxford (1977)
7.14   Kenny, A.: The Aristotelean Ethics. A Study of the Relationship between the
        Eudemian and Nicomachean Ethics of Aristotle. Oxford (1978)
7.15   Compare Raphael, D.D. (ed.): British Moralists (1650–1800) I–II. Oxford (1969)
7.16   Paton, H.J.: The Categorical Imperative. A Study in Kant's Moral Philosophy.
        London (1947); Williams, B.A.D.: The Concept of the Categorical Imperative. A
        Study of the Place of the Categorical Imperative in Kant's Ethical Theory. Oxford
        (1968)
7.17   Mill, J.S.: Utilitarianism. London (1861); Rescher, N.: Distributive Justice. A Con-
        structive Critique of the Utilitarian Theory of Distribution. Indianapolis (1966)
7.18   Rawls, J.: A Theory of Justice. London (1972)
7.19   Simon, J.: Friedrich Nietzsche. In: Höffe, O. (ed.): Klassiker der Philosophie II.
        München (1981) 203–224, 482–484
7.20   Heidegger, M.: Die Technik und die Kehre. Pfullingen (1962); Loscerbo, J.: Being
        and Technology. A Study in the Philosophy of Martin Heidegger. The Hague
        (1981)
7.21   Jonas, H.: Das Prinzip Verantwortung. Versuch einer Ethik für die technologische
        Zivilisation. Suhrkamp: Frankfurt (1979)
7.22   Lenk, H.: Zwischen Wissenschaft und Ethik. Suhrkamp: Frankfurt (1992); Mit-
        telstraß, J.: Leonardo Welt. Über Wissenschaft, Forschung und Verantwortung.
        Suhrkamp: Frankfurt (1992)
7.23   Kant, I.: Kritik der reinen Vernunft (1787), B 833

Subject Index

Name Index

Springer
and the
environment

At Springer we firmly believe that an international science publisher has a special obligation to the environment, and our corporate policies consistently reflect this conviction.
We also expect our business partners – paper mills, printers, packaging manufacturers, etc. – to commit themselves to using materials and production processes that do not harm the environment. The paper in this book is made from low- or no-chlorine pulp and is acid free, in conformance with international standards for paper permanency.

Springer

Printing: Mercedesdruck, Berlin
Binding: Buchbinderei Lüderitz & Bauer, Berlin